Ordinary Differential Equations and Linear Algebra

Ordinary Differential Equations and Linear Algebra
A Systems Approach

Todd Kapitula
Calvin College
Grand Rapids, Michigan

Society for Industrial and Applied Mathematics
Philadelphia

Copyright © 2015 by the Society for Industrial and Applied Mathematics

10 9 8 7 6 5 4 3 2 1

All rights reserved. Printed in the United States of America. No part of this book may be reproduced, stored, or transmitted in any manner without the written permission of the publisher. For information, write to the Society for Industrial and Applied Mathematics, 3600 Market Street, 6th Floor, Philadelphia, PA 19104-2688 USA.

The Google™ search engine and PageRank™ algorithm are trademarks of Google Inc. PRChecker.info is not affiliated with Google Inc. but provides only publicly available information about PageRank values of web sites. PRChecker.info provides our services on "as is" and "as available" bases, and we do not provide any guarantees regarding this service's stability and/or availability.

MATLAB is a registered trademark of The MathWorks, Inc. For MATLAB product information, please contact The MathWorks, Inc., 3 Apple Hill Drive, Natick, MA 01760-2098 USA, 508-647-7000, Fax: 508-647-7001, *info@mathworks.com*, *www.mathworks.com*.

WolframAlpha is a registered trademark of Wolfram Alpha LLC.

Publisher	David Marshall
Acquisitions Editor	Elizabeth Greenspan
Developmental Editor	Gina Rinelli
Managing Editor	Kelly Thomas
Production Editor	Lisa Briggeman
Copy Editor	Bruce Owens
Production Manager	Donna Witzleben
Production Coordinator	Cally Shrader
Compositor	Techsetters, Inc.
Graphic Designer	Lois Sellers

Library of Congress Cataloging-in-Publication Data

Kapitula, Todd.
  Ordinary differential equations and linear algebra : a systems approach / Todd Kapitula, Calvin College, Grand Rapids, Michigan.
      pages cm. -- (Other titles in applied mathematics ; 145)
  Includes bibliographical references and index.
  ISBN 978-1-611974-08-9
  1. Differential equations. 2. Algebras, Linear. I. Title.
  QA372.K2155 2015
  515'.352--dc23   2015030491

 is a registered trademark.

Contents

Preface   vii

1   Essentials of linear algebra   1
    1.1   Solving linear systems 2
    1.2   Vector algebra and matrix/vector multiplication 12
    1.3   Matrix algebra: Addition, subtraction, and multiplication . . . 18
    1.4   Matrix algebra: The inverse of a square matrix 20
    1.5   The structure of the solution 23
    1.6   Sets of linear combinations of vectors 29
    1.7   Subspaces 40
    1.8   Basis and dimension 46
    1.9   Equivalence results 55
    1.10   The determinant of a square matrix 60
    1.11   Linear algebra with complex-valued numbers, vectors,
              and matrices 65
    1.12   Eigenvalues and eigenvectors 71
    1.13   Case studies 85
    Group projects 97

2   Scalar first-order linear differential equations   101
    2.1   Motivating problems 101
    2.2   General theory 104
    2.3   The structure of the solution 111
    2.4   The homogeneous solution 115
    2.5   The particular and general solution 116
    2.6   Case studies 126
    Group projects 133

3   Systems of first-order linear differential equations   135
    3.1   Motivating problems 135
    3.2   Scalar equations and equivalent systems 142
    3.3   Existence and uniqueness theory 144
    3.4   The structure of the solution 144
    3.5   The homogeneous solution 149
    3.6   The particular solution 175
    3.7   Case studies 183
    Group projects 197

4	**Scalar higher-order linear differential equations**	**201**
4.1	The homogeneous solution	202
4.2	The particular solution	209
4.3	Case studies	218
	Group projects	233

5	**Discontinuous forcing and the Laplace transform**	**237**
5.1	The Heaviside (unit) step function and the (Dirac) delta function	237
5.2	The Laplace transform	240
5.3	Application: Solving second-order scalar ODEs	247
5.4	Case studies	251
5.5	The transfer function, convolutions, and variation of parameters	261
	Group projects	269

6	**Odds and ends**	**273**
6.1	Separation of variables	273
6.2	Nonlinear scalar autonomous ODEs and the phase line	276
6.3	Case studies	279
6.4	Power series solutions	283
6.5	Case studies	287

Bibliography   **295**

Index   **299**

Preface

This book arose from lecture notes that I began to develop in 2010–2011 for a first course in ordinary differential equations (ODEs). At Calvin College, the students in this course are primarily engineers. In our engineering program, the only (formal) linear algebra the students are required to see throughout their undergraduate career is what is presented in the ODE course. This is not unusual, as the ABET Accreditation Criteria of 2014–2015 do not explicitly require a course devoted to the study of linear algebra alone. Since, in my opinion, the amount of material on linear algebra covered in, e.g., the classical text of Boyce and DiPrima [9] is insufficient if that is all you will see in your academic career, I found it necessary to supplement the ODE course with my own notes on linear algebra. Eventually, it became clear that in order to have a seamless transition between the linear algebra and ODEs, there needed to be one text. This is not a new idea; for example, two recent texts that have a substantive linear algebra component are by Boelkins et al. [6] and Edwards and Penney [15].

Because there is a substantive linear algebra component in this text, I—and, more important, the students—find it much easier, when discussing the solutions of linear systems of ODEs, to focus more on the ODE aspects of the problems and less on the underlying algebraic manipulations. I have found that doing the linear algebra first allows for a more extensive exploration of linear systems of ODEs. Most important, it becomes possible to study and solve much more interesting examples and applications. The inclusion of more modeling and model analysis is extremely important; indeed, it is precisely what is recommended in the 2013 report by the National Academy of Sciences on the current state and future of the mathematical sciences.

The applications presented in this text are labeled "Case Studies." I chose this moniker because I want to convey to the reader that in solving particular problems, we were going to do more than simply find a solution; instead, we were going to take time to determine what the solution was telling us about the dynamical behavior for the given physical system. There are 20 case studies presented herein. Some are classical, e.g., damped mass-spring systems and mixing problems (compartment models), but several are not typically found in a text such as this. Nonclassical examples include a discrete susceptible-infected-recovered (SIR) model, a study of the effects on the body of lead ingestion, strongly damped systems (which can be recast as a singular perturbation problem), and a (simple) problem in the mathematics of climate. It is (probably) not possible to present all of these case studies in a one-semester course. On the other hand, the large number allows the instructor to choose a subset that will be of particular interest to his or her class.

The book is formatted as follows. In Chapter 1, we discuss not only the basics of linear algebra that will be needed for solving systems of linear ODEs, e.g.,

Gaussian elimination, matrix algebra, and eigenvalues/eigenvectors, but also such foundational material as subspaces, dimension, etc. While the latter material is not necessary to solve ODEs, I find that this is a natural time to introduce students to these more abstract linear algebra concepts. Moreover, since linear algebra is such important foundational material for a mathematical understanding of all of the sciences, I feel that it is essential that the students learn as much as they reasonably can in the short amount of time that is available. It is typically the case that the material in Chapter 1 can be covered in about 12–15 class periods. Primarily because of time constraints, when presenting this material, I focus primarily on the case of the vector space $\mathbb{R}^n$. The culminating section in the chapter is that on eigenvalues and eigenvectors. Here, I especially emphasize the utility of writing a given vector as a linear combination of eigenvectors. The closing section consists of four case studies: three determine the large "time" behavior associated with discrete dynamical systems, and one is a study of digital filters. If the reader and/or instructor wishes to have a supplementary text for this chapter, the free book by Hefferon [20][1] is an excellent companion.

Once the linear algebra has been mastered, in Chapter 2 we begin the study of ODEs by solving scalar first-order linear ODEs. We briefly discuss the general existence/uniqueness theory as well as the numerical solution. When solving ODEs numerically, we use the MATLAB programs `dfield8b.m` and `pplane8b.m` developed by J. Polking (provided on the book website at `http://www.siam.org/books/ot145`). These MATLAB programs have accompanying Java applets DFIELD and PPLANE.[2] My experience is that these software tools are more than sufficient to numerically solve the problems discussed in this chapter. We next analytically find the homogeneous and particular solutions to the linear problem. In this construction, we do three things:

(a) derive and write the homogeneous solution formula in such a way that the later notion of a homogeneous solution being thought of as the product of a matrix-valued solution and a constant vector is a natural extension,

(b) derive and write the variation-of-parameters solution formula in such a manner that the ideas easily generalize to systems, and

(c) develop the technique of undetermined coefficients.

The chapter closes with a careful analysis of, first, the one-tank mixing problem under the assumption that the incoming concentration varies periodically in time and, second, a mathematical finance problem. The idea here is to

(a) show students that understanding is not achieved with a solution formula; instead, it is necessary that the formula be written "correctly" so that as much physical information as possible can be gleaned from it;

(b) introduce students to the ideas of amplitude plots and phase plots; and

(c) set students up for the later analysis of the periodically forced mass spring.

As a final note, in many (if not almost all) texts, there is typically in this type of chapter an extensive discussion on nonlinear ODEs. I chose to provide only a cursory treatment of this topic at the end of this book because of

---

[1]`http://joshua.smcvt.edu/linearalgebra/`.
[2]`http://math.rice.edu/~dfield/dfpp.html`.

(a) my desire for my students to understand and focus on linearity and its consequences, and

(b) the fact that we at Calvin College teach a follow-up course on nonlinear dynamics using the wonderful text by Strogatz [37].

In Chapter 3, we study systems of linear ODEs. We start with five physical examples, three of which are mathematically equivalent in that they are modeled by a second-order scalar ODE. We show that nth-order scalar ODEs are equivalent to first-order systems and thus (we hope) convince the student that it is acceptable to skip (for the moment) a direct study of these higher-order scalar problems. We almost immediately focus on the case of the homogeneous problem being constant coefficient and derive the homogeneous solution via an expansion in terms of eigenvectors. From a pedagogical perspective, I find (and my students seem to agree) that this is a natural way to see how the eigenvalues and eigenvectors of a matrix play a key role in the construction of the homogeneous solution and in particular how using a particular basis may greatly simplify a given problem. Moreover, I find that this approach serves as an indirect introduction to the notion of Fourier expansions, which is, of course, used extensively in a successor course on linear partial differential equations (PDEs). After we construct the homogeneous solutions, we discuss the associated phase plane. As for the particular solution, we mimic the discussion of the previous chapter and simply show the few modifications that must be made in order for the previous scalar results to be valid for systems. My experience has been that the manner in which things were done in the previous chapter helps the student see that it is not the case that we are learning something entirely new and different; instead, we are just expanding on an already understood concept. The chapter closes with a careful analysis of three problems: a two-tank mixing problem in which the incoming concentration into one of the tanks is assumed to vary periodically in time, a study of the effect of lead ingestion, and an SIR model associated with zoonotic (animal-to-human) bacterial infections. As in the previous chapter, the goal is to not only construct the solution to the mathematical model but also understand how the solution helps us understand the dynamics of the given physical system.

In Chapter 4, we solve higher-order scalar ODEs. Because all of the theoretical work has already been done in the previous chapter, it is not necessary to spend too much time on this particular task. In particular, there is a relatively short presentation on how one can use the systems theory to solve the scalar problem. The variation of parameters formula is not re-derived; instead, it is presented just as a special case of the formula for systems. We conclude with a careful study of several problems: the undamped and damped mass-spring systems, a (linear) pendulum driven by a constant torque, and a coupled mass-spring system. Nice illustrative Java applets treat the forced and damped oscillations of a spring pendulum[3] and coupled oscillators.[4] There are also illustrative movies that are generated by MATLAB (provided on the book website, http://www.siam.org/books/ot145).

In Chapter 5, we solve scalar ODEs using the Laplace transform. The focus here is to solve only those problems for which the forcing term is a linear combination of Heaviside functions and delta functions. In my opinion, any other type of forcing term can be more easily handled with the method of either undetermined

---
[3]http://www.walter-fendt.de/ph14e/resonance.htm.
[4]http://www.lon-capa.org/%7emmp/applist/coupled/osc2.htm.

coefficients or variation of parameters. Moreover, we focus on using the Laplace transform as a method to find the particular solution with zero initial data. The understanding here is that we can find the homogeneous solution using the ideas and techniques from previous chapters. Because of the availability of SAGE (or WolframAlpha or any other CAS), we spend little time on partial fraction expansions and the inversion of the Laplace transform. The subsequent case studies are somewhat novel. We start by providing a physical interpretation of delta function forcing. We then find a way to stop the oscillations for an undamped mass-spring system. For another problem, we study a one-tank mixing problem in which the incoming concentration varies periodically in time. The injection strategy is modeled as an infinite sum of delta functions. The last case study involves the analysis of a strongly damped mass-spring problem. We show that this system can be thought of as a singular perturbation problem that is (formally) mathematically equivalent to a one-tank mixing problem. We next discuss how the Laplace transform generates transfer function and the manner in which this function is a model for a physical system. Finally, we show that the convolution integral leads to a variation-of-parameters formula for the particular solution.

In Chapter 6, we cover topics that are not infrequently discussed if time permits: separation of variables, phase-line analysis, and series solutions. Each topic is only briefly touched on, but enough material is presented herein for the student to get a good idea of what each one is about. For the latter two topics, I present case studies that could lead to a more detailed examination of the topic (using outside resources) if the student and/or instructor wishes to study it more.

Almost every section concludes with a set of homework problems. Moreover, there is a section at the end of each chapter that is labeled "Group Projects." The problems contained in these sections are more challenging, and I find it to be the case that students have a better chance of understanding and solving them if they work together in groups of 2–4 people. My experience is that students truly enjoy working on these problems and very much appreciate working collaboratively. I typically assign 1–3 of these types of problems per semester.

None of the homework problems have attached to them a solution that is included in this book. I suspect that many (if not most) students will find this lack of solved problems troubling. Two relatively cheap (potentially supplemental) texts that address this issue are Lipschutz and Lipson [26] for the linear algebra material and Bronson and Costa [10] for the ODE material. Of course, other books, e.g., [5, 12, 18, 32], can be found simply by going to the library and looking through the (perhaps) dozens of appropriate books.

Throughout this text, we expect students to use a CAS to do some of the intermediate calculations. I generally use SAGE[5] and/or WolframAlpha.[6] There are several advantages to using these particular CAS:

(a) It is not necessary to learn a programming language to use WolframAlpha.

(b) The commands are intuitive.

(c) Online help is readily available.

(d) Both SAGE and WolframAlpha are easily accessible and free (as of June 2015).

---

[5] http://www.sagemath.org/.
[6] http://www.wolframalpha.com/.

Of course, one can use other CAS packages. Since there is currently no universal agreement (even within my department!) as to which package is best to use, I do not want to limit this text to a particular system. Consequently, I do not include output of any particular CAS in the text. Within the text, I use the moniker SAGE to indicate that I have done a calculation using a CAS. Moreover, to help the students become acquainted with how a CAS is used in the intermediate calculations, on the book website there is a SAGE worksheet that lists the commands used to do a particular referenced calculation. My expectations here are

(a) that the student will use whichever CAS he or she wants, and

(b) that interested students who have some experience with a particular CAS will quickly learn how to use it.

Finally, in this text, a CAS is *never* used to completely solve a given problem, as it is important that the student thoroughly understand what intermediate calculations are needed to find the solution. As seen in the Case Studies, having a mathematical solution to the problem (which is the purview of the CAS) solves only half the problem. More work, in which a CAS may or may not be helpful, is needed in order to extract from the solution the desired answer to the given problem. My goal is to help students learn how to provide a physical interpretation of the solution.

All figures appear in black and white in the print version of this book. Figures are reproduced in full color in the ebook version, available for purchase from http://www.siam.org/books/ot145.

Acknowledgments. I am indebted to Kate Ardinger, Tom Jager, Jeff Humpherys, Michael Kapitula, Keith Promislow, Thomas Scofield, Matt Walhout, and anonymous reviewers for discussions about and a careful reading of this manuscript. The implementation of their suggestions and comments greatly improved the text. Any and all mistakes are mine. Finally, I am thankful for the support, encouragement, and love of my wife, Laura.

For the glory of the most high God alone,
And for my neighbour to learn from.
— *J. S. Bach*

Chapter 1
Essentials of linear algebra

> *Mathematics is the art of reducing any problem to linear algebra.*
>
> — William Stein
>
> *To many, mathematics is a collection of theorems. For me, mathematics is a collection of examples; a theorem is a statement about a collection of examples and the purpose of proving theorems is to classify and explain the examples.*
>
> — John Conway

The average college student knows how to solve two equations in two unknowns in an elementary way: the method of substitution. For example, consider the system of equations
$$2x + y = 6, \quad 2x + 4y = 5.$$
Solving the first equation for y gives $y = 6 - 2x$, and substituting this expression into the second equation yields
$$2x + 4(6 - 2x) = 5 \quad \rightsquigarrow \quad x = \frac{19}{6}.$$
Substitution into either of the equations gives the value of y; namely, $y = -1/3$. For systems of three or more equations, this algorithm is algebraically unwieldy. Furthermore, it is inefficient, as it is often the case not very clear as to which variable(s) should be substituted into which equation(s). Thus, at the very least, we should develop an efficient algorithm for solving large systems of equations. Perhaps more troubling (at least to the mathematician!) is the fact that the method of substitution does not yield any insight into the structure of the solution set. An analysis and understanding of this structure is the topic of linear algebra. As we will see, not only will we gain a much better understanding of how to solve linear algebraic systems, but, by considering the problem more abstractly, we will better understand how to solve linear systems of ordinary differential equations (ODEs).

This chapter is organized in the following manner. We begin our discussion of linear systems of equations by developing an efficient solution algorithm: Gaussian elimination. We then consider the problem using matrices and vectors and spend considerable time and energy trying to understand the solution structure via these objects. In particular, we show that that the solution is composed of two pieces. One piece is intrinsically associated with the matrix alone, and the other piece reflects an interaction between the matrix and nonhomogeneous term. We conclude the chapter by looking at special vectors associated with square matrices: the eigenvectors. These vectors have the special algebraic property that the matrix multiplied by an eigenvector is simply a scalar multiple of that eigenvector (this scalar is known as the associated eigenvalue). As we will see, the eigenvalues and eigenvectors are the key objects associated with a matrix that allow us to easily and explicitly write down and understand the solution to linear dynamical systems (both discrete and continuous).

1.1 • Solving linear systems

1.1.1 • Notation and terminology

A *linear equation* in n variables is an algebraic equation of the form

$$a_1 x_1 + a_2 x_2 + \cdots + a_n x_n = b. \tag{1.1.1}$$

The (possibly complex-valued) numbers $a_1, a_2, \ldots, a_n$ are the *coefficients*, and the unknowns to be solved for are the *variables* $x_1, \ldots, x_n$. The variables are also sometimes called *unknowns*. An example in two variables is

$$2x_1 - 5x_2 = 7,$$

and an example in three variables is

$$x_1 - 3x_2 + 9x_3 = -2.$$

A *system of linear equations* is a collection of m linear equations (1.1.1) and can be written as

$$\begin{aligned} a_{11} x_1 + a_{12} x_2 + \cdots + a_{1n} x_n &= b_1 \\ a_{21} x_1 + a_{22} x_2 + \cdots + a_{2n} x_n &= b_2 \\ &\vdots \\ a_{m1} x_1 + a_{m2} x_2 + \cdots + a_{mn} x_n &= b_m. \end{aligned} \tag{1.1.2}$$

The coefficient a_{jk} is associated with the variable x_k in the jth equation. An example of two equations in three variables is

$$\begin{aligned} x_1 - 4x_2 &= 6 \\ 3x_1 + 2x_2 - 5x_3 &= 2. \end{aligned} \tag{1.1.3}$$

1.1. Solving linear systems

> Until we get to our discussion of eigenvalues and eigenvectors in section 1.12, we will assume that the coefficients and variables are real numbers, i.e., $a_{jk}, x_j \in \mathbb{R}$. This is done solely for the sake of pedagogy and exposition. It cannot be stressed too much, however, that *everything* we do preceding section 1.12 still works even if we remove this restriction and allow these numbers to be complex (have nonzero imaginary part).

When there is a large number of equations and/or variables, it is awkward to write down a linear system in the form of (1.1.2). It is more convenient instead to use a *matrix* formulation. A matrix is a rectangular array of numbers with m rows and n columns, and such a matrix is said to be an $m \times n$ (read "m by n") matrix. If $m = n$, the matrix is said to be a *square matrix*.

> For an $m \times n$ matrix with real entries, we will say $A \in \mathcal{M}_{m \times n}(\mathbb{R})$. If the matrix is square, i.e., $m = n$, then we will write $A \in \mathcal{M}_n(\mathbb{R})$. The $\mathbb{R}$ is there to emphasize that all of the entries are real numbers. If the entries are allowed to be complex, we will write $A \in \mathcal{M}_{m \times n}(\mathbb{C})$, or $A \in \mathcal{M}_n(\mathbb{C})$.

The *coefficient matrix* for the linear system (1.1.2) is given by

$$A = \begin{pmatrix} a_{11} & a_{12} & \cdots & a_{1n} \\ a_{21} & a_{22} & \cdots & a_{2n} \\ \vdots & \vdots & \cdots & \vdots \\ a_{m1} & a_{m2} & \cdots & a_{mn} \end{pmatrix}, \tag{1.1.4}$$

and the coefficient a_{jk}, which is associated with the variable x_k in the jth equation, is in the jth row and kth column. For example, the coefficient matrix for the system (1.1.3) is given by

$$A = \begin{pmatrix} 1 & -4 & 0 \\ 3 & 2 & -5 \end{pmatrix} \in \mathcal{M}_{2 \times 3}(\mathbb{R}),$$

with

$$a_{11} = 1,\ a_{12} = -4,\ a_{13} = 0,\ a_{21} = 3,\ a_{22} = 2,\ a_{23} = -5.$$

A *vector*, say $v \in \mathcal{M}_{m \times 1}(\mathbb{R})$, is a matrix with only one column. A vector is sometimes called a *column vector* or *m-vector*. To clearly distinguish between vectors and matrices, we will say that if $v \in \mathcal{M}_{m \times 1}(\mathbb{R})$, then $v \in \mathbb{R}^m$. The variables in the system (1.1.2) will be written as the vector

$$x = \begin{pmatrix} x_1 \\ x_2 \\ \vdots \\ x_n \end{pmatrix} \in \mathbb{R}^n,$$

and the variables on the right-hand side will be written as the vector

$$b = \begin{pmatrix} b_1 \\ b_2 \\ \vdots \\ b_m \end{pmatrix} \in \mathbb{R}^m.$$

The zero vector, $0 \in \mathbb{R}^m$, *is the vector with a zero in each entry.*

In conclusion, for the system (1.1.2), there are three matrix-valued quantities: the coefficient matrix A, the vector of unknowns x, and the right-hand-side vector b. We will represent the linear system (1.1.2)

$$Ax = b. \tag{1.1.5}$$

We will later see what it means to multiply a matrix and a vector. The linear system is said to be *homogeneous* if $b = 0$; otherwise, the system is said to be *nonhomogeneous*.

1.1.2 ▪ Solutions of linear systems

A *solution* to the linear system (1.1.5) (or, equivalently, (1.1.2)) is a vector x that satisfies all m equations simultaneously. For example, consider the linear system of three equations in three unknowns for which

$$A = \begin{pmatrix} 1 & 0 & -1 \\ 3 & 1 & 0 \\ 1 & -1 & -1 \end{pmatrix}, \quad b = \begin{pmatrix} 0 \\ 1 \\ -4 \end{pmatrix}, \tag{1.1.6}$$

i.e.,

$$x_1 - x_3 = 0, \quad 3x_1 + x_2 = 1, \quad x_1 - x_2 - x_3 = -4.$$

It is not difficult to check that a solution is given by

$$x = \begin{pmatrix} -1 \\ 4 \\ -1 \end{pmatrix} \rightsquigarrow x_1 = -1, \ x_2 = 4, \ x_3 = -1.$$

A system of linear equations with at least one solution is said to be *consistent*; otherwise, it is *inconsistent*.

How many solutions does a linear system have? Consider the system given by

$$2x_1 - x_2 = -2, \quad -x_1 + 3x_2 = 11.$$

The first equation represents a line in the $x_1 x_2$-plane with slope 2, and the second equation represents a line with slope $1/3$. Since lines with different slopes intersect at a unique point, the system is consistent with a unique solution. It is not difficult to check that the solution is given by $(x_1, x_2) = (1, 4)$. Next, consider the system given by

$$2x_1 - x_2 = -2, \quad -4x_1 + 2x_2 = 8.$$

Each equation represents a line with slope 2, so that the lines are parallel. Since the second equation is a multiple of the first equation, the system is consistent.

1.1. Solving linear systems

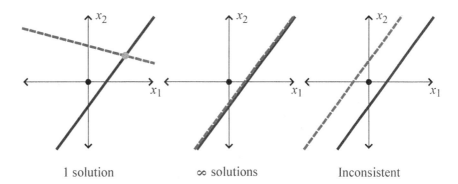

Figure 1.1. *A graphical depiction of the three possibilities for linear systems of two equations in two unknowns. The left panel shows the case when the corresponding lines are not parallel, and the other two panels show the cases when the lines are parallel.*

Moreover, each point on the line is a solution, so there are an infinite number of solutions. On the other hand, the system

$$2x_1 - x_2 = -2, \quad -4x_1 + 2x_2 = 7$$

is inconsistent, as the second equation is no longer a scalar multiple of the first equation. See Figure 1.1 for graphical representations of these three cases.

We see that a linear system with two equations and two unknowns is consistent with a unique solution, is consistent with an infinite number of solutions, or is inconsistent. It is not difficult to show that this fact holds for linear systems with three unknowns. Each linear equation in the system represents a plane in $x_1 x_2 x_3$-space. Given any two planes, we know that they either are parallel or intersect along a line. Thus, if the system has two equations, then it will be either consistent with an infinite number of solutions or inconsistent. Suppose that the system with two equations is consistent and add a third linear equation. Further, suppose that the original two planes intersect along a line. This new plane either is parallel to the line or intersects it at precisely one point. If the original two planes are the same, then the new plane either is parallel to both or intersects it along a line. In conclusion, for a system of equations with three variables, there is either a unique solution, an infinite number of solutions, or no solution.

For systems with four or more variables, this geometric argument is difficult to visualize. However, using the row-reduced echelon form of a matrix (see Definition 1.4), one can show the following algebraically:

> **Theorem 1.1.** *If the linear system (1.1.2) is consistent, then there is either a unique solution or an infinite number of solutions.*

Remark 1.2. *Theorem 1.1 does not hold for nonlinear systems. For example, the nonlinear system*

$$x_1^2 + x_2^2 = 2, \quad x_1 + x_2 = 0$$

is consistent and has the two solutions $(-1, 1), (1, -1)$.

It is often the case that if a linear system is consistent, then more cannot be said about the number of solutions without directly solving the system. However, in the argument leading up to Theorem 1.1, we did see that for a system of two equations in three unknowns, if the system was consistent, then there were necessarily an infinite number of solutions. This result holds in general:

> **Corollary 1.3.** *Suppose that the linear system is such that $m < n$; i.e., there are fewer equations than unknowns (the system is underdetermined). If the system is consistent, then there are an infinite number of solutions.*

1.1.3 ▪ Solving by Gaussian elimination

We now need to understand how to systematically solve the linear system (1.1.5),

$$Ax = b.$$

While the method of substitution works fine for two equations in two unknowns, it quickly breaks down as a practical method when there are three or more variables involved in the system. We need to come up with something else.

The simplest linear system to solve for two equations in two unknowns is

$$x_1 = b_1, \quad x_2 = b_2.$$

The coefficient matrix is

$$I_2 = \begin{pmatrix} 1 & 0 \\ 0 & 1 \end{pmatrix} \in \mathscr{M}_2(\mathbb{R}),$$

which is known as the *identity matrix*. The unique solution to this system is $x = b$. The simplest linear system to solve for three equations in three unknowns is

$$x_1 = b_1, \quad x_2 = b_2, \quad x_3 = b_3.$$

The coefficient matrix is now

$$I_3 = \begin{pmatrix} 1 & 0 & 0 \\ 0 & 1 & 0 \\ 0 & 0 & 1 \end{pmatrix} \in \mathscr{M}_3(\mathbb{R}),$$

The identity matrix, I_n, is a square matrix with ones on the diagonal and zeros everywhere else. The subscript refers to the size of the matrix.

which is the 3×3 identity matrix. The unique solution to this system is again $x = b$. Continuing in this fashion, the simplest linear system for n equations in n unknowns to solve is

$$x_1 = b_1, \, x_2 = b_2, \, x_3 = b_3, \ldots, x_n = b_n.$$

The coefficient matrix associated with this system is I_n, and the solution is $x = b$.

Suppose that the number of equations is not equal to the number of unknowns. For example, a particularly simple system to solve is given by

$$x_1 - 3x_3 + 4x_4 = 2, \quad x_2 + x_3 - 6x_4 = 5. \tag{1.1.7}$$

1.1. Solving linear systems

The coefficient matrix for this system is

$$A = \begin{pmatrix} 1 & 0 & -3 & 4 \\ 0 & 1 & 1 & -6 \end{pmatrix} \in \mathcal{M}_{2\times 4}(\mathbb{R}).$$

Solving the system for the first two variables in terms of the latter two yields

$$x_1 = 3x_3 - 4x_4 + 2, \quad x_2 = -x_3 + 6x_4 + 5.$$

On setting $x_3 = s$ and $x_4 = t$, where the dummy variables $s, t \in \mathbb{R}$ are arbitrary, we see that the solution to this system is

$$x_1 = 2+3s-t, \; x_2 = 5-s+6t, \; x_3 = s, \; x_4 = t \rightsquigarrow x = \begin{pmatrix} 2+3s-4t \\ 5-s+6t \\ s \\ t \end{pmatrix}, \quad s, t \in \mathbb{R}.$$

Since s and t are arbitrary, there is an infinite number of solutions. This was expected, for, as we saw in Corollary 1.3, consistent underdetermined systems will have an infinite number of solutions.

The coefficient matrices for the problems considered so far share a common feature, which is detailed below:

RREF

Definition 1.4. *A matrix is said to be in row-reduced echelon form (RREF) if*

(a) *all nonzero rows are above any zero row,*

(b) *the first nonzero entry in a row (the leading entry) is a one, and*

(c) *every other entry in a column with a leading one is zero.*

Those columns with a leading entry are known as pivot columns, and the leading entries are called pivot positions.

The RREF of a given matrix is unique [39].

Example 1.5. Consider the matrix in RREF given by

$$A = \begin{pmatrix} 1 & 0 & -3 & 0 & 7 \\ 0 & 1 & -3 & 0 & 2 \\ 0 & 0 & 0 & 1 & -4 \\ 0 & 0 & 0 & 0 & 0 \end{pmatrix} \in \mathcal{M}_{4\times 5}(\mathbb{R}).$$

As a rule of thumb, when putting an augmented matrix into RREF, the idea is to place 1's on the diagonal and 0's everywhere else (as much as possible).

The first, second, and fourth columns are the pivot columns, and the pivot positions are the first entry in the first row, the second entry in the second row, and the fourth entry in the third row. ∎

If a coefficient matrix is in RREF, then the linear system is particularly easy to solve. Thus, our goal is to take a given linear system with its attendant coefficient matrix and then perform allowable algebraic operations so that the new system has a coefficient matrix that is in RREF. The allowable algebraic operations for solving a linear system are

(a) multiply any equation by a constant,

(b) add/subtract equations, and

(c) switch the ordering of equations.

On performing these operations, the resulting system is not the same as the original; however, the new system is equivalent to the old in that, for consistent systems, the solution values remain unchanged. If the original system is inconsistent, then so will any new system resulting from performing the above operations.

In order to do these operations most efficiently using matrices, it is best to work with the *augmented matrix* associated with the linear system $Ax = b$, namely, the matrix $(A|b)$. The augmented matrix is formed by adding a column, namely, the vector b, to the coefficient matrix. For example, for the linear system associated with (1.1.6), the augmented matrix is given by

$$(A|b) = \begin{pmatrix} 1 & 0 & -1 & | & 0 \\ 3 & 1 & 0 & | & 1 \\ 1 & -1 & -1 & | & -4 \end{pmatrix}, \qquad (1.1.8)$$

and the augmented matrix for the linear system (1.1.7) is

$$(A|b) = \begin{pmatrix} 1 & 0 & -3 & 4 & | & 2 \\ 0 & 1 & 1 & -6 & | & 5 \end{pmatrix}.$$

The allowable operations on the individual equations in the linear system correspond to operations on the *rows* of the augmented matrix. In particular, when doing *Gaussian elimination* on an augmented matrix in order to put it into RREF, we are allowed to

(a) multiply any row by a constant,

(b) add/subtract rows, and

(c) switch the ordering of the rows.

Once we have performed Gaussian elimination on an augmented matrix in order to put it into RREF, we can easily solve the resultant system.

Example 1.6. Consider the linear system associated with the augmented matrix in (1.1.8). We will henceforth let ρ_j denote the jth row of a matrix. The operation "$a\rho_j + b\rho_k$" will be taken to mean multiply the jth row by a, multiply the kth row by b, add the two resultant rows together, and replace the kth row with this sum. With this notation in mind, performing Gaussian elimination yields

$$(A|b) \xrightarrow{-3\rho_1+\rho_2} \begin{pmatrix} 1 & 0 & -1 & | & 0 \\ 0 & 1 & 3 & | & 1 \\ 1 & -1 & -1 & | & -4 \end{pmatrix} \xrightarrow{-\rho_1+\rho_3} \begin{pmatrix} 1 & 0 & -1 & | & 0 \\ 0 & 1 & 3 & | & 1 \\ 0 & -1 & 0 & | & -4 \end{pmatrix}$$

$$\xrightarrow{\rho_2+\rho_3} \begin{pmatrix} 1 & 0 & -1 & | & 0 \\ 0 & 1 & 3 & | & 1 \\ 0 & 0 & 3 & | & -3 \end{pmatrix} \xrightarrow{(1/3)\rho_3} \begin{pmatrix} 1 & 0 & -1 & | & 0 \\ 0 & 1 & 3 & | & 1 \\ 0 & 0 & 1 & | & -1 \end{pmatrix}$$

$$\xrightarrow{-3\rho_3+\rho_2} \begin{pmatrix} 1 & 0 & -1 & | & 0 \\ 0 & 1 & 0 & | & 4 \\ 0 & 0 & 1 & | & -1 \end{pmatrix} \xrightarrow{\rho_3+\rho_1} \begin{pmatrix} 1 & 0 & 0 & | & -1 \\ 0 & 1 & 0 & | & 4 \\ 0 & 0 & 1 & | & -1 \end{pmatrix}.$$

The new linear system is

$$x_1 = -1,\ x_2 = 4,\ x_3 = -1 \quad \leadsto \quad x = \begin{pmatrix} -1 \\ 4 \\ -1 \end{pmatrix},$$

which is also immediately seen to be the solution. ∎

Example 1.7. Consider the linear system

$$\begin{aligned} x_1 - 2x_2 - x_3 &= 0 \\ 3x_1 + x_2 + 4x_3 &= 7 \\ 2x_1 + 3x_2 + 5x_3 &= 7. \end{aligned}$$

Performing Gaussian elimination on the augmented matrix yields

$$\begin{pmatrix} 1 & -2 & -1 & | & 0 \\ 3 & 1 & 4 & | & 7 \\ 2 & 3 & 5 & | & 7 \end{pmatrix} \xrightarrow{-3\rho_1 + \rho_2} \begin{pmatrix} 1 & -2 & -1 & | & 0 \\ 0 & 7 & 7 & | & 7 \\ 2 & 3 & 5 & | & 7 \end{pmatrix}$$

$$\xrightarrow{-2\rho_1 + \rho_3} \begin{pmatrix} 1 & -2 & -1 & | & 0 \\ 0 & 7 & 7 & | & 7 \\ 0 & 7 & 7 & | & 7 \end{pmatrix} \xrightarrow{-\rho_2 + \rho_3} \begin{pmatrix} 1 & -2 & -1 & | & 0 \\ 0 & 7 & 7 & | & 7 \\ 0 & 0 & 0 & | & 0 \end{pmatrix}$$

$$\xrightarrow{(1/7)\rho_2} \begin{pmatrix} 1 & -2 & -1 & | & 0 \\ 0 & 1 & 1 & | & 1 \\ 0 & 0 & 0 & | & 0 \end{pmatrix} \xrightarrow{2\rho_2 + \rho_1} \begin{pmatrix} 1 & 0 & 1 & | & 2 \\ 0 & 1 & 1 & | & 1 \\ 0 & 0 & 0 & | & 0 \end{pmatrix}.$$

The new linear system to be solved is given by

$$x_1 + x_3 = 2, \quad x_2 + x_3 = 1, \quad 0x_1 + 0x_2 + 0x_3 = 0.$$

Ignoring the last equation, this is a system of two equations with three unknowns; consequently, since the system is consistent, it must be the case that there are an infinite number of solutions. The variables x_1 and x_2 are associated with leading entries in the RREF form of the augmented matrix. As for the variable x_3, which is associated with the third column, which in turn is not a pivot column, we say the following:

> **Free variable**
>
> **Definition 1.8.** *A free variable of a linear system is a variable that is associated with a column in the RREF matrix that is not a pivot column.*

Since x_3 is a free variable, it can be arbitrarily chosen. On setting $x_3 = t$, where $t \in \mathbb{R}$, the other variables are

$$x_1 = 2 - t, \quad x_2 = 1 - t.$$

The solution is then

$$x = \begin{pmatrix} 2-t \\ 1-t \\ t \end{pmatrix}, \quad t \in \mathbb{R}. \quad \blacksquare$$

Example 1.9. Consider a linear system that is a variant of the one given above, namely,

$$x_1 - 2x_2 - x_3 = 0$$
$$3x_1 + x_2 + 4x_3 = 7$$
$$2x_1 + 3x_2 + 5x_3 = 8.$$

On doing Gaussian elimination of the augmented matrix, we see that

$$\begin{pmatrix} 1 & -2 & -1 & | & 0 \\ 3 & 1 & 4 & | & 7 \\ 2 & 3 & 5 & | & 8 \end{pmatrix} \xrightarrow{\text{RREF}} \begin{pmatrix} 1 & 0 & 1 & | & 2 \\ 0 & 1 & 1 & | & 1 \\ 0 & 0 & 0 & | & 1 \end{pmatrix}.$$

The new linear system to be solved is

$$x_1 + x_3 = 2, \quad x_2 + x_3 = 1, \quad 0x_1 + 0x_2 + 0x_3 = 1.$$

Since the last equation clearly does not have a solution, the system is inconsistent. ∎

Example 1.10. Consider a linear system for which the coefficient matrix and non-homogeneous term are

$$A = \begin{pmatrix} 1 & 2 & 3 \\ 4 & 5 & 6 \\ 7 & 8 & 2 \end{pmatrix}, \quad b = \begin{pmatrix} -1 \\ 4 \\ -7 \end{pmatrix}.$$

Using SAGE, we find

$$(A|b) \xrightarrow{\text{RREF}} \begin{pmatrix} 1 & 0 & 0 & | & 139/21 \\ 0 & 1 & 0 & | & -152/21 \\ 0 & 0 & 1 & | & 48/21 \end{pmatrix}.$$

The solution is the last column of the augmented matrix,

$$x = \begin{pmatrix} 139/21 \\ -152/21 \\ 48/21 \end{pmatrix} \sim \begin{pmatrix} 6.62 \\ -7.24 \\ 2.29 \end{pmatrix}. \quad \blacksquare$$

Exercises

Exercise 1.1.1. *Solve each system of equations or explain why no solution exists.*

(a) $x_1 + 2x_2 = 4$, $-2x_1 + 3x_2 = -1$

(b) $x_1 + 2x_2 = 4$, $x_1 + 2x_2 = -1$

(c) $x_1 + 2x_2 = 4$, $4x_1 + 8x_2 = 15$

Exercise 1.1.2. *Each of the linear systems below is represented by an augmented matrix in RREF. If the system is consistent, express the solution in vector form.*

1.1. Solving linear systems

(a) $\begin{pmatrix} 1 & 0 & 0 & | & -3 \\ 0 & 1 & 1 & | & 2 \\ 0 & 0 & 0 & | & 5 \end{pmatrix}$

(b) $\begin{pmatrix} 1 & 0 & 0 & | & -4 \\ 0 & 1 & 2 & | & 7 \\ 0 & 0 & 0 & | & 0 \end{pmatrix}$

(c) $\begin{pmatrix} 1 & 0 & 0 & | & 3 \\ 0 & 1 & 0 & | & 4 \\ 0 & 0 & 0 & | & -2 \end{pmatrix}$

(d) $\begin{pmatrix} 1 & 0 & 0 & 3 & | & -1 \\ 0 & 1 & 1 & 4 & | & 3 \\ 0 & 0 & 0 & 0 & | & 0 \end{pmatrix}$

Exercise 1.1.3. Determine all value(s) of r that make each augmented matrix correspond to a consistent linear system. For each such r, express the solution to the corresponding linear system in vector form.

(a) $\begin{pmatrix} 1 & 4 & | & -3 \\ -2 & -8 & | & r \end{pmatrix}$

(b) $\begin{pmatrix} 1 & 4 & | & -3 \\ 2 & r & | & -6 \end{pmatrix}$

(c) $\begin{pmatrix} 1 & 4 & | & -3 \\ -3 & r & | & -9 \end{pmatrix}$

(d) $\begin{pmatrix} 1 & r & | & -3 \\ -3 & r & | & 8 \end{pmatrix}$

Exercise 1.1.4. The augmented matrix for a linear system is given by

$$\begin{pmatrix} 1 & 1 & 3 & | & 2 \\ 1 & 2 & 4 & | & 3 \\ 1 & 3 & a & | & b \end{pmatrix}.$$

(a) For what value(s) of a and b will the system have infinitely many solutions?

(b) For what value(s) of a and b will the system be inconsistent?

Exercise 1.1.5. Solve each linear system and express the solution in vector form.

(a) $3x_1 + 2x_2 = 16$, $-2x_1 + 3x_2 = 11$

(b) $3x_1 + 2x_2 - x_3 = -2$, $-3x_1 - x_2 + x_3 = 5$, $3x_1 + 2x_2 + x_3 = 2$

(c) $2x_1 + x_2 = -1$, $x_1 - x_3 = -2$, $-x_1 + 3x_2 + 7x_3 = 11$

(d) $x_1 + x_2 - x_3 = 0$, $2x_1 - 3x_2 + 5x_3 = 0$, $4x_1 - x_2 + 3x_3 = 0$

(e) $x_2 + x_3 - x_4 = 0$, $x_1 + x_2 + x_3 + x_4 = 6$
$2x_1 + 4x_2 + x_3 - 2x_4 = -1$, $3x_1 + x_2 - 2x_3 + 2x_4 = 3$

Exercise 1.1.6. *If the coefficient matrix satisfies $A \in \mathcal{M}_{9 \times 6}(\mathbb{R})$ and if the RREF of the augmented matrix $(A|b)$ has three zero rows, is the solution unique? Why or why not?*

Exercise 1.1.7. *If the coefficient matrix satisfies $A \in \mathcal{M}_{5 \times 7}(\mathbb{R})$ and if the linear system $Ax = b$ is consistent, is the solution unique? Why or why not?*

Exercise 1.1.8. *Find a quadratic polynomial $p(t) = a_0 + a_1 t + a_2 t^2$ that passes through the points $(-2, 12), (1, 6), (2, 18)$. (Hint: $p(1) = 6$ implies that $a_0 + a_1 + a_2 = 6$.)*

Exercise 1.1.9. *Find a cubic polynomial $p(t) = a_0 + a_1 t + a_2 t^2 + a_3 t^3$ that passes through the points $(-1, -3), (0, 1), (1, 3), (2, 17)$.*

Exercise 1.1.10. *Determine if each of the following statements is true or false. Provide an explanation for your answer.*

(a) *A system of four linear equations in three unknowns can have exactly five solutions.*

(b) *If a system has a free variable, then there will be an infinite number of solutions.*

(c) *If a system is consistent, then there is a free variable.*

(d) *If the RREF of the augmented matrix has four zero rows and if the system is consistent, then there will be an infinite number of solutions.*

(e) *If the RREF of the augmented matrix has no zero rows, then the system is consistent.*

1.2 • Vector algebra and matrix/vector multiplication

Now that we have an efficient algorithm to solve the linear system $Ax = b$, we need to next understand what it means from a geometric perspective to solve the system. For example, if the system is consistent, how does the vector b relate to the coefficients of the coefficient matrix A? In order to answer this question, we need to make sense of the expression Ax (matrix/vector multiplication).

1.2.1 • Linear combinations of vectors

We begin by considering the addition/subtraction of vectors and the product of a scalar with a vector. We will define the addition/subtraction of two n-vectors to be exactly what is expected, and the same will hold true for the multiplication of a vector by a scalar, namely,

$$x \pm y = \begin{pmatrix} x_1 \pm y_1 \\ x_2 \pm y_2 \\ \vdots \\ x_n \pm y_n \end{pmatrix}, \quad cx = \begin{pmatrix} cx_1 \\ cx_2 \\ \vdots \\ cx_n \end{pmatrix}.$$

1.2. Vector algebra and matrix/vector multiplication

Vector addition and subtraction are done component by component, and scalar multiplication of a vector means that each component of the vector is multiplied by the scalar. For example,

$$\begin{pmatrix} -2 \\ 5 \end{pmatrix} + \begin{pmatrix} 3 \\ -1 \end{pmatrix} = \begin{pmatrix} 1 \\ 4 \end{pmatrix}, \quad 3\begin{pmatrix} 2 \\ -3 \end{pmatrix} = \begin{pmatrix} 6 \\ -9 \end{pmatrix}.$$

These are *linear operations*. Combining these two operations, we have more generally the following:

Linear combination

Definition 1.11. *A linear combination of the n-vectors $a_1,\ldots,a_k$ is given by the vector b, where*

$$b = x_1 a_1 + x_2 a_2 + \cdots + x_k a_k = \sum_{j=1}^{k} x_j a_j.$$

The scalars $x_1,\ldots,x_k$ are known as weights.

With this notion of linear combinations of vectors, we can rewrite linear systems of equations in vector notation. For example, consider the linear system

$$\begin{aligned} x_1 - x_2 + x_3 &= -1 \\ 3x_1 + 2x_2 + 8x_3 &= 7 \\ x_1 + 2x_2 + 4x_3 &= 5. \end{aligned} \qquad (1.2.1)$$

Since two vectors are equal if and only if all of their coefficients are equal, we can write (1.2.1) in vector form as

$$\begin{pmatrix} x_1 - x_2 + x_3 \\ 3x_1 + 2x_2 + 8x_3 \\ x_1 + 2x_2 + 4x_3 \end{pmatrix} = \begin{pmatrix} -1 \\ 7 \\ 5 \end{pmatrix}.$$

Using linearity, we can write the vector on the left-hand side as

$$\begin{pmatrix} x_1 - x_2 + x_3 \\ 3x_1 + 2x_2 + 8x_3 \\ x_1 + 2x_2 + 4x_3 \end{pmatrix} = x_1 \begin{pmatrix} 1 \\ 3 \\ 1 \end{pmatrix} + x_2 \begin{pmatrix} -1 \\ 2 \\ 2 \end{pmatrix} + x_3 \begin{pmatrix} 1 \\ 8 \\ 4 \end{pmatrix},$$

so the system (1.2.1) is equivalent to

$$x_1 \begin{pmatrix} 1 \\ 3 \\ 1 \end{pmatrix} + x_2 \begin{pmatrix} -1 \\ 2 \\ 2 \end{pmatrix} + x_3 \begin{pmatrix} 1 \\ 8 \\ 4 \end{pmatrix} = \begin{pmatrix} -1 \\ 7 \\ 5 \end{pmatrix}.$$

After setting

$$a_1 = \begin{pmatrix} 1 \\ 3 \\ 1 \end{pmatrix}, \quad a_2 = \begin{pmatrix} -1 \\ 2 \\ 2 \end{pmatrix}, \quad a_3 = \begin{pmatrix} 1 \\ 8 \\ 4 \end{pmatrix}, \quad b = \begin{pmatrix} -1 \\ 7 \\ 5 \end{pmatrix},$$

the linear system can then be rewritten as the linear combination of vectors

$$x_1 a_1 + x_2 a_2 + x_3 a_3 = b. \tag{1.2.2}$$

In conclusion, asking for solutions to the linear system (1.2.1) can instead be thought of as asking if the vector b is a linear combination of the vectors a_1, a_2, a_3. It can be checked that after Gaussian elimination,

$$\begin{pmatrix} 1 & -1 & 1 & | & -1 \\ 3 & 2 & 8 & | & 7 \\ 1 & 2 & 4 & | & 5 \end{pmatrix} \xrightarrow{\text{RREF}} \begin{pmatrix} 1 & 0 & 2 & | & 1 \\ 0 & 1 & 1 & | & 2 \\ 0 & 0 & 0 & | & 0 \end{pmatrix}.$$

The free variable is x_3, so the solution to the linear system (1.2.1) can be written as

$$x_1 = 1 - 2t, \ x_2 = 2 - t, \ x_3 = t; \quad t \in \mathbb{R}. \tag{1.2.3}$$

In vector form, this form of the solution is

$$x = \begin{pmatrix} 1 - 2t \\ 2 - t \\ t \end{pmatrix} = \begin{pmatrix} 1 \\ 2 \\ 0 \end{pmatrix} + t \begin{pmatrix} -2 \\ -1 \\ 1 \end{pmatrix}, \quad t \in \mathbb{R}.$$

The vector b is a linear combination of the vectors a_1, a_2, a_3, and the weights are given in (1.2.3):

$$b = (1 - 2t) a_1 + (2 - t) a_2 + t a_3, \quad t \in \mathbb{R}.$$

1.2.2 ▪ Matrix/vector multiplication

With this observation in mind, we now define the multiplication of a matrix and a vector so that the resultant corresponds to a linear system. For the linear system of (1.2.1), let A be the coefficient matrix,

$$A = (a_1 \ a_2 \ a_3) \in \mathcal{M}_3(\mathbb{R}).$$

Here, each column of A is thought of as a vector. If for

$$x = \begin{pmatrix} x_1 \\ x_2 \\ x_3 \end{pmatrix}$$

we define

$$Ax := x_1 a_1 + x_2 a_2 + x_3 a_3,$$

then by using (1.2.2) we have that the linear system is given by

$$Ax = b. \tag{1.2.4}$$

(Compare with (1.1.5).) In other words, by writing the linear system in the form of (1.2.4), we really mean the linear combinations of (1.2.2), which in turn is equivalent to the original system (1.2.1).

1.2. Vector algebra and matrix/vector multiplication

Matrix/vector multiplication

Definition 1.12. *Suppose that $A = (a_1 \ a_2 \ \ldots \ a_n)$, where each vector $a_j \in \mathbb{R}^m$ is an m-vector. For $x \in \mathbb{R}^n$, we define matrix/vector multiplication as*

$$Ax = x_1 a_1 + x_2 a_2 + \cdots + x_n a_n = \sum_{j=1}^{n} x_j a_j.$$

Note that $A \in \mathcal{M}_{m \times n}(\mathbb{R})$ and $x \in \mathbb{R}^n$, so by definition

$$\underbrace{A}_{\mathbb{R}^{m \times n}} \underbrace{x}_{\mathbb{R}^{n \times 1}} = \underbrace{b}_{\mathbb{R}^{m \times 1}}.$$

In order for a matrix/vector multiplication to make sense, the number of columns in the matrix A must the be same as the number of entries in the vector x. The product will be a vector in which the number of entries is equal to the number of rows in A.

Example 1.13. We have

$$\begin{pmatrix} 1 & 2 \\ 3 & 4 \end{pmatrix} \begin{pmatrix} -3 \\ 5 \end{pmatrix} = -3 \begin{pmatrix} 1 \\ 3 \end{pmatrix} + 5 \begin{pmatrix} 2 \\ 4 \end{pmatrix} = \begin{pmatrix} 7 \\ 11 \end{pmatrix}$$

and

$$\begin{pmatrix} 1 & 2 & 5 \\ 3 & 4 & 6 \end{pmatrix} \begin{pmatrix} 2 \\ -1 \\ 3 \end{pmatrix} = 2 \begin{pmatrix} 1 \\ 3 \end{pmatrix} - \begin{pmatrix} 2 \\ 4 \end{pmatrix} + 3 \begin{pmatrix} 5 \\ 6 \end{pmatrix} = \begin{pmatrix} 15 \\ 20 \end{pmatrix}.$$

Note that in the first example, a 2×2 matrix multiplied a 2×1 matrix in order to get a 2×1 matrix, whereas in the second example, a 2×3 matrix multiplied a 3×1 matrix in order to get a 2×1 matrix. ∎

The multiplication of a matrix and a vector is a *linear operation*, as it satisfies the property that the product of a matrix with a linear combination of vectors is the same thing as first taking the individual matrix/vector products and then taking the appropriate linear combination of the resultant two vectors:

Lemma 1.14. *If $A \in \mathcal{M}_{m \times n}(\mathbb{R})$ with $x, y \in \mathbb{R}^n$, then*

$$A(cx + dy) = cAx + dAy.$$

Proof. Writing $A = (a_1 \ a_2 \ \cdots \ a_n)$ and using the fact that

$$cx + dy = \begin{pmatrix} cx_1 + dy_1 \\ cx_2 + dy_2 \\ \vdots \\ cx_n + dy_n \end{pmatrix},$$

we have

$$\begin{aligned}
A(cx+dy) &= (cx_1+dy_1)a_1 + (cx_2+dy_2)a_2 + \cdots + (cx_n+dy_n)a_n \\
&= [cx_1a_1 + cx_2a_2 + \cdots + cx_na_n] + [dy_1a_1 + dy_2a_2 + \cdots dy_na_n] \\
&= c[x_1a_1 + x_2a_2 + \cdots + x_na_n] + d[y_1a_1 + y_2a_2 + \cdots + y_na_n] \\
&= cAx + dAy. \quad \square
\end{aligned}$$

Remark 1.15. *We are already familiar with linear operators, which are simply operators that satisfy the linearity property of Lemma 1.14, in other contexts. For example, if D represents differentiation, i.e., $D[f(t)] = f'(t)$, then we know from Calculus I that*

$$D[af(t) + bg(t)] = af'(t) + bg'(t) = aD[f(t)] + bD[g(t)].$$

Similarly, if $\mathscr{I}$ represents definite integration, e.g., $\mathscr{I}[f(t)] = \int_0^1 f(t)\,dt$, then we again know from Calculus I that

$$\mathscr{I}[af(t) + bg(t)] = a\int_0^1 f(t)\,dt + b\int_0^1 g(t)\,dt = a\mathscr{I}[f(t)] + b\mathscr{I}[g(t)].$$

While we will not explore this issue too deeply in this text, the implication of this fact is that much of what we study about the actions of matrices on the set of vectors also applies to operations such as differentiation and integration on the set of functions.

For a simple example of a nonlinear operator, *i.e., an operator that is not linear, consider $\mathscr{F}(x) = x^2$. We have*

$$\mathscr{F}(ax) = a^2x^2, \quad a\mathscr{F}(x) = ax^2.$$

These two quantities are clearly equal for all x if and only if $a \in \{0, 1\}$. Consequently, $\mathscr{F}$ cannot be a linear operator.

─────────── **Exercises** ───────────

Exercise 1.2.1. *For each of the problems below, compute the product Ax when it is well defined. If the product cannot be computed, explain why.*

(a) $A = \begin{pmatrix} 1 & -3 \\ -3 & 2 \end{pmatrix}$, $x = \begin{pmatrix} -4 \\ 2 \end{pmatrix}$

(b) $A = \begin{pmatrix} 1 & -2 & 5 \\ 2 & 0 & -3 \end{pmatrix}$, $x = \begin{pmatrix} 2 \\ -1 \\ 7 \end{pmatrix}$

(c) $A = \begin{pmatrix} 1 & -2 \\ 5 & 2 \\ 0 & -3 \end{pmatrix}$, $x = \begin{pmatrix} 2 \\ -1 \\ 7 \end{pmatrix}$

(d) $A = \begin{pmatrix} 2 & -1 & -3 \end{pmatrix}$, $x = \begin{pmatrix} 1 \\ 6 \\ -4 \end{pmatrix}$

1.2. Vector algebra and matrix/vector multiplication

Exercise 1.2.2. Let

$$a_1 = \begin{pmatrix} -1 \\ 2 \\ 1 \end{pmatrix}, \quad a_2 = \begin{pmatrix} 3 \\ 1 \\ 1 \end{pmatrix}, \quad a_3 = \begin{pmatrix} 1 \\ 5 \\ 3 \end{pmatrix}, \quad b = \begin{pmatrix} -3 \\ 1 \\ 5 \end{pmatrix}.$$

Is b a linear combination of a_1, a_2, a_3? If so, are the weights unique?

Exercise 1.2.3. Let

$$A = \begin{pmatrix} 2 & 5 \\ -3 & -1 \end{pmatrix}, \quad b = \begin{pmatrix} 5 \\ 6 \end{pmatrix}.$$

Is the linear system $Ax = b$ consistent? If so, what particular linear combination(s) of the columns of A give the vector b?

Exercise 1.2.4. Find all of the solutions to the homogeneous problem $Ax = 0$ when

(a) $A = \begin{pmatrix} 1 & -3 & 6 \\ 2 & 0 & 7 \end{pmatrix}$

(b) $A = \begin{pmatrix} 1 & -3 & -4 \\ -2 & 4 & -12 \\ 0 & 2 & -4 \end{pmatrix}$

(c) $A = \begin{pmatrix} 2 & 3 & 6 \\ -3 & 5 & -1 \\ 1 & -1 & 1 \end{pmatrix}$

Exercise 1.2.5. Let

$$A = \begin{pmatrix} 2 & -1 \\ -6 & 3 \end{pmatrix}, \quad b = \begin{pmatrix} b_1 \\ b_2 \end{pmatrix}.$$

Describe the set of all vectors b for which $Ax = b$ is consistent.

Exercise 1.2.6. Determine if each of the following statements is true or false. Provide an explanation for your answer.

(a) The homogeneous system $Ax = 0$ is consistent.

(b) If b is a linear combination of a_1, a_2, then there exist unique scalars x_1, x_2 such that $b = x_1 a_1 + x_2 a_2$.

(c) If $Ax = b$ is consistent, then b is a linear combination of the rows of A.

(d) A linear combination of five vectors in $\mathbb{R}^3$ produces a vector in $\mathbb{R}^5$.

(e) In order to compute Ax, the vector x must have the same number of entries as the number of rows in A.

1.3 ▪ Matrix algebra: Addition, subtraction, and multiplication

Now that we have defined vector algebra and matrix/vector multiplication, we briefly consider the algebra of matrices, in particular, addition, subtraction, and multiplication. Division will be discussed later in section 1.4. Just like for vectors, the addition and subtraction are straightforward, as is scalar multiplication. If we denote two matrices as $A = (a_{jk}) \in \mathcal{M}_{m \times n}(\mathbb{R})$ and $B = (b_{jk}) \in \mathcal{M}_{m \times n}(\mathbb{R})$, then it is the case that

$$A \pm B = (a_{jk} \pm b_{jk}), \quad cA = (ca_{jk}).$$

In other words, we add/subtract two matrices of the same size component by component, and if we multiply a matrix by a scalar, then we multiply each component by that scalar. This is exactly what we do in the addition/subtraction of vectors and the multiplication of a vector by a scalar. For example, if

$$A = \begin{pmatrix} 1 & 2 \\ -1 & -3 \end{pmatrix}, \quad B = \begin{pmatrix} 2 & 1 \\ 4 & 3 \end{pmatrix},$$

then

$$A + B = \begin{pmatrix} 3 & 3 \\ 3 & 0 \end{pmatrix}, \quad 3A = \begin{pmatrix} 3 & 6 \\ -3 & -9 \end{pmatrix}.$$

Regarding the multiplication of two matrices, we simply generalize the matrix/vector multiplication. For a given $A \in \mathcal{M}_{m \times n}(\mathbb{R})$, recall that for $b \in \mathbb{R}^n$,

$$Ab = b_1 a_1 + b_2 a_2 + \cdots + b_n a_n, \quad A = (a_1 \, a_2 \, \cdots \, a_n).$$

If $B = (b_1 \, b_2 \, \cdots \, b_\ell) \in \mathcal{M}_{n \times \ell}(\mathbb{R})$ (note that each column $b_j \in \mathbb{R}^n$), we then define the multiplication of A and B by

$$\underbrace{A}_{\mathcal{M}_{m \times n}(\mathbb{R})} \cdot \underbrace{B}_{\mathcal{M}_{n \times \ell}(\mathbb{R})} = \underbrace{(Ab_1 \, Ab_2 \, \cdots \, Ab_\ell)}_{\mathcal{M}_{m \times \ell}(\mathbb{R})}.$$

The number of columns of A must match the number of rows of B in order for the operation to make sense. Furthermore, the number of rows of the product is the number of rows of A, and the number of columns of the product is the number of columns of B. For example, if

$$A = \begin{pmatrix} 1 & 2 & 3 \\ -1 & -3 & 2 \end{pmatrix}, \quad B = \begin{pmatrix} 2 & 1 \\ 4 & 3 \\ 6 & 4 \end{pmatrix},$$

then

$$AB = \left(A \begin{pmatrix} 2 \\ 4 \\ 6 \end{pmatrix} \quad A \begin{pmatrix} 1 \\ 3 \\ 4 \end{pmatrix} \right) = \begin{pmatrix} 28 & 19 \\ -2 & -2 \end{pmatrix} \in \mathcal{M}_2(\mathbb{R}),$$

and

$$BA = \left(B \begin{pmatrix} 1 \\ -1 \end{pmatrix} \quad B \begin{pmatrix} 2 \\ -3 \end{pmatrix} \quad B \begin{pmatrix} 3 \\ 2 \end{pmatrix} \right) = \begin{pmatrix} 1 & 1 & 8 \\ 1 & -1 & 18 \\ 2 & 0 & 26 \end{pmatrix} \in \mathcal{M}_3(\mathbb{R}).$$

1.3. Matrix algebra: Addition, subtraction, and multiplication

As the above example illustrates, it may *not* necessarily be the case that $AB = BA$. In this example, changing the order of multiplication leads to resultant matrices of different sizes. However, even if the resultant matrices are the same size, they need not be the same. Suppose that

$$A = \begin{pmatrix} 1 & 2 \\ -1 & -3 \end{pmatrix}, \quad B = \begin{pmatrix} 2 & 1 \\ 4 & 3 \end{pmatrix}.$$

We have

$$AB = \begin{pmatrix} A\begin{pmatrix} 2 \\ 4 \end{pmatrix} & A\begin{pmatrix} 1 \\ 3 \end{pmatrix} \end{pmatrix} = \begin{pmatrix} 10 & 7 \\ -14 & -10 \end{pmatrix} \in \mathcal{M}_2(\mathbb{R})$$

and

$$BA = \begin{pmatrix} B\begin{pmatrix} 1 \\ -1 \end{pmatrix} & B\begin{pmatrix} 2 \\ -3 \end{pmatrix} \end{pmatrix} = \begin{pmatrix} 1 & 1 \\ 1 & -1 \end{pmatrix} \in \mathcal{M}_2(\mathbb{R}).$$

These are clearly not the same matrix. Thus, in general we cannot expect matrix multiplication to be commutative.

On the other hand, even though matrix multiplication is not necessarily commutative, it is associative, i.e.,

$$A(B + C) = AB + AC.$$

This fact follows from the fact that matrix/vector multiplication is a linear operation (recall Lemma 1.14) and the definition of matrix/matrix multiplication through matrix/vector multiplication. In particular, if we write

$$B = (b_1 \, b_2 \cdots b_\ell), \quad C = (c_1 \, c_2 \cdots c_\ell),$$

then on writing

$$B + C = (b_1 + c_1 \, b_2 + c_2 \cdots b_\ell + c_\ell),$$

we have

$$\begin{aligned} A(B + C) &= A(b_1 + c_1 \, b_2 + c_2 \cdots b_\ell + c_\ell) \\ &= (A(b_1 + c_1) \, A(b_2 + c_2) \cdots A(b_\ell + c_\ell)) \\ &= (Ab_1 + Ac_1 \, Ab_2 + Ac_2 \cdots Ab_\ell + Ac_\ell) \\ &= (Ab_1 \, Ab_2 \cdots Ab_\ell) + (Ac_1 \, Ac_2 \cdots Ac_\ell) \\ &= AB + AC. \end{aligned}$$

Indeed, while we will not discuss the details here, it is a fact that just like matrix/vector multiplication, matrix/matrix multiplication is a linear operation:

$$A(bB + cC) = bAB + cAC.$$

(See Exercise 1.3.1.)

There is a special matrix that plays the role of the scalar 1 in matrix multiplication: the identity matrix I_n. If $A \in \mathcal{M}_{m \times n}(\mathbb{R})$, then it is straightforward to check that

$$AI_n = A, \quad I_m A = A.$$

(See Exercise 1.3.4.) In particular, if $x \in \mathbb{R}^n$, then it is true that $I_n x = x$. For an explicit example of this fact, if $n = 3$,

$$\begin{pmatrix} 1 & 0 & 0 \\ 0 & 1 & 0 \\ 0 & 0 & 1 \end{pmatrix} \begin{pmatrix} x_1 \\ x_2 \\ x_3 \end{pmatrix} = x_1 \begin{pmatrix} 1 \\ 0 \\ 0 \end{pmatrix} + x_2 \begin{pmatrix} 0 \\ 1 \\ 0 \end{pmatrix} + x_3 \begin{pmatrix} 0 \\ 0 \\ 1 \end{pmatrix} = \begin{pmatrix} x_1 \\ x_2 \\ x_3 \end{pmatrix}.$$

---------- **Exercises** ----------

Exercise 1.3.1. *Let*

$$A = \begin{pmatrix} 1 & -2 & -5 \\ -2 & 3 & -7 \end{pmatrix}, \quad B = \begin{pmatrix} 2 & 0 \\ -2 & 3 \\ 1 & 5 \end{pmatrix}, \quad C = \begin{pmatrix} 1 & 2 \\ 5 & -4 \\ 3 & -1 \end{pmatrix}.$$

Compute the prescribed algebraic operation if it is well defined. If it cannot be done, explain why.

(a) $3B - 2C$

(b) $4A + 2B$

(c) AB

(d) CA

Exercise 1.3.2. *Suppose that $A \in \mathcal{M}_{m \times n}(\mathbb{R})$ and $B \in \mathcal{M}_{n \times k}(\mathbb{R})$ with $m \neq n$ and $n \neq k$ (i.e., neither matrix is square).*

(a) *What is the size of AB?*

(b) *Can m, k be chosen so that BA is well defined? If so, what is the size of BA?*

(c) *Is it possible for $AB = BA$? Explain.*

Exercise 1.3.3. *Suppose that $A \in \mathcal{M}_{m \times n}(\mathbb{R})$ and that $B, C \in \mathcal{M}_{n \times k}(\mathbb{R})$. Show that*

$$A(bB + cC) = bAB + cAC.$$

Exercise 1.3.4. *Suppose that $A \in \mathcal{M}_{m \times n}(\mathbb{R})$. Show that $AI_n = A$ and that $I_m A = A$.*

1.4 ▪ Matrix algebra: The inverse of a square matrix

We learned how to do matrix addition/subtraction and multiplication in section 1.3: how about matrix division? If such a thing exists, then we can (formally) write the solution to linear systems as

$$Ax = b \quad \leadsto \quad x = \frac{1}{A} b.$$

As currently written, this calculation makes no sense. However, using the analogy that $1/2$ is the unique number such that $1/2 \cdot 2 = 1$, we could define $1/A$ to be

1.4. Matrix algebra: The inverse of a square matrix

that matrix such that $1/A \cdot A = I_n$. It is not clear that for a given matrix A, the corresponding matrix $1/A$ must exist. Continuing with the analogy, there is no number $c \in \mathbb{R}$ such that $c \cdot 0 = 1$. Moreover, even if $1/A$ does exist, it is not at all clear as to how it should be computed.

When solving the linear system as above, we are implicitly assuming that

(a) a solution exists for any b, and

(b) the solution is unique.

As we will see in section 1.9.4, these two conditions can be satisfied only if the matrix is square (the linear system has the same number of equations as unknowns). Consequently, for the rest of the discussion, we will consider only square matrices, and we will call $1/A$ the *inverse* of a square matrix $A \in \mathcal{M}_n(\mathbb{R})$. If it exists, it will be denoted by A^{-1} (think $1/2 = 2^{-1}$), and it will have the property that

$$A^{-1}A = AA^{-1} = I_n. \tag{1.4.1}$$

Assuming that the inverse exists, it allows us to solve the linear system $Ax = b$ via a matrix/vector multiplication, namely,

$$Ax = b \rightsquigarrow A^{-1}Ax = A^{-1}b \rightsquigarrow I_n x = A^{-1}b \rightsquigarrow x = A^{-1}b.$$

Lemma 1.16. *Consider the linear system $Ax = b$, where $A \in \mathcal{M}_n(\mathbb{R})$ is invertible, i.e., A^{-1} exists. The solution to the linear system is given by*

$$x = A^{-1}b.$$

How do we compute the inverse? Denote $A^{-1} = (a_1^{-1}\ a_2^{-1}\ \cdots\ a_n^{-1})$ and let e_j denote the jth column of I_n, i.e., $I_n = (e_1\ e_2\ \cdots\ e_n)$. Using (1.4.1),

$$(e_1\ e_2\ \cdots\ e_n) = I_n = AA^{-1} = (Aa_1^{-1}\ Aa_2^{-1}\ \cdots\ Aa_n^{-1}).$$

Equating columns gives $Aa_j^{-1} = e_j$ for $j = 1, \ldots, n$. Thus, the jth column of A^{-1} is the solution to $Ax = e_j$. If the inverse exists, each of these linear systems must have a unique solution. This is possible if and only if the RREF of A is I_n (e.g., see the discussion starting in section 1.9.4). As for the augmented matrix,

$$(A|e_j) \xrightarrow{\text{RREF}} (I_n|a_j^{-1}), \quad j = 1, \ldots, n. \tag{1.4.2}$$

We now consider the collection of linear systems (1.4.2) through a different lens. First, consider a general collection of linear systems with the same coefficient matrix:

$$Ax_1 = b_1,\ Ax_2 = b_2,\ \ldots,\ Ax_m = b_m.$$

Using the definition of matrix/matrix multiplication, this collection of linear systems can be written more compactly as $AX = B$, where

$$X = (x_1\ x_2\ \cdots\ x_m), \quad B = (b_1\ b_2\ \cdots\ b_m).$$

Solving this new system is accomplished by forming the augmented matrix $(A|B)$ and then row reducing.

Now, (1.4.2) is equivalent to solving n linear systems:
$$Ax = e_j, \quad j = 1, \ldots, n.$$

Using the above, this collection of linear systems can be written more compactly as
$$AX = I_n.$$

Forming the augmented matrix $(A|I_n)$, we find the inverse of A via
$$(A|I_n) \xrightarrow{\text{RREF}} (I_n|A^{-1}).$$

Lemma 1.17. *The square matrix $A \in \mathcal{M}_n(\mathbb{R})$ is invertible if and only if the RREF of A is I_n. The inverse is computed via*
$$(A|I_n) \xrightarrow{\text{RREF}} (I_n|A^{-1}).$$

Example 1.18. Suppose that
$$A = \begin{pmatrix} 1 & 2 \\ 3 & 5 \end{pmatrix}, \quad b = \begin{pmatrix} 2 \\ -6 \end{pmatrix}.$$

We have
$$(A|I_2) \xrightarrow{\text{RREF}} \begin{pmatrix} 1 & 0 & -5 & 2 \\ 0 & 1 & 3 & -1 \end{pmatrix} \rightsquigarrow A^{-1} = \begin{pmatrix} -5 & 2 \\ 3 & -1 \end{pmatrix}.$$

Consequently, the solution to the linear system $Ax = b$ is given by
$$x = A^{-1}b = \begin{pmatrix} -22 \\ 12 \end{pmatrix}. \blacksquare$$

Example 1.19. Suppose that
$$A = \begin{pmatrix} 1 & 2 \\ 3 & 6 \end{pmatrix}.$$

We have
$$(A|I_2) \xrightarrow{\text{RREF}} \begin{pmatrix} 1 & 2 & 1 & 0 \\ 0 & 0 & -3 & 1 \end{pmatrix};$$

consequently, since the left-hand side of the augmented matrix cannot be row reduced to I_2, A^{-1} does not exist. $\blacksquare$

Example 1.20. Of course, the inverse can also be computed using SAGE. For example,
$$A = \begin{pmatrix} 1 & 2 & 3 \\ -1 & 2 & -3 \\ 5 & 6 & 7 \end{pmatrix} \rightsquigarrow A^{-1} = \frac{1}{8} \begin{pmatrix} -8 & -1 & 3 \\ 2 & 2 & 0 \\ 4 & -1 & -1 \end{pmatrix}. \blacksquare$$

Exercises

Exercise 1.4.1. *Find the inverse, if it exists, of the following matrices:*

(a) $\begin{pmatrix} 3 & 7 \\ -1 & 4 \end{pmatrix}$

(b) $\begin{pmatrix} -2 & 3 \\ 4 & -6 \end{pmatrix}$

(c) $\begin{pmatrix} 5 & 0 & 0 \\ 0 & 6 & 4 \\ 0 & -2 & -1 \end{pmatrix}$

Exercise 1.4.2. *Use A^{-1}, if it exists, to solve the linear system $Ax = b$. If A^{-1} does not exist, find all solutions to the system if it is consistent.*

(a) $A = \begin{pmatrix} 3 & 7 \\ -1 & 4 \end{pmatrix}$, $b = \begin{pmatrix} 5 \\ -6 \end{pmatrix}$

(b) $A = \begin{pmatrix} -2 & 3 \\ 4 & -6 \end{pmatrix}$, $b = \begin{pmatrix} -4 \\ 8 \end{pmatrix}$

(c) $A = \begin{pmatrix} 5 & 0 & 0 \\ 0 & 6 & 4 \\ 0 & -2 & -1 \end{pmatrix}$, $b = \begin{pmatrix} 3 \\ 1 \\ 9 \end{pmatrix}$

Exercise 1.4.3. *Let $A = \begin{pmatrix} 2 & -5 \\ 3 & 4 \end{pmatrix}$, $b_1 = \begin{pmatrix} 7 \\ -8 \end{pmatrix}$, $b_2 = \begin{pmatrix} 0 \\ 6 \end{pmatrix}$. Use A^{-1} to solve the systems $Ax = b_1$ and $Ax = b_2$.*

Exercise 1.4.4. *Suppose that $A, B \in \mathcal{M}_n(\mathbb{R})$ are invertible matrices. Show that*
$$(AB)^{-1} = B^{-1}A^{-1}.$$

Exercise 1.4.5. *Determine if each of the following statements is true or false. Provide an explanation for your answer.*

(a) *If A has a pivot in every row, then the matrix is invertible.*

(b) *If $Ax = b$ has a unique solution, then A is invertible.*

(c) *If $A, B \in \mathcal{M}_n(\mathbb{R})$ are invertible, then $(AB)^{-1} = A^{-1}B^{-1}$.*

(d) *If A is a square matrix whose RREF has one zero row, then A is invertible.*

(e) *If the RREF of A has no zero rows, then the matrix is invertible.*

1.5 ▪ The structure of the solution

We now show that we can decompose the solution to the consistent linear system,
$$Ax = b, \qquad (1.5.1)$$
into two distinct sets. The general solution will be a sum of a vector from each set.

1.5.1 • Homogeneous solutions

There is always a solution to a homogeneous system, as $A \cdot 0 = 0$.

A class of linear systems that are important to solve arises when $b = 0$:

> **Homogeneous systems**
>
> **Definition 1.21.** *A homogeneous linear system is given by $Ax = 0$.*

Homogeneous linear systems have the important property that linear combinations of solutions are solutions.

> **Lemma 1.22.** *Suppose that x_1, x_2 are two solutions to the homogeneous linear system $Ax = 0$. Then $x = c_1 x_1 + c_2 x_2$ for any $c_1, c_2 \in \mathbb{R}$ is also a solution to the homogeneous linear system.*

Proof. The result follows immediately from the linearity of matrix/vector multiplication (see Lemma 1.14). In particular, we have

$$A(c_1 x_1 + c_2 x_2) = c_1 A x_1 + c_2 A x_2 = c_1 0 + c_2 0 = 0. \quad \square$$

If $x_1, x_2, \ldots, x_k$ are solutions to a homogeneous linear system, then by Lemma 1.22, any finite linear combination of these solutions,

$$x_\text{h} = c_1 x_1 + c_2 x_2 + \cdots + c_k x_k, \qquad (1.5.2)$$

is also a solution to the homogeneous system. The subscript "h" denotes homogeneous.

A general homogeneous solution is not unique.

> **Homogeneous solution**
>
> **Definition 1.23.** *We will say that x_h is the (general) homogeneous solution if any solution to the homogeneous system can be written as in (1.5.2) for some values for the constants $c_1, c_2, \ldots, c_k$.*

When solving a homogeneous system, we row reduce the augmented matrix $(A|0)$. However, when doing the row reduction, it is enough to simply row reduce the matrix A. This is due to the fact that row reducing the augmented matrix $(A|0)$ yields no additional information, as the right-hand side of the augmented matrix will remain the zero vector no matter the particular row operation.

Example 1.24. Consider the homogeneous system $Ax = 0$, where

$$A = \begin{pmatrix} 2 & -3 \\ -4 & 6 \end{pmatrix}.$$

Recall that in order to solve the linear system, it is enough to put A into RREF. It is straightforward to check that

$$A \xrightarrow{\text{RREF}} \begin{pmatrix} 1 & -3/2 \\ 0 & 0 \end{pmatrix},$$

1.5. The structure of the solution

which provides the equivalent linear system:

$$x_1 - \frac{3}{2}x_2 = 0.$$

On setting $x_2 = t$, the homogeneous solution is

$$x_h = \begin{pmatrix} 3/2\,t \\ t \end{pmatrix} = t \begin{pmatrix} 3/2 \\ 1 \end{pmatrix}, \quad t \in \mathbb{R}. \quad \blacksquare$$

As we see in this example (and all following examples), there will be nontrivial (nonzero) homogeneous solutions if and only if there are free variables for the linear system. Moreover, the number of vectors needed to construct the homogeneous solution will be the number of free variables associated with the linear system.

Example 1.25. Consider the homogeneous system $Ax = 0$, where

$$A = \begin{pmatrix} 1 & -1 & 1 & 0 \\ -2 & 1 & -5 & -1 \\ 3 & -3 & 3 & 0 \end{pmatrix}.$$

Using Gaussian elimination,

$$A \xrightarrow{\text{RREF}} \begin{pmatrix} 1 & 0 & 4 & 1 \\ 0 & 1 & 3 & 1 \\ 0 & 0 & 0 & 0 \end{pmatrix},$$

which provides the equivalent linear system:

$$x_1 + 4x_3 + x_4 = 0, \quad x_2 + 3x_3 + x_4 = 0.$$

On setting $x_3 = s$, $x_4 = t$, the homogeneous solution is

$$x_h = \begin{pmatrix} -4s - t \\ -3s - t \\ s \\ t \end{pmatrix} = s \begin{pmatrix} -4 \\ -3 \\ 1 \\ 0 \end{pmatrix} + t \begin{pmatrix} -1 \\ -1 \\ 0 \\ 1 \end{pmatrix}, \quad s, t \in \mathbb{R}. \quad \blacksquare$$

Example 1.26. Consider the homogeneous linear system $Ax = 0$, where

$$A = \begin{pmatrix} 3 & 4 & 7 & -1 \\ 2 & 6 & 8 & -4 \\ -5 & 3 & -2 & -8 \\ 7 & -2 & 5 & 9 \end{pmatrix}.$$

Using SAGE to do the row reduction, we find

$$A \xrightarrow{\text{RREF}} \begin{pmatrix} 1 & 0 & 1 & 1 \\ 0 & 1 & 1 & -1 \\ 0 & 0 & 0 & 0 \\ 0 & 0 & 0 & 0 \end{pmatrix}.$$

The homogeneous linear system associated with the RREF of A is

$$x_1 + x_3 + x_4 = 0, \quad x_2 + x_3 - x_4 = 0.$$

The free variables are x_3 and x_4, so the homogeneous solution is

$$x_\text{h} = s \begin{pmatrix} -1 \\ 1 \\ 0 \\ 1 \end{pmatrix} + t \begin{pmatrix} -1 \\ -1 \\ 1 \\ 0 \end{pmatrix}, \quad s, t \in \mathbb{R}. \quad \blacksquare$$

1.5.2 • Particular solutions

Again consider the homogeneous equation $Ax = 0$, which is always consistent with homogeneous solution x_h. Let a *particular solution* to the nonhomogeneous problem (1.5.1) ($b \neq 0$) be designated as x_p. As a consequence of the linearity of matrix/vector multiplication, we have

$$A(x_\text{h} + x_\text{p}) = Ax_\text{h} + Ax_\text{p} = 0 + b = b.$$

In other words, the sum of homogeneous and particular solutions, $x_\text{h} + x_\text{p}$, is a solution to the linear system (1.5.1). Indeed, any solution can be written in such a manner simply by writing a solution x as $x = x_\text{h} + (x - x_\text{h})$ and designating $x_\text{p} = x - x_\text{h}$.

A variant of this result will be foundational for solving linear ordinary differential equations.

> **Theorem 1.27.** *All solutions to the linear system (1.5.1) are of the form*
>
> $$x = x_\text{h} + x_\text{p},$$
>
> *where the homogeneous solution x_h is independent of b and a particular solution x_p depends on b.*

Recall by Lemma 1.22 that if x_1 is a solution to a homogeneous linear system, then so is $c_1 x_1$ for any constant $c_1 \in \mathbb{R}$. This property does not hold for particular solutions. Indeed, since

$$A(cx_\text{p}) = cAx_\text{p} = cb,$$

we have that cx_p is a particular solution if and only if $c = 1$.

Example 1.28. Consider a linear system for which

$$A = \begin{pmatrix} 1 & 3 & 4 & -1 \\ -1 & 4 & 3 & -6 \\ 2 & -6 & -4 & 10 \\ 0 & 5 & 5 & -5 \end{pmatrix}, \quad b = \begin{pmatrix} -2 \\ -5 \\ 8 \\ -5 \end{pmatrix}.$$

On performing Gaussian elimination, the RREF of the augmented matrix is given by

$$(A|b) \xrightarrow{\text{RREF}} \left(\begin{array}{cccc|c} 1 & 0 & 1 & 2 & 1 \\ 0 & 1 & 1 & -1 & -1 \\ 0 & 0 & 0 & 0 & 0 \\ 0 & 0 & 0 & 0 & 0 \end{array} \right).$$

1.5. The structure of the solution

The original linear system is then equivalent to the system

$$x_1 + x_3 + 2x_4 = 1, \quad x_2 + x_3 - x_4 = -1. \tag{1.5.3}$$

The free variables for this system are x_3, x_4, so by setting $x_3 = s$ and $x_4 = t$, we get the solution to be

$$x = \begin{pmatrix} -s - 2t + 1 \\ -s + t - 1 \\ s \\ t \end{pmatrix} = s \begin{pmatrix} -1 \\ -1 \\ 1 \\ 0 \end{pmatrix} + t \begin{pmatrix} -2 \\ 1 \\ 0 \\ 1 \end{pmatrix} + \begin{pmatrix} 1 \\ -1 \\ 0 \\ 0 \end{pmatrix}.$$

The claim is that for the solution written in this form,

$$x_h = s \begin{pmatrix} -1 \\ -1 \\ 1 \\ 0 \end{pmatrix} + t \begin{pmatrix} -2 \\ 1 \\ 0 \\ 1 \end{pmatrix}, \quad x_p = \begin{pmatrix} 1 \\ -1 \\ 0 \\ 0 \end{pmatrix}.$$

It is easy to check that x_p is a particular solution:

$$Ax_p = A \begin{pmatrix} 1 \\ -1 \\ 0 \\ 0 \end{pmatrix} = \begin{pmatrix} 1 \\ -1 \\ 2 \\ 0 \end{pmatrix} - \begin{pmatrix} 3 \\ 4 \\ -6 \\ 5 \end{pmatrix} = \begin{pmatrix} -2 \\ -5 \\ 8 \\ -5 \end{pmatrix} = b.$$

Note that x_p is the last column of the RREF of the augmented matrix $(A|b)$. Similarly, in order to see that x_h is a homogeneous solution, use the linearity of matrix/vector multiplication and check that

$$Ax_h = A(s \begin{pmatrix} -1 \\ -1 \\ 1 \\ 0 \end{pmatrix} + t \begin{pmatrix} -2 \\ 1 \\ 0 \\ 1 \end{pmatrix}) = sA \begin{pmatrix} -1 \\ -1 \\ 1 \\ 0 \end{pmatrix} + tA \begin{pmatrix} -2 \\ 1 \\ 0 \\ 1 \end{pmatrix} = 0. \quad \blacksquare$$

Example 1.29. Consider the linear system $Ax = b$, where

$$A = \begin{pmatrix} 3 & 4 & -7 & 2 \\ 2 & 6 & 9 & -2 \\ -5 & 3 & 2 & -13 \\ 7 & -2 & 5 & 16 \end{pmatrix}, \quad b = \begin{pmatrix} 5 \\ 27 \\ 11 \\ -1 \end{pmatrix}.$$

We will use SAGE to assist in finding the homogeneous and particular solutions. Since

$$(A|b) \xrightarrow{\text{RREF}} \begin{pmatrix} 1 & 0 & 0 & 2 & | & 0 \\ 0 & 1 & 0 & -1 & | & 3 \\ 0 & 0 & 1 & 0 & | & 1 \\ 0 & 0 & 0 & 0 & | & 0 \end{pmatrix},$$

an equivalent linear system is

$$x_1 + 2x_4 = 0, \quad x_2 - x_4 = 3, \quad x_3 = 1.$$

The variable x_4 is the free variable, and the solution is given by

$$x = \begin{pmatrix} -2t \\ t+3 \\ 1 \\ t \end{pmatrix} = t \begin{pmatrix} -2 \\ 1 \\ 0 \\ 1 \end{pmatrix} + \begin{pmatrix} 0 \\ 3 \\ 1 \\ 0 \end{pmatrix}, \quad t \in \mathbb{R}.$$

The homogeneous solution is that with a free parameter:

$$x_h = t \begin{pmatrix} -2 \\ 1 \\ 0 \\ 1 \end{pmatrix}, \quad t \in \mathbb{R}.$$

A particular solution is what remains:

$$x_p = \begin{pmatrix} 0 \\ 3 \\ 1 \\ 0 \end{pmatrix}.$$

Note that the chosen particular solution is the last column of the RREF of $(A|b)$. ∎

Exercises

Exercise 1.5.1. *For each matrix A, find the homogeneous solution.*

(a) $A = \begin{pmatrix} 1 & 2 \\ 3 & 1 \end{pmatrix}$

(b) $A = \begin{pmatrix} 1 & 3 \\ -2 & -6 \end{pmatrix}$

(c) $A = \begin{pmatrix} 1 & 2 \\ 4 & 8 \end{pmatrix}$

(d) $A = \begin{pmatrix} 1 & 2 & 3 & -2 \\ 3 & 1 & 3 & 0 \end{pmatrix}$

(e) $A = \begin{pmatrix} 1 & 2 & 1 \\ 3 & 1 & 8 \\ -1 & -3 & 0 \end{pmatrix}$

(f) $A = \begin{pmatrix} 1 & 3 & 8 \\ -1 & 2 & 2 \\ 3 & -4 & -2 \end{pmatrix}$

(g) $A = \begin{pmatrix} -2 & 1 & -5 & -6 \\ 3 & -2 & 7 & 11 \\ 4 & 5 & 17 & -16 \end{pmatrix}$

Exercise 1.5.2. For each matrix A and vector b, write the solution to $Ax = b$ as $x = x_h + x_p$, where x_h is the homogeneous solution and x_p is a particular solution. Explicitly identify x_h and x_p.

(a) $A = \begin{pmatrix} 1 & 3 \\ -2 & -6 \end{pmatrix}$, $b = \begin{pmatrix} -4 \\ 8 \end{pmatrix}$

(b) $A = \begin{pmatrix} 1 & 3 & 8 \\ -1 & 2 & 2 \\ 3 & -4 & -2 \end{pmatrix}$, $b = \begin{pmatrix} 10 \\ 0 \\ 4 \end{pmatrix}$

(c) $A = \begin{pmatrix} -2 & 1 & -5 & -6 \\ 3 & -2 & 7 & 11 \\ 4 & 5 & 17 & -16 \end{pmatrix}$, $b = \begin{pmatrix} -8 \\ 13 \\ 2 \end{pmatrix}$

Exercise 1.5.3. Given the RREF of $(A|b)$, find the general solution. Identify the homogeneous solution, x_h, and particular solution, x_p.

(a) $\left(\begin{array}{cc|c} 1 & 0 & 7 \\ 0 & 1 & 5 \end{array} \right)$

(b) $\left(\begin{array}{ccc|c} 1 & 0 & 2 & -3 \\ 0 & 1 & -4 & 2 \\ 0 & 0 & 0 & 0 \end{array} \right)$

(c) $\left(\begin{array}{ccc|c} 0 & 1 & 0 & 4 \\ 0 & 0 & 1 & 6 \end{array} \right)$

(d) $\left(\begin{array}{cccc|c} 1 & 0 & -3 & 5 & 7 \\ 0 & 1 & 1 & -2 & 9 \end{array} \right)$

(e) $\left(\begin{array}{cccc|c} 1 & 0 & 0 & -6 & 2 \\ 0 & 0 & 1 & 2 & -5 \\ 0 & 0 & 0 & 0 & 0 \end{array} \right)$

Exercise 1.5.4. Suppose that $A \in \mathcal{M}_{m \times n}(\mathbb{R})$.

(a) Show that if $m < n$, then there are necessarily nontrivial homogeneous solutions.

(b) Give examples to show that if $m \geq n$, then there may or may not be nontrivial homogeneous solutions.

1.6 ▪ Sets of linear combinations of vectors

Now that we know how to do matrix algebra and algebraically solve linear systems using Gaussian elimination, we need to consider more carefully the geometry associated with linear systems of equations. The linear system

$$Ax = b, \quad A = (a_1 \, a_2 \, \cdots \, a_k)$$

can by the definition of matrix/vector multiplication be written as

$$x_1 a_1 + x_2 a_2 + \cdots + x_k a_k = b.$$

The linear system is consistent if and only if the vector b is some linear combination of the vectors $a_1, a_2, \ldots, a_k$. We first study the set of *all* linear combinations of these vectors. Once this set has been properly described, we will consider the problem of determining which (and how many) of the original set of vectors are needed in order to adequately describe it.

1.6.1 ▪ Span of a set of vectors

A particular linear combination of the vectors $a_1, a_2, \ldots, a_k$ is given by $x_1 a_1 + \cdots + x_k a_k$. The collection of all possible linear combinations of these vectors is known as the *span* of the vectors.

Span

Definition 1.30. *Let $S = \{a_1, a_2, \ldots, a_k\}$ be a set of n-vectors. The span of S,*

$$\mathrm{Span}(S) = \mathrm{Span}\{a_1, a_2, \ldots, a_k\},$$

is the collection of all linear combinations. In other words, $b \in \mathrm{Span}(S)$ if and only if for some $x \in \mathbb{R}^k$,

$$b = x_1 a_1 + x_2 a_2 + \cdots + x_k a_k.$$

The span of a collection of vectors has geometric meaning. First, suppose that $a_1 \in \mathbb{R}^3$. Recall that lines in $\mathbb{R}^3$ are defined parametrically by

$$r(t) = r_0 + t v,$$

where v is a vector parallel to the line and r_0 corresponds to a point on the line. Since

$$\mathrm{Span}\{a_1\} = \{t a_1 : t \in \mathbb{R}\},$$

this set is the line through the origin that is parallel to a_1.

Now suppose that $a_1, a_2 \in \mathbb{R}^3$ are not parallel, i.e., $a_2 \neq c a_1$ for some $c \in \mathbb{R}$. Set $v = a_1 \times a_2$, so v is a 3-vector that is perpendicular to both a_1 and a_2. The linearity of the dot product and the fact that $v \cdot a_1 = v \cdot a_2 = 0$ yields

$$v \cdot (x_1 a_1 + x_2 a_2) = x_1 v \cdot a_1 + x_2 v \cdot a_2 = 0.$$

Thus,

$$\mathrm{Span}\{a_1, a_2\} = \{x_1 a_1 + x_2 a_2 : x_1, x_2 \in \mathbb{R}\} \subset \{y \in \mathbb{R}^3 : y \cdot v = 0\}.$$

Conversely, the equality $y \cdot v = 0$ is a homogeneous linear equation in the three variables y_1, y_2, y_3:

$$v_1 y_1 + v_2 y_2 + v_3 y_3 = 0.$$

This homogeneous equation has two free variables, so a homogeneous solution will be a linear combination of two vectors that are not parallel. Without loss of generality, we can assume that those two vectors are a_1 and a_2. In conclusion, $\mathrm{Span}\{a_1, a_2\}$

1.6. Sets of linear combinations of vectors

is the collection of all vectors that are perpendicular to v, which means the set is the plane through the origin that is perpendicular to v. There are higher-dimensional analogues, but unfortunately they are difficult to visualize.

Now let us consider the computation that must be done in order to determine if $b \in \text{Span}(S)$. By definition, $b \in \text{Span}(S)$; i.e., b is a linear combination of the vectors $a_1, \ldots, a_k$, if and only if there exist constants $x_1, x_2, \ldots, x_k$ such that

$$x_1 a_1 + x_2 a_2 + \cdots + x_k a_k = b.$$

On setting

$$A = (a_1 \, a_2 \, \cdots \, a_k), \quad x = \begin{pmatrix} x_1 \\ x_2 \\ \vdots \\ x_k \end{pmatrix},$$

by using the Definition 1.12 of matrix/vector multiplication, we have that this condition is equivalent to solving the linear system $Ax = b$. This yields the following:

Lemma 1.31. *Suppose that $S = \{a_1, a_2, \ldots, a_k\}$ and set $A = (a_1 \, a_2 \, \cdots \, a_k)$. The vector $b \in \text{Span}(S)$ if and only if the linear system $Ax = b$ is consistent.*

Example 1.32. Letting

$$a_1 = \begin{pmatrix} 1 \\ 2 \end{pmatrix}, \quad a_2 = \begin{pmatrix} 1 \\ 1 \end{pmatrix}, \quad b = \begin{pmatrix} -1 \\ 2 \end{pmatrix},$$

let us determine if $b \in \text{Span}\{a_1, a_2\}$. As we have seen in Lemma 1.31, this question is equivalent to determining if the linear system $Ax = b$ is consistent. Since after Gaussian elimination

$$(A|b) \xrightarrow{\text{RREF}} \begin{pmatrix} 1 & 0 & \bigm| & 3 \\ 0 & 1 & \bigm| & -4 \end{pmatrix},$$

the linear system $Ax = b$ is equivalent to

$$x_1 = 3, \quad x_2 = -4,$$

which is easily solved. Thus, not only is $b \in \text{Span}\{a_1, a_2\}$, but it is the case that $b = 3a_1 - 4a_2$. ∎

Example 1.33. Letting

$$a_1 = \begin{pmatrix} 1 \\ 2 \\ -4 \end{pmatrix}, \quad a_2 = \begin{pmatrix} 3 \\ -1 \\ 5 \end{pmatrix}, \quad b = \begin{pmatrix} 7 \\ -7 \\ r \end{pmatrix},$$

let us determine those value(s) of r for which $b \in \text{Span}\{a_1, a_2\}$. As we have seen in Lemma 1.31, this question is equivalent to determining if the linear system $Ax = b$ is consistent. Since after Gaussian elimination

$$(A|b) \xrightarrow{\text{RREF}} \begin{pmatrix} 1 & 0 & \bigm| & -2 \\ 0 & 1 & \bigm| & 3 \\ 0 & 0 & \bigm| & r-23 \end{pmatrix},$$

the linear system is consistent if and only if $r = 23$. In this case, $x_1 = -2, x_2 = 3$, so that $b \in \text{Span}\{a_1, a_2\}$ with $b = -2a_1 + 3a_2$. ∎

Spanning set

Definition 1.34. *Let $S = \{a_1, a_2, \ldots, a_k\}$, where each vector $a_j \in \mathbb{R}^n$. We say that S is a spanning set for $\mathbb{R}^n$ if each $b \in \mathbb{R}^n$ is realized as a linear combination of the vectors in S:*
$$b = x_1 a_1 + x_2 a_2 + \cdots + x_k a_k.$$
In other words, S is a spanning set if the linear system,
$$Ax = b, \quad A = (a_1 \, a_2 \, \cdots \, a_k),$$
is consistent for any b.

Example 1.35. For
$$a_1 = \begin{pmatrix} 1 \\ 2 \\ 1 \end{pmatrix}, \quad a_2 = \begin{pmatrix} 3 \\ -4 \\ 2 \end{pmatrix}, \quad a_3 = \begin{pmatrix} 4 \\ -2 \\ 3 \end{pmatrix}, \quad a_4 = \begin{pmatrix} 4 \\ 7 \\ -5 \end{pmatrix},$$

let us determine if $S = \{a_1, a_2, a_3, a_4\}$ is a spanning set for $\mathbb{R}^3$. Using Lemma 1.31, we need to know if $Ax = b$ is consistent for *any* $b \in \mathbb{R}^3$, where $A = (a_1 \, a_2 \, a_3 \, a_4)$. In order for this to be the case, the RREF of the augmented matrix $(A|b)$ must always correspond to a consistent system; in particular, the coefficient side of the RREF of the augmented matrix must have no zero rows. Thus, in order to answer the question, it is sufficient to consider the RREF of A. Since

$$A \xrightarrow{\text{RREF}} \begin{pmatrix} 1 & 0 & 1 & 0 \\ 0 & 1 & 1 & 0 \\ 0 & 0 & 0 & 1 \end{pmatrix},$$

which has no zero rows, the linear system will always be consistent. The set S is a spanning set for $\mathbb{R}^3$. ∎

1.6.2 ▪ Linear independence of a set of vectors

We now consider the question of how many of the vectors $a_1, a_2, \ldots, a_k$ are needed to completely describe $\text{Span}(\{a_1, a_2, \ldots, a_k\})$. For example, let $S = \{a_1, a_2, a_3\}$, where

$$a_1 = \begin{pmatrix} 1 \\ -1 \\ 0 \end{pmatrix}, \quad a_2 = \begin{pmatrix} 1 \\ 0 \\ 1 \end{pmatrix}, \quad a_3 = \begin{pmatrix} 5 \\ -2 \\ 3 \end{pmatrix},$$

and consider $\text{Span}(S)$. If $b \in \text{Span}(S)$, then on using Definition 1.30, we know there exist constants x_1, x_2, x_3 such that

$$b = x_1 a_1 + x_2 a_2 + x_3 a_3.$$

1.6. Sets of linear combinations of vectors

Now, it can be checked that

$$a_3 = 2a_1 + 3a_2 \quad \rightsquigarrow \quad a_1 + 3a_2 - a_3 = 0, \tag{1.6.1}$$

so the vector a_3 is a linear combination of a_1 and a_2. The original linear combination can then be rewritten as

$$b = x_1 a_1 + x_2 a_2 + x_3(2a_1 + 3a_2) = (x_1 + 2x_3)a_1 + (x_2 + 3x_3)a_2.$$

In other words, the vector b is a linear combination of a_1 and a_2 alone. Thus, the addition of a_3 in the definition of $\text{Span}(S)$ is superfluous, so we can write

$$\text{Span}(S) = \text{Span}\{a_1, a_2\}.$$

Since $a_2 \neq c a_1$ for some $c \in \mathbb{R}$, we cannot reduce the collection of vectors comprising the spanning set any further.

We say that if some nontrivial linear combination of some set of vectors produces the zero vector, such as in (1.6.1), then the following holds:

In the preceding example, the set $\{a_1, a_2, a_3\}$ is linearly dependent, whereas the set $\{a_1, a_2\}$ is linearly independent.

Linear dependence

Definition 1.36. *The set of vectors $S = \{a_1, a_2, \ldots, a_k\}$ is linearly dependent if there is a nontrivial vector $x \neq 0 \in \mathbb{R}^k$ such that*

$$x_1 a_1 + x_2 a_2 + \cdots + x_k a_k = 0. \tag{1.6.2}$$

Otherwise, the set of vectors is linearly independent.

If the set of vectors is linearly dependent, then (at least) one vector in the collection can be written as a linear combination of the other vectors (again see (1.6.1)). In particular, two vectors will be linearly dependent if and only if one is a multiple of the other. An examination of (1.6.2) through the lens of matrix/vector multiplication reveals the left-hand side is Ax. Consequently, we determine if a set of vectors is linearly dependent or independent by solving the homogeneous linear system:

$$Ax = 0, \quad A = (a_1 \; a_2 \; \cdots \; a_k).$$

If there is a nontrivial solution, then the vectors will be linearly dependent; otherwise, they will be independent.

Lemma 1.37. *Let $S = \{a_1, a_2, \ldots, a_k\}$ be a set of n-vectors and set*

$$A = (a_1 \; a_2 \; \cdots \; a_k) \in \mathcal{M}_{n \times k}(\mathbb{R}).$$

The vectors are linearly dependent if and only if the linear system $Ax = 0$ has a nontrivial solution. Alternatively, the vectors are linearly independent if and only if the only solution to $Ax = 0$ is $x = 0$.

Recall that a homogeneous system will have nontrivial (nonzero) solutions to the homogeneous problem if and only if there are free variables. Moreover, we assign free variables to those columns that are not pivot columns. In particular, if all of the columns of A are pivot columns, then the vectors must be linearly independent. With this observation in mind, we can restate Lemma 1.37:

Corollary 1.38. *Let $S = \{a_1, a_2, \ldots, a_k\}$ be a set of n-vectors and set*

$$A = (a_1\ a_2\ \cdots\ a_k) \in \mathcal{M}_{n \times k}(\mathbb{R}).$$

The vectors are linearly independent if and only if all of the columns of A are pivot columns.

Example 1.39. Let

$$a_1 = \begin{pmatrix} 1 \\ 0 \\ 1 \end{pmatrix}, \quad a_2 = \begin{pmatrix} 3 \\ -1 \\ 4 \end{pmatrix}, \quad a_3 = \begin{pmatrix} -1 \\ 1 \\ -2 \end{pmatrix}, \quad a_4 = \begin{pmatrix} -3 \\ 3 \\ -2 \end{pmatrix},$$

and consider the sets

$$S_1 = \{a_1, a_2\}, \quad S_2 = \{a_1, a_2, a_3\}, \quad S_3 = \{a_1, a_2, a_3, a_4\}.$$

For each set of vectors, we wish to determine if they are linearly independent. If they are not, then we will write down a linear combination of the vectors that yields the zero vector.

Forming the augmented matrix and performing Gaussian elimination gives the RREF of each given matrix to be

$$A_1 = (a_1\ a_2) \xrightarrow{\text{RREF}} \begin{pmatrix} 1 & 0 \\ 0 & 1 \\ 0 & 0 \end{pmatrix}, \quad A_2 = (a_1\ a_2\ a_3) \xrightarrow{\text{RREF}} \begin{pmatrix} 1 & 0 & 2 \\ 0 & 1 & -1 \\ 0 & 0 & 0 \end{pmatrix},$$

and

$$A_3 = (a_1\ a_2\ a_3\ a_4) \xrightarrow{\text{RREF}} \begin{pmatrix} 1 & 0 & 2 & 0 \\ 0 & 1 & -1 & 0 \\ 0 & 0 & 0 & 1 \end{pmatrix}.$$

By Corollary 1.38, the vectors in S_1 are linearly independent. However, the same cannot be said for the latter two sets.

The homogeneous linear system associated with the RREF of A_2 is

$$x_1 + 2x_3 = 0, \quad x_2 - x_3 = 0.$$

Since x_3 is a free variable, a solution is

$$x_1 = -2t,\ x_2 = t,\ x_3 = t \quad \rightsquigarrow \quad x = \begin{pmatrix} -2 \\ 1 \\ 1 \end{pmatrix}, \quad t \in \mathbb{R}.$$

Using the definition of matrix/vector multiplication, we conclude the relationship:

$$0 = A_2 \begin{pmatrix} -2 \\ 1 \\ 1 \end{pmatrix} = -2a_1 + a_2 + a_3.$$

1.6. Sets of linear combinations of vectors

Moreover,

$$a_3 = 2a_1 - a_2 \rightsquigarrow \text{Span}\{a_1, a_2, a_3\} = \text{Span}\{a_1, a_2\}.$$

The homogeneous linear system associated with the RREF of A_3 is

$$x_1 + 2x_3 = 0, \quad x_2 - x_3 = 0, \quad x_4 = 0.$$

Since x_3 is still a free variable, a solution is

$$x_1 = -2t, \ x_2 = t, \ x_3 = t, \ x_4 = 0 \rightsquigarrow x = \begin{pmatrix} -2 \\ 1 \\ 1 \\ 0 \end{pmatrix}, \quad t \in \mathbb{R}.$$

Using the definition of matrix/vector multiplication, we conclude the relationship as before:

$$0 = A_3 \begin{pmatrix} -2 \\ 1 \\ 1 \\ 0 \end{pmatrix} = -2a_1 + a_2 + a_3.$$

Moreover,

$$a_3 = 2a_1 - a_2 \rightsquigarrow \text{Span}\{a_1, a_2, a_3, a_4\} = \text{Span}\{a_1, a_2, a_4\}. \blacksquare$$

Example 1.40. Suppose that $S = \{a_1, a_2, a_3, a_4, a_5\}$, where each $a_j \in \mathbb{R}^4$. Further suppose that the RREF of A is

$$A = (a_1 \ a_2 \ a_3 \ a_4 \ a_5) \xrightarrow{\text{RREF}} \begin{pmatrix} 1 & 0 & 2 & 0 & -3 \\ 0 & 1 & -1 & 1 & 2 \\ 0 & 0 & 0 & 0 & 0 \\ 0 & 0 & 0 & 0 & 0 \end{pmatrix}.$$

The first two columns of A are the pivot columns, and the remaining columns are associated with free variables. The homogeneous system associated with the RREF of A is

$$x_1 + 2x_3 - 3x_5 = 0, \quad x_2 - x_3 + x_4 + 2x_5 = 0.$$

Since x_3, x_4, x_5 are free variables, in vector form the solution to the homogeneous system is

$$x = r \begin{pmatrix} -2 \\ 1 \\ 1 \\ 0 \\ 0 \end{pmatrix} + s \begin{pmatrix} 0 \\ -1 \\ 0 \\ 1 \\ 0 \end{pmatrix} + t \begin{pmatrix} 3 \\ -2 \\ 0 \\ 0 \\ 1 \end{pmatrix} \quad r, s, t \in \mathbb{R}.$$

We then have the relationships

$$0 = A \begin{pmatrix} -2 \\ 1 \\ 1 \\ 0 \\ 0 \end{pmatrix} = -2a_1 + a_2 + a_3,$$

$$0 = A \begin{pmatrix} 0 \\ -1 \\ 0 \\ 1 \\ 0 \end{pmatrix} = -a_2 + a_4,$$

and

$$0 = A \begin{pmatrix} 3 \\ -2 \\ 0 \\ 0 \\ 1 \end{pmatrix} = 3a_1 - 2a_2 + a_5.$$

The last three vectors are each a linear combination of the first two,

$$a_3 = 2a_1 - a_2, \quad a_4 = a_2, \quad a_5 = -3a_1 + 2a_2,$$

so

$$\text{Span}\{a_1, a_2, a_3, a_4, a_5\} = \text{Span}\{a_1, a_2\}. \quad \blacksquare$$

1.6.3 • Linear independence of a set of functions

When discussing linear dependence, we can use Definition 1.36 in a more general sense. Suppose that $\{f_1, f_2, \ldots, f_k\}$ is a set of real-valued functions, each of which has at least $k-1$ continuous derivatives. We say that these functions are linearly dependent on the interval $a < t < b$ if there is a nontrivial *constant* vector $x \in \mathbb{R}^k$ such that

$$x_1 f_1(t) + x_2 f_2(t) + \cdots + x_k f_k(t) \equiv 0, \quad a < t < b.$$

How do we determine if this set of functions is linearly dependent? The problem is unlike the previous examples in that it is not clear how to formulate this problem as a homogeneous linear system.

We overcome this difficulty in the following manner. Since the linear combination of the functions is identically zero, it will be the case that a derivative of the linear combination will also be identically zero, i.e.,

$$x_1 f_1'(t) + x_2 f_2'(t) + \cdots + x_k f_k'(t) \equiv 0, \quad a < t < b.$$

We can take a derivative of the above to then get

$$x_1 f_1''(t) + x_2 f_2''(t) + \cdots + x_k f_k''(t) \equiv 0, \quad a < t < b,$$

and continuing in the fashion, we have for $j = 0, \ldots, k-1$,

$$x_1 f_1^{(j)}(t) + x_2 f_2^{(j)}(t) + \cdots + x_k f_k^{(j)}(t) \equiv 0, \quad a < t < b.$$

1.6. Sets of linear combinations of vectors

We have now derived a system of k linear equations that is given by

$$W(t)x \equiv 0, \quad W(t) := \begin{pmatrix} f_1(t) & f_2(t) & \cdots & f_k(t) \\ f_1'(t) & f_2'(t) & \cdots & f_k'(t) \\ f_1''(t) & f_2''(t) & \cdots & f_k''(t) \\ \vdots & \vdots & \vdots & \vdots \\ f_1^{(k-1)}(t) & f_2^{(k-1)}(t) & \cdots & f_k^{(k-1)}(t) \end{pmatrix}.$$

The matrix $W(t)$ is known as the *Wronskian* for the set of functions $\{f_1(t), f_2(t), \ldots, f_k(t)\}$.

We now see that the functions will be linearly dependent if there is a nontrivial vector x that does *not* depend on t, such that $W(t)x = 0$ for each $a < t < b$. This is a t-dependent homogeneous linear system. In order for there to be a nontrivial solution, the linear system must have at least one free variable for each t. Since the Wronskian is a square matrix, this means that if the functions are linearly dependent, then the RREF of $W(t)$ must not be I_k for each t. Conversely, the functions will be linearly independent if there is (at least) one value of t_0 such that the RREF of $W(t_0)$ is the identity matrix I_k.

Example 1.41. Consider the set $\{1, t, t^2, t^3\}$ on the interval $-\infty < t < +\infty$. The Wronskian associated with this set of functions is

$$W(t) = \begin{pmatrix} 1 & t & t^2 & t^3 \\ 0 & 1 & 2t & 3t^2 \\ 0 & 0 & 2 & 6t \\ 0 & 0 & 0 & 6 \end{pmatrix}.$$

It is clear that

$$W(0) = \begin{pmatrix} 1 & 0 & 0 & 0 \\ 0 & 1 & 0 & 0 \\ 0 & 0 & 2 & 0 \\ 0 & 0 & 0 & 6 \end{pmatrix} \xrightarrow{\text{RREF}} I_4,$$

so the set of functions is linearly independent. ∎

Example 1.42. Consider the set $\{\sin(t), \cos(t)\}$ on the interval $0 \leq t \leq 2\pi$. The Wronskian for this set of functions is

$$W(t) = \begin{pmatrix} \sin(t) & \cos(t) \\ \cos(t) & -\sin(t) \end{pmatrix}.$$

Since

$$W(\pi/2) = \begin{pmatrix} 1 & 0 \\ 0 & -1 \end{pmatrix} \xrightarrow{\text{RREF}} I_2,$$

the set of functions is linearly independent. ∎

Exercises

Exercise 1.6.1. *Determine if $b \in \text{Span}\{a_1, \ldots, a_\ell\}$ for the following vectors. If the answer is yes, give the linear combination(s) which makes it true.*

(a) $b = \begin{pmatrix} 1 \\ 3 \end{pmatrix}$, $a_1 = \begin{pmatrix} 2 \\ 3 \end{pmatrix}$, $a_2 = \begin{pmatrix} 3 \\ -5 \end{pmatrix}$

(b) $b = \begin{pmatrix} -2 \\ 5 \end{pmatrix}$, $a_1 = \begin{pmatrix} 4 \\ 3 \end{pmatrix}$, $a_2 = \begin{pmatrix} 2 \\ 1 \end{pmatrix}$

(c) $b = \begin{pmatrix} -5 \\ -4 \\ 15 \end{pmatrix}$, $a_1 = \begin{pmatrix} 1 \\ -1 \\ 6 \end{pmatrix}$, $a_2 = \begin{pmatrix} 2 \\ 1 \\ -3 \end{pmatrix}$, $a_3 = \begin{pmatrix} 4 \\ -1 \\ -9 \end{pmatrix}$

(d) $b = \begin{pmatrix} 1 \\ -2 \\ 4 \end{pmatrix}$, $a_1 = \begin{pmatrix} 1 \\ 3 \\ 0 \end{pmatrix}$, $a_2 = \begin{pmatrix} 3 \\ -1 \\ 5 \end{pmatrix}$, $a_3 = \begin{pmatrix} 1 \\ -1 \\ 1 \end{pmatrix}$

Exercise 1.6.2. *Find the equation of the line in $\mathbb{R}^2$ that corresponds to* Span $\{v_1\}$, *where*
$$v_1 = \begin{pmatrix} 2 \\ -5 \end{pmatrix}.$$

Exercise 1.6.3. *Find the equation of the plane in $\mathbb{R}^3$ that corresponds to* Span $\{v_1, v_2\}$, *where*
$$v_1 = \begin{pmatrix} 1 \\ -2 \\ -1 \end{pmatrix}, \quad v_2 = \begin{pmatrix} 3 \\ 0 \\ 4 \end{pmatrix}.$$

Exercise 1.6.4. *For the given set S, determine whether the set is linearly dependent or linearly independent.*

(a) $S = \{v_1, v_2\}$, *where*
$$v_1 = \begin{pmatrix} 3 \\ -4 \end{pmatrix}, \quad v_2 = \begin{pmatrix} 1 \\ 2 \end{pmatrix}.$$

(b) $S = \{v_1, v_2\}$, *where*
$$v_1 = \begin{pmatrix} 2 \\ -5 \end{pmatrix}, \quad v_2 = \begin{pmatrix} -4 \\ 10 \end{pmatrix}.$$

(c) $S = \{v_1, v_2, v_3\}$, *where*
$$v_1 = \begin{pmatrix} 2 \\ 3 \\ -1 \end{pmatrix}, \quad v_2 = \begin{pmatrix} 1 \\ 2 \\ -6 \end{pmatrix}, \quad v_3 = \begin{pmatrix} 9 \\ 15 \\ 7 \end{pmatrix}.$$

(d) $S = \{v_1, v_2, v_3, v_4\}$, *where*
$$v_1 = \begin{pmatrix} 1 \\ -2 \\ 0 \end{pmatrix}, \quad v_2 = \begin{pmatrix} 3 \\ -2 \\ 5 \end{pmatrix}, \quad v_3 = \begin{pmatrix} -7 \\ 1 \\ 4 \end{pmatrix}, \quad v_4 = \begin{pmatrix} 0 \\ -5 \\ 8 \end{pmatrix}.$$

1.6. Sets of linear combinations of vectors

Exercise 1.6.5. *Set*

$$v_1 = \begin{pmatrix} 1 \\ -2 \\ -5 \end{pmatrix}, \quad v_2 = \begin{pmatrix} -2 \\ 1 \\ 3 \end{pmatrix}, \quad v_3 = \begin{pmatrix} 8 \\ -7 \\ r \end{pmatrix}.$$

(a) *For which value(s) of r are the vectors v_1, v_2, v_3 linearly independent?*

(b) *For which value(s) of r is $v_3 \in \text{Span}\{v_1, v_2\}$?*

(c) *How are (a) and (b) related?*

Exercise 1.6.6. *Is the set of vectors*

$$S = \left\{ \begin{pmatrix} 2 \\ -1 \\ 4 \\ 6 \end{pmatrix}, \begin{pmatrix} 1 \\ -1 \\ 6 \\ 8 \end{pmatrix}, \begin{pmatrix} 0 \\ 3 \\ 2 \\ -5 \end{pmatrix}, \begin{pmatrix} -1 \\ 1 \\ 0 \\ 7 \end{pmatrix} \right\}$$

a spanning set for $\mathbb{R}^4$? Why or why not?

Exercise 1.6.7. *Determine if the set of vectors is linearly independent. If the answer is YES, give the weights for the linear combination that results in the zero vector.*

(a) $a_1 = \begin{pmatrix} 1 \\ -4 \end{pmatrix}, \quad a_2 = \begin{pmatrix} -3 \\ 12 \end{pmatrix}$

(b) $a_1 = \begin{pmatrix} 2 \\ 3 \end{pmatrix}, \quad a_2 = \begin{pmatrix} -1 \\ 5 \end{pmatrix}$

(c) $a_1 = \begin{pmatrix} 1 \\ 0 \\ 0 \end{pmatrix}, \quad a_2 = \begin{pmatrix} 3 \\ 2 \\ 3 \end{pmatrix}, \quad a_3 = \begin{pmatrix} 3 \\ 2 \\ 0 \end{pmatrix}$

(d) $a_1 = \begin{pmatrix} 1 \\ 3 \\ -2 \end{pmatrix}, \quad a_2 = \begin{pmatrix} -3 \\ -5 \\ 6 \end{pmatrix}, \quad a_3 = \begin{pmatrix} 0 \\ 5 \\ -6 \end{pmatrix}$

(e) $a_1 = \begin{pmatrix} 2 \\ -1 \\ 4 \end{pmatrix}, \quad a_2 = \begin{pmatrix} 3 \\ 4 \\ 2 \end{pmatrix}, \quad a_3 = \begin{pmatrix} 0 \\ -11 \\ 8 \end{pmatrix}$

Exercise 1.6.8. *Show that the following sets of functions are linearly independent:*

(a) $\{e^t, e^{2t}, e^{3t}\}$, *where* $-\infty < t < +\infty$

(b) $\{1, \cos(t), \sin(t)\}$, *where* $-\infty < t < +\infty$

(c) $\{e^t, te^t, t^2 e^t, t^3 e^t\}$, *where* $-\infty < t < +\infty$

(d) $\{1, t, t^2, \ldots, t^k\}$ *for any $k \geq 4$, where* $-\infty < t < +\infty$

(e) $\{e^{at}, e^{bt}\}$ *for $a \neq b$, where* $-\infty < t < +\infty$

Exercise 1.6.9. *Determine if each of the following statements is true or false. Provide an explanation for your answer.*

(a) *The span of any two nonzero vectors in $\mathbb{R}^3$ can be viewed as a plane through the origin in $\mathbb{R}^3$.*

(b) *If $Ax = b$ is consistent, then $b \in \text{Span}\{a_1, a_2, \ldots, a_n\}$ for $A = (a_1 \, a_2 \, \cdots \, a_n)$.*

(c) *The number of free variables for a linear system is the same as the number of pivot columns for the coefficient matrix.*

(d) *The span of a single nonzero vector in $\mathbb{R}^2$ can be viewed as a line through the origin in $\mathbb{R}^2$.*

(e) *If $A \in \mathcal{M}_{m \times n}(\mathbb{R})$ with $m > n$, then the columns of A must be linearly independent.*

(f) *If $A \in \mathcal{M}_{m \times n}(\mathbb{R})$ has a pivot in every column, then the columns of A span $\mathbb{R}^m$.*

(g) *If $A \in \mathcal{M}_{m \times n}(\mathbb{R})$ with $m \neq n$, then it is possible for the columns of A to both span $\mathbb{R}^m$ and be linearly independent.*

1.7 • Subspaces

Recall by Lemma 1.22 that solutions to the homogeneous linear system $Ax = 0$ satisfy the linearity property that if x_1, x_2 are homogeneous solutions, then so is $c_1 x_1 + c_2 x_2$ for any $c_1, c_2 \in \mathbb{R}$. We will label sets that have this linearity property as follows:

If S is a subspace, then $0 \in S$.

> **Subspace**
>
> **Definition 1.43.** *A nonempty set $S \subset \mathbb{R}^n$ is a subspace if*
>
> $$x_1, x_2 \in S \quad \rightsquigarrow \quad c_1 x_1 + c_2 x_2 \in S, \quad c_1, c_2 \in \mathbb{R}.$$

1.7.1 • Vector spaces

The set $\mathbb{R}^n$ is an example of a *vector space*. A *real* vector space, V, is a collection of elements, called vectors, on which are defined two operations: addition and scalar multiplication by real numbers. If $x, y \in V$, then $c_1 x + c_2 y \in V$ for any real scalars c_1, c_2. The following axioms must also be satisfied:

(a) commutativity of vector addition: $x + y = y + x$;

(b) associativity of vector addition: $(x + y) + z = x + (y + z)$;

(c) existence of an additive identity: there is a $0 \in V$ such that $x + 0 = x$;

(d) existence of an additive inverse: for each x, there is a y such that $x + y = 0$;

(e) existence of a multiplicative identity: $1 \cdot x = x$;

1.7. Subspaces

(f) first multiplicative distributive law: $c(x+y) = cx + cy$;

(g) second multiplicative distributive law: $(c_1 + c_2)x = c_1 x + c_2 x$; and

(h) relation to ordinary multiplication: $(c_1 c_2)x = c_1(c_2 x) = c_2(c_1 x)$.

Examples of vector spaces include

(a) the set of n-vectors, $\mathbb{R}^n$;

(b) the set of matrices $\mathcal{M}_{m \times n}(\mathbb{R})$; and

(c) the set of all polynomials of degree n.

1.7.2 ▪ Subspaces and span

The set of all vectors with complex-valued coefficients, $\mathbb{C}^n$, is a vector space (see section 1.11 if you are not familiar with complex numbers). In this case, the constants c_1 and c_2 are complex-valued.

Going back to subspaces of $\mathbb{R}^n$, which is all we will (primarily) be concerned with in this text, by using Definition 1.43, we see that the span of a collection of vectors is a subspace:

> **Lemma 1.44.** *The set $S = \text{Span}\{a_1, a_2, \ldots, a_k\}$ is a subspace.*

Proof. Suppose that $b_1, b_2 \in S$. By Lemma 1.31, there exist vectors x_1, x_2 such that for $A = (a_1 \, a_2 \, \cdots \, a_k)$,
$$A x_1 = b_1, \quad A x_2 = b_2.$$
We must now show that for the vector $b = c_1 b_1 + c_2 b_2$, there is a vector x such that $Ax = b$, as it will then be true that $b \in S$. However, if we choose $x = c_1 x_1 + c_2 x_2$, then by the linearity of matrix/vector multiplication, we have that
$$Ax = A(c_1 x_1 + c_2 x_2) = c_1 A x_1 + c_2 A x_2 = c_1 b_1 + c_2 b_2 = b. \quad \square$$

The converse also holds. It can be shown that any subspace of $\mathbb{R}^n$ is realized as the span of some *finite* collection of vectors in $\mathbb{R}^n$ (see Lemma 1.67). In other words, in the vector space $\mathbb{R}^n$, there are no other subspaces other than those given in Lemma 1.44.

> **Theorem 1.45.** *$S \subset \mathbb{R}^n$ is a subspace if and only if there is a collection of vectors $\{a_1, a_2, \ldots, a_k\}$ such that $S = \text{Span}\{a_1, a_2, \ldots, a_k\}$.*

Example 1.46. Suppose that
$$S = \left\{ \begin{pmatrix} x_1 + 2x_2 \\ -3x_2 \\ 4x_1 + x_2 \end{pmatrix} : x_1, x_2 \in \mathbb{R} \right\}.$$
Since
$$\begin{pmatrix} x_1 + 2x_2 \\ -3x_2 \\ 4x_1 + x_2 \end{pmatrix} = x_1 \begin{pmatrix} 1 \\ 0 \\ 4 \end{pmatrix} + x_2 \begin{pmatrix} 2 \\ -3 \\ 1 \end{pmatrix} = \text{Span}\left\{ \begin{pmatrix} 1 \\ 0 \\ 4 \end{pmatrix}, \begin{pmatrix} 2 \\ -3 \\ 1 \end{pmatrix} \right\},$$
by Lemma 1.44, the set is a subspace. ∎

Example 1.47. Suppose that

$$S = \left\{ \begin{pmatrix} x_1 + 2x_2 \\ 1 - 3x_2 \\ 4x_1 + x_2 \end{pmatrix} : x_1, x_2 \in \mathbb{R} \right\}.$$

We have that $b \in S$ if and only if

$$b = \begin{pmatrix} 0 \\ 1 \\ 0 \end{pmatrix} + x_1 \begin{pmatrix} 1 \\ 0 \\ 4 \end{pmatrix} + x_2 \begin{pmatrix} 2 \\ -3 \\ 1 \end{pmatrix}.$$

If S is a subspace, then it must contain the zero vector. Writing

$$0 = \begin{pmatrix} 0 \\ 1 \\ 0 \end{pmatrix} + x_1 \begin{pmatrix} 1 \\ 0 \\ 4 \end{pmatrix} + x_2 \begin{pmatrix} 2 \\ -3 \\ 1 \end{pmatrix} = \begin{pmatrix} 0 \\ 1 \\ 0 \end{pmatrix} + \begin{pmatrix} 1 & 2 \\ 0 & -3 \\ 4 & 1 \end{pmatrix} \begin{pmatrix} x_1 \\ x_2 \end{pmatrix},$$

on using

$$\begin{pmatrix} 1 & 2 & | & 0 \\ 0 & -3 & | & -1 \\ 4 & 1 & | & 0 \end{pmatrix} \xrightarrow{\text{RREF}} \begin{pmatrix} 1 & 0 & | & 0 \\ 0 & 1 & | & 0 \\ 0 & 0 & | & 1 \end{pmatrix},$$

we see the linear system is inconsistent. Since $0 \notin S$, the set is not a subspace. ∎

1.7.3 • The null space

Associated with a matrix, there will be two important subspaces. The first is the following:

Null(A) *is a nonempty set, as $A \cdot 0 = 0$ implies $\{0\} \subset$ Null(A).*

> **Null space**
>
> **Definition 1.48.** *The null space of A, denoted by* Null(A), *is the set of all solutions to a homogeneous linear system:*
>
> $$\text{Null}(A) := \{x : Ax = 0\}.$$

As we saw in Lemma 1.22, if $x_1, x_2 \in$ Null(A), then $c_1 x_1 + c_2 x_2 \in$ Null(A). Consequently, by Definition 1.43 the null space is a subspace. The vector space in which it lives is determined by the number of columns in the matrix.

> **Lemma 1.49.** *If $A \in \mathcal{M}_{m \times n}(\mathbb{R})$, then* Null($A$) $\subset \mathbb{R}^n$ *is a subspace.*

Example 1.50. Suppose that

$$\begin{pmatrix} -4 \\ 2 \\ 2 \end{pmatrix}, \begin{pmatrix} -1 \\ 4 \\ 0 \end{pmatrix} \in \text{Null}(A).$$

1.7. Subspaces

Since Null(A) is a subspace,

$$c_1\begin{pmatrix} -4 \\ 2 \\ 2 \end{pmatrix} + c_2\begin{pmatrix} -1 \\ 4 \\ 0 \end{pmatrix} \in \text{Null}(A),$$

so

$$\text{Span}\left\{\begin{pmatrix} -4 \\ 2 \\ 2 \end{pmatrix}, \begin{pmatrix} -1 \\ 4 \\ 0 \end{pmatrix}\right\} \subset \text{Null}(A). \blacksquare$$

Example 1.51. Let us find Null(A) for

$$A = \begin{pmatrix} 1 & 2 & 3 & -3 \\ 2 & 1 & 3 & 0 \\ 1 & -1 & 0 & 3 \\ -3 & 2 & -1 & -7 \end{pmatrix}.$$

We solve the homogeneous linear system, $Ax = 0$. Using SAGE to perform the row reduction,

$$A \xrightarrow{\text{RREF}} \begin{pmatrix} 1 & 0 & 1 & 1 \\ 0 & 1 & 1 & -2 \\ 0 & 0 & 0 & 0 \\ 0 & 0 & 0 & 0 \end{pmatrix},$$

so the equivalent linear system is

$$x_1 + x_3 + x_4 = 0, \quad x_2 + x_3 - 2x_4 = 0.$$

Setting $x_3 = s$, $x_4 = t$, the homogeneous solution is

$$x_\text{h} = \begin{pmatrix} -s-t \\ -s+2t \\ s \\ t \end{pmatrix} = s\begin{pmatrix} -1 \\ -1 \\ 1 \\ 0 \end{pmatrix} + t\begin{pmatrix} -1 \\ 2 \\ 0 \\ 1 \end{pmatrix}, \quad s, t \in \mathbb{R},$$

so

$$\text{Null}(A) = \text{Span}\left\{\begin{pmatrix} -1 \\ -1 \\ 1 \\ 0 \end{pmatrix}, \begin{pmatrix} -1 \\ 2 \\ 0 \\ 1 \end{pmatrix}\right\}. \blacksquare$$

1.7.4 · The column space

Another important example of a subspace that is directly associated with a matrix is the column space:

The column space is also known as the range of A, $R(A)$.

> **Column space**
>
> **Definition 1.52.** *The column space of a matrix A, $\text{Col}(A)$, is the set of all linear combinations of the columns of A.*

Setting $A = (a_1\ a_1\ \cdots\ a_k)$, we can rewrite the column space as

$$\mathrm{Col}(A) = \{x_1 a_1 + x_1 a_2 + \cdots + x_k a_k : x_1, x_2, \ldots, x_k \in \mathbb{R}\}$$
$$= \mathrm{Span}\{a_1, a_2, \ldots, a_k\}.$$

By Lemma 1.44, $\mathrm{Col}(A)$ is a subspace. Furthermore, if $b \in \mathrm{Col}(A)$, then for some weights $x_1, \ldots, x_k$,

$$b = x_1 a_1 + x_2 a_2 + \cdots + x_k a_k \quad \leadsto \quad b = Ax.$$

The formulation on the right follows from the definition of matrix/vector multiplication. This gives us the following:

Lemma 1.53. $\mathrm{Col}(A)$ *is a subspace, and the column space has the equivalent definition*

$$\mathrm{Col}(A) = \{b : Ax = b \text{ is consistent}\}.$$

Example 1.54. Suppose that

$$A = \begin{pmatrix} -2 & 3 & 5 \\ 1 & 5 & 4 \\ -3 & 1 & 4 \end{pmatrix}, \quad b = \begin{pmatrix} -3 \\ 8 \\ -8 \end{pmatrix}.$$

We wish to determine if $b \in \mathrm{Col}(A)$ and, if so, how can it be written as a linear combination of the columns of A. The first question is equivalent to asking if the linear system $Ax = b$ is consistent. Since

$$(A|b) \xrightarrow{\text{RREF}} \begin{pmatrix} 1 & 0 & -1 & | & 3 \\ 0 & 1 & 1 & | & 1 \\ 0 & 0 & 0 & | & 0 \end{pmatrix},$$

the linear system is consistent. As for the second question, since the solution to the linear system is

$$x = \begin{pmatrix} 3 \\ 1 \\ 0 \end{pmatrix} + t \begin{pmatrix} 1 \\ -1 \\ 1 \end{pmatrix}, \quad t \in \mathbb{R},$$

we have, e.g., $b = 3a_1 + a_2$. ∎

Exercises

Exercise 1.7.1. Set

$$S = \left\{ \begin{pmatrix} 2s - 3t \\ -s + 4t \\ 7t \end{pmatrix} : s, t \in \mathbb{R} \right\} \subset \mathbb{R}^3.$$

Is S a subspace? If so, determine vectors $x_1, x_2, \ldots, x_k$ such that $S = \mathrm{Span}\{x_1, x_2, \ldots, x_k\}$. Otherwise, explain why S is not a subspace.

1.7. Subspaces

Exercise 1.7.2. Set

$$S = \left\{ \begin{pmatrix} 4s+2t \\ 1-3s-t \\ s+9t \end{pmatrix} : s,t \in \mathbb{R} \right\} \subset \mathbb{R}^3.$$

Is S a subspace? If so, determine vectors $x_1, x_2, \ldots, x_k$ such that $S = \text{Span}\{x_1, x_2, \ldots, x_k\}$. Otherwise, explain why S is not a subspace.

Exercise 1.7.3. Find $\text{Null}(A)$ and write it as a span of a collection of vectors.

(a) $A = \begin{pmatrix} 3 & 1 & 1 \\ 4 & -2 & 8 \end{pmatrix}$

(b) $A = \begin{pmatrix} -2 & 3 & 5 \\ 1 & 5 & 4 \\ -3 & 1 & 4 \end{pmatrix}$

(c) $A = \begin{pmatrix} 1 & -2 & 1 \\ 0 & -1 & -1 \\ -3 & 4 & -5 \\ -1 & 5 & -2 \end{pmatrix}$

(d) $A = \begin{pmatrix} -2 & 3 & 1 & -8 \\ 5 & -1 & 4 & 7 \\ 6 & 2 & 8 & 2 \end{pmatrix}$

Exercise 1.7.4. If

$$A = \begin{pmatrix} -4 & 7 & 0 \\ 3 & 5 & 1 \end{pmatrix}, \quad b = \begin{pmatrix} -6 \\ 5 \end{pmatrix},$$

is $b \in \text{Col}(A)$? Explain your answer.

Exercise 1.7.5. If

$$A = \begin{pmatrix} 3 & -1 \\ 2 & 7 \\ -4 & 9 \end{pmatrix}, \quad b = \begin{pmatrix} 2 \\ 4 \\ -1 \end{pmatrix},$$

is $b \in \text{Col}(A)$? Explain your answer.

Exercise 1.7.6. If

$$A = \begin{pmatrix} 5 & 2 & -3 \\ -4 & 9 & 0 \\ 2 & 6 & -7 \end{pmatrix}, \quad b = \begin{pmatrix} -5 \\ 8 \\ 2 \end{pmatrix},$$

is $b \in \text{Col}(A)$? Explain your answer.

Exercise 1.7.7. Set

$$A = \begin{pmatrix} 1 & -3 \\ 2 & 5 \\ -1 & 4 \end{pmatrix}, \quad u = \begin{pmatrix} 2 \\ -4 \\ 7 \end{pmatrix}, \quad v = \begin{pmatrix} -3 \\ 16 \\ 5 \end{pmatrix}.$$

(a) *Is $u \in \text{Col}(A)$? Explain.*

(b) *Is $v \in \text{Col}(A)$? Explain.*

(c) *Describe all vectors that belong to $\text{Col}(A)$ as the span of a finite set of vectors.*

Exercise 1.7.8. *Suppose that $A \in \mathcal{M}_{m \times n}(\mathbb{R})$. For what relationship between m and n will it be necessarily true that*

(a) *$\text{Null}(A)$ is nontrivial and*

(b) *the columns of A do not span $\mathbb{R}^m$.*

Exercise 1.7.9. *Determine if each of the following statements is true or false. Provide an explanation for your answer.*

(a) *If $A \in \mathcal{M}_{3 \times 5}(\mathbb{R})$, it is possible that*

$$\text{Col}(A) = \left\{ \begin{pmatrix} 6s+t \\ s-2t \\ 4+3s \end{pmatrix} : s,t \in \mathbb{R} \right\} \subset \mathbb{R}^3.$$

(b) *It is possible for $\text{Null}(A)$ to be the empty set.*

(c) *If for some $A \in \mathcal{M}_{6 \times 3}(\mathbb{R})$,*

$$\begin{pmatrix} -4 \\ 5 \\ 9 \end{pmatrix} \in \text{Null}(A),$$

then the columns of A are linearly independent.

Exercise 1.7.10. *Let $A, B \subset \mathbb{R}^n$ be subspaces, and define*

$$A + B = \{x : x = a + b, \ a \in A, \ b \in B\}.$$

Show that $A + B$ is a subspace.

1.8 • Basis and dimension

1.8.1 • Basis

The next question to consider is the "size" of a subspace. The number that we assign to the "size" should reflect the intuition that a plane in $\mathbb{R}^3$ is bigger than a line in $\mathbb{R}^3$, and hence the number assigned to a plane should be larger than the number associated with the line. Regarding a given plane going through the origin in $\mathbb{R}^3$, while the geometric object itself is unique, there are many ways to describe it. For example, in Calculus we learned that it can be described as being the set of all vectors that are perpendicular to a certain vector. Conversely, we could describe it as the span of a collection of vectors that lie in the plane. The latter notion is the one that we will use, as it more easily generalizes to the spaces $\mathbb{R}^n$ for $n \geq 4$.

1.8. Basis and dimension

One way to determine the "size" of the subspace is to then count the number of spanning vectors. Because an arbitrarily high number of vectors could be used as the spanning set, in order to uniquely determine the size of the space, we must restrict the possible number of spanning vectors as much as possible. This restriction requires that we use only the *linearly independent* vectors (see Definition 1.36) in the spanning set. We first label these vectors:

Basis

Definition 1.55. *A set $B = \{a_1, a_2, \ldots, a_k\}$ is a basis for a subspace S if*

(a) *the vectors $a_1, a_2, \ldots, a_k$ are linearly independent, and*

(b) $S = \text{Span}\{a_1, a_2, \ldots, a_k\}$.

In other words, the set of vectors is a basis if

(a) any vector in S can be written as a linear combination of the basis vectors, and

(b) there are not so many vectors in the set that (at least) one of them can be written as a linear combination of the others.

Example 1.56. We wish to find a basis for the column space, $\text{Col}(A)$, of the matrix

$$A = \begin{pmatrix} -2 & 3 & -1 \\ 3 & -5 & 1 \\ 6 & -7 & 5 \end{pmatrix}.$$

Since

$$A \xrightarrow{\text{RREF}} \begin{pmatrix} 1 & 0 & 2 \\ 0 & 1 & 1 \\ 0 & 0 & 0 \end{pmatrix},$$

we have

$$\text{Null}(A) = \text{Span}\left\{ \begin{pmatrix} -2 \\ -1 \\ 1 \end{pmatrix} \right\} \rightsquigarrow a_3 = 2a_1 + a_2.$$

Consequently, regarding the column space, we have

$$\text{Col}(A) = \text{Span}\left\{ \begin{pmatrix} -2 \\ 3 \\ 6 \end{pmatrix}, \begin{pmatrix} 3 \\ -5 \\ -7 \end{pmatrix} \right\}.$$

These first two columns are clearly linearly independent. In conclusion, these two pivot columns are a basis for $\text{Col}(A)$. ∎

The previous example, as well as the several examples we did in section 1.6, point us toward a general truth. In all these examples, a set of linearly independent vectors in the set of column vectors, $\{a_1, a_2, \ldots, a_k\}$, which form a spanning set for $\text{Col}(A)$, were found by removing from the original set those columns that correspond to free variables for associated homogeneous linear systems. In other words, a set of linearly independent vectors that span the column space is the collection of pivot columns. In general, it is a relatively straightforward exercise (see Exercise 1.8.11) to show that each column of a matrix A that is not a pivot column can be written as a linear combination of the pivot columns. Thus, by the definition of basis, we have the following:

Lemma 1.57. *The pivot columns of $A \in \mathcal{M}_{m \times n}(\mathbb{R})$ form a basis for the column space, $\mathrm{Col}(A)$.*

Example 1.58. Let

$$a_1 = \begin{pmatrix} 1 \\ -1 \\ 2 \end{pmatrix}, \quad a_2 = \begin{pmatrix} 3 \\ 4 \\ 7 \end{pmatrix}, \quad a_3 = \begin{pmatrix} 0 \\ -7 \\ -1 \end{pmatrix}$$

and set $S = \{a_1, a_2, a_3\}$. It can be checked that

$$A = (a_1 \; a_2 \; a_3) \xrightarrow{\text{RREF}} \begin{pmatrix} 1 & 0 & 3 \\ 0 & 1 & -1 \\ 0 & 0 & 0 \end{pmatrix}.$$

The first and second columns of A are the pivot columns, so

$$\mathrm{Span}\{a_1, a_2, a_3\} = \mathrm{Span}\{a_1, a_2\}.$$

Since

$$\mathrm{Null}(A) = \mathrm{Span}\left\{ \begin{pmatrix} -3 \\ 1 \\ 1 \end{pmatrix} \right\},$$

we have for the remaining vector

$$-3a_1 + a_2 + a_3 = 0 \quad \rightsquigarrow \quad a_3 = 3a_1 - a_2. \quad \blacksquare$$

Example 1.59. Suppose that $S = \{a_1, a_2, a_3, a_4\}$, where each $a_j \in \mathbb{R}^5$. Further suppose that the RREF of A is

$$A = (a_1 \; a_2 \; a_3 \; a_4) \xrightarrow{\text{RREF}} \begin{pmatrix} 1 & 1 & 0 & 3 \\ 0 & 0 & 1 & 4 \\ 0 & 0 & 0 & 0 \\ 0 & 0 & 0 & 0 \\ 0 & 0 & 0 & 0 \end{pmatrix}.$$

The first and third columns of A are the pivot columns, so

$$\mathrm{Span}\{a_1, a_2, a_3, a_4\} = \mathrm{Span}\{a_1, a_3\}.$$

Since

$$\mathrm{Null}(A) = \mathrm{Span}\left\{ \begin{pmatrix} -1 \\ 1 \\ 0 \\ 0 \end{pmatrix}, \begin{pmatrix} -3 \\ 0 \\ -4 \\ 1 \end{pmatrix} \right\},$$

1.8. Basis and dimension

for the other two vectors, we have the relationships

$$a_2 = a_1, \quad a_4 = 3a_1 + 4a_3. \quad \blacksquare$$

A basis for a subspace is not unique. For example,

$$B_1 = \left\{ \begin{pmatrix} 1 \\ 0 \\ 0 \end{pmatrix}, \begin{pmatrix} 0 \\ 1 \\ 0 \end{pmatrix} \right\}, \quad B_2 = \left\{ \begin{pmatrix} 1 \\ 1 \\ 0 \end{pmatrix}, \begin{pmatrix} 1 \\ -1 \\ 0 \end{pmatrix} \right\},$$

are each a basis for the $x_1 x_2$-plane in $\mathbb{R}^3$. However, we do have the intuitive result that the *number* of basis vectors for a subspace is unique.

Lemma 1.60. *If $A = \{a_1, \ldots, a_k\}$ and $B = \{b_1, \ldots, b_m\}$ are two bases for a subspace S, then $k = m$. In other words, all bases for a subspace have the same number of vectors.*

Proof. The result is geometrically intuitive. The mathematical proof is as follows. Start by forming the matrices A and B via

$$A = (a_1 \, a_2 \, \cdots \, a_k), \quad B = (b_1 \, b_2 \, \cdots \, b_m).$$

The columns of each matrix are linearly independent, so by Lemma 1.37, the null space of each matrix is trivial,

$$\text{Null}(A) = \{0\}, \quad \text{Null}(B) = \{0\}.$$

Since A is a basis, each vector in B is a linear combination of the vectors in A; in particular, for each b_j, there is a vector c_j such that

$$b_j = Ac_j, \quad j = 1, \ldots, m.$$

If we set

$$C = (c_1 \, c_2 \, \cdots \, c_m) \in \mathcal{M}_{k \times m}(\mathbb{R}),$$

the matrices A and B are then related by

$$B = (b_1 \, b_2 \, \cdots \, b_m) = (Ac_1 \, Ac_2 \, \cdots \, Ac_m) = AC.$$

Suppose that $s \in S$. Since A and B are each a basis, there exist unique vectors x_A and x_B such that

$$s = Ax_A = Bx_B.$$

But the relation $B = AC$ implies that

$$Ax_A = (AC)x_B = A(Cx_B) \quad \leadsto \quad A(x_A - Cx_B) = 0.$$

The implicated equality follows from the linearity of matrix/matrix multiplication. Recalling that $\text{Null}(A) = \{0\}$, x_A and x_B are related via

$$A(x_A - Cx_B) = 0 \quad \leadsto \quad x_A - Cx_B = 0 \quad \leadsto \quad x_A = Cx_B.$$

Finally, consider the linear system $Cx = y$. For a given $y = x_A$, there is a unique solution $x = x_B$. This follows simply from the fact that for a given vector $s \in S$, there are the corresponding vectors x_A and x_B, and these are related through matrix/vector multiplication, $x_A = Cx_B$. Since the linear system always has a unique solution, the matrix C must be square. Thus, $k = m$, which is the desired result. □

1.8.2 ▪ Dimension and rank

Because the number of vectors in a basis of a subspace is fixed, this quantity gives a good way to describe the "size" of a subspace.

> **Dimension**
>
> **Definition 1.61.** *If $B = \{a_1, a_2, \ldots, a_k\}$ is a basis for a subspace S, then the dimension of S, $\dim[S]$, is the number of basis vectors:*
>
> $$\dim[S] = k.$$

$\dim[\{0\}] = 0$

Example 1.62. Let e_j for $j = 1, \ldots, n$ denote the jth column vector in the identity matrix I_n. Since I_n is in RREF, by Lemma 1.57, the pivot columns, $\{e_1, e_2, \ldots, e_n\}$, are linearly independent and form a basis for $\text{Col}(I_n)$. Since the matrix is square and the RREF has no zero rows, the columns form a spanning set for $\mathbb{R}^n$. In conclusion, the columns of the identity matrix form a basis for $\mathbb{R}^n$. By Definition 1.61, we then have the familiar result that $\dim[\mathbb{R}^n] = n$. ∎

Regarding the dimension of the column space, we use the following moniker:

> **Rank**
>
> **Definition 1.63.** *The dimension of the column space of a matrix is known as the rank:*
>
> $$\text{rank}(A) = \dim[\text{Col}(A)].$$

We now relate the column space to the null space through their dimensions. The pivot columns of the RREF of A are a basis for $\text{Col}(A)$ (see Lemma 1.57), so

$$\text{rank}(A) = \text{\# of pivot columns}.$$

In addition, the number of free variables is the number of linearly independent vectors that form a spanning set for $\text{Null}(A)$. Consequently, we can say that

$$\dim[\text{Null}(A)] = \text{\# of free variables}.$$

Since a column of the RREF of A either is a pivot column or is associated with a free variable (see Figure 1.2), on using the fact that the sum of the number of pivot columns and the number of free variables is the total number of columns, we get the following:

1.8. Basis and dimension

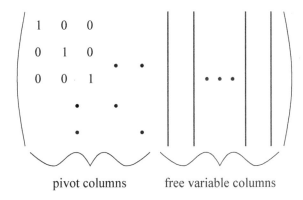

pivot columns   free variable columns

Figure 1.2. *A cartoon of a matrix in RREF. A vertical line represents a column associated with a free variable. The remaining columns are pivot columns.*

Lemma 1.64. *For the matrix $A \in \mathcal{M}_{m \times n}(\mathbb{R})$,*

$$\mathrm{rank}(A) + \dim[\mathrm{Null}(A)] = n.$$

Example 1.65. Suppose that

$$A = \begin{pmatrix} 1 & 3 & -2 & 1 \\ 1 & -1 & 2 & 1 \\ 3 & 4 & -1 & 1 \end{pmatrix} \xrightarrow{\mathrm{RREF}} \begin{pmatrix} 1 & 0 & 1 & 0 \\ 0 & 1 & -1 & 0 \\ 0 & 0 & 0 & 1 \end{pmatrix}.$$

Since the pivot columns are the first, second, and fourth columns of the RREF of A, a basis for $\mathrm{Col}(A)$ is given by the set

$$\left\{ \begin{pmatrix} 1 \\ 1 \\ 3 \end{pmatrix}, \begin{pmatrix} 3 \\ -1 \\ 4 \end{pmatrix}, \begin{pmatrix} 1 \\ 1 \\ 1 \end{pmatrix} \right\}.$$

For this matrix, $\mathrm{rank}(A) = 3$, and $\dim[\mathrm{Null}(A)] = 1 > 0$. ∎

Example 1.66. We now compute $\mathrm{Col}(A)$ and $\mathrm{Null}(A)$ with the assistance of SAGE. Here $A \in \mathcal{M}_4(\mathbb{R})$ is given by

$$A = \begin{pmatrix} 1 & 2 & 3 & 4 \\ -1 & 2 & 1 & 0 \\ 5 & 6 & 11 & 16 \\ 2 & 4 & 6 & 8 \end{pmatrix} \xrightarrow{\mathrm{RREF}} \begin{pmatrix} 1 & 0 & 1 & 2 \\ 0 & 1 & 1 & 1 \\ 0 & 0 & 0 & 0 \\ 0 & 0 & 0 & 0 \end{pmatrix}.$$

Since there are two pivot columns and two columns associated with free variables, $\mathrm{rank}(A) = 2$, and $\dim[\mathrm{Null}(A)] = 2$. A basis for $\mathrm{Col}(A)$ is the pivot columns of A, which are the first two columns:

$$\mathrm{Col}(A) = \mathrm{Span}\left\{ \begin{pmatrix} 1 \\ -1 \\ 5 \\ 2 \end{pmatrix}, \begin{pmatrix} 2 \\ 2 \\ 6 \\ 4 \end{pmatrix} \right\}.$$

As for the null space, the homogeneous linear system associated with the RREF of A is

$$x_1 + x_3 + 2x_4 = 0, \quad x_2 + x_3 + x_4 = 0.$$

Since x_3 and x_4 are the free variables, the homogeneous solution is

$$x_h = \begin{pmatrix} -s - 2t \\ -s - t \\ s \\ t \end{pmatrix} = s \begin{pmatrix} -1 \\ -1 \\ 1 \\ 0 \end{pmatrix} + t \begin{pmatrix} -2 \\ -1 \\ 0 \\ 1 \end{pmatrix}.$$

The null space is then

$$\text{Null}(A) = \text{Span} \left\{ \begin{pmatrix} -1 \\ -1 \\ 1 \\ 0 \end{pmatrix}, \begin{pmatrix} -2 \\ -1 \\ 0 \\ 1 \end{pmatrix} \right\}. \quad \blacksquare$$

Finally, we can now complete the proof of Theorem 1.45 and make the definitive statement that any subspace of $\mathbb{R}^n$ can be realized as the span of a finite set of vectors.

Lemma 1.67. *Let $S \subset \mathbb{R}^n$ be a subspace. For some $k \leq n$, there is a finite collection of linearly independent vectors, $a_1, a_2, \ldots, a_k$, such that $S = \text{Span}\{a_1, a_2, \ldots, a_k\}$.*

Proof. By Exercise 1.8.13, we can write $S = \text{Span}(S)$, so S is the span of some collection of vectors. We now need to provide an upper bound for the dimension of the subspace. Let $\{a_1, a_2, \ldots, a_k\} \subset S$ be a collection of linearly independent vectors. Set

$$A = (a_1 \; a_2 \; \cdots \; a_k) \in \mathcal{M}_{n \times k}(\mathbb{R}).$$

Using the result of Exercise 1.8.14 we have $\text{Col}(A) \subset S$. Since the columns are linearly independent, $\text{rank}(A) = k$. In order for every column to be a pivot column, we need $k \leq n$. In other words, the subspace can contain no more than n linearly independent vectors, so $\dim[S] \leq n$. The result follows. $\square$

Exercises

Exercise 1.8.1. *For each of the following matrices, not only find a basis for $\text{Col}(A)$ and $\text{Null}(A)$ but also determine $\text{rank}(A)$ and $\dim[\text{Null}(A)]$.*

(a) $A = \begin{pmatrix} 1 & 3 & 2 \\ 2 & 1 & 4 \\ 4 & 7 & 8 \end{pmatrix}$

(b) $A = \begin{pmatrix} -3 & 1 & 3 & 4 \\ 1 & 2 & -1 & -2 \\ -3 & 8 & 4 & 2 \end{pmatrix}$

1.8. Basis and dimension

(c) $A = \begin{pmatrix} 1 & 3 & -2 & 1 \\ 2 & 1 & 3 & 2 \\ 3 & 4 & 5 & 6 \end{pmatrix}$

Exercise 1.8.2. Set
$$A = \begin{pmatrix} 1 & -3 & -3 \\ 2 & 5 & -4 \end{pmatrix}, \quad v = \begin{pmatrix} 2 \\ -4 \\ 7 \end{pmatrix}.$$

(a) *Is $v \in \mathrm{Null}(A)$? Explain.*

(b) *Describe all vectors that belong to $\mathrm{Null}(A)$ as the span of a finite set of vectors.*

(c) *What is $\dim[\mathrm{Null}(A)]$?*

Exercise 1.8.3. *Suppose that $A \in \mathcal{M}_{7 \times 8}(\mathbb{R})$. If*

(a) *the RREF of A has two zero rows, what is $\mathrm{rank}(A)$?*

(b) *A has five pivot columns, what is $\dim[\mathrm{Null}(A)]$?*

Exercise 1.8.4. *Can a set of eight vectors be a basis for $\mathbb{R}^7$? Explain.*

Exercise 1.8.5. *Can a set of five vectors be a basis for $\mathbb{R}^6$? Explain.*

Exercise 1.8.6. *Is the set*
$$S = \left\{ \begin{pmatrix} 1 \\ 5 \end{pmatrix}, \begin{pmatrix} -2 \\ 3 \end{pmatrix} \right\}$$
a basis for $\mathbb{R}^2$? Explain. If not, find a basis for $\mathrm{Span}(S)$ and determine $\dim[\mathrm{Span}(S)]$.

Exercise 1.8.7. *Is the set*
$$S = \left\{ \begin{pmatrix} 1 \\ 5 \\ -2 \end{pmatrix}, \begin{pmatrix} -2 \\ 3 \\ 0 \end{pmatrix}, \begin{pmatrix} 3 \\ 1 \\ -5 \end{pmatrix} \right\}$$
a basis for $\mathbb{R}^3$? Explain. If not, find a basis for $\mathrm{Span}(S)$ and determine $\dim[\mathrm{Span}(S)]$.

Exercise 1.8.8. *Is the set*
$$S = \left\{ \begin{pmatrix} 1 \\ 1 \\ 3 \end{pmatrix}, \begin{pmatrix} -2 \\ 1 \\ 4 \end{pmatrix}, \begin{pmatrix} 4 \\ 1 \\ 2 \end{pmatrix} \right\}$$
a basis for $\mathbb{R}^3$? Explain. If not, find a basis for $\mathrm{Span}(S)$ and determine $\dim[\mathrm{Span}(S)]$.

Exercise 1.8.9. Set
$$A = \begin{pmatrix} 1 & 2 & 5 \\ -1 & 5 & 2 \\ 2 & -7 & -1 \end{pmatrix}, \quad b = \begin{pmatrix} 8 \\ 13 \\ -17 \end{pmatrix}.$$

(a) *Find a basis for* Col(A).

(b) *What is* rank(A)?

(c) *The vector* $b \in$ Col(A). *Write the vector as a linear combination of the basis vectors chosen in* (a).

Exercise 1.8.10. *Set*

$$A = \begin{pmatrix} 1 & -3 & -2 & 0 \\ 2 & -6 & 1 & 5 \\ -1 & 3 & 3 & 1 \\ -3 & 9 & 1 & -5 \end{pmatrix}, \quad b = \begin{pmatrix} -5 \\ 5 \\ 8 \\ 0 \end{pmatrix}.$$

(a) *Find a basis for* Col(A).

(b) *What is* rank(A)?

(c) *The vector* $b \in$ Col(A). *Write the vector as a linear combination of the basis vectors chosen in* (a).

Exercise 1.8.11. *Let* $S = \{a_1, a_2, \ldots, a_k\}$ *be a set of vectors and set* $A = (a_1 \ a_2 \ \cdots \ a_k)$.

(a) *Show that each column of A that is not a pivot column can be written as a linear combination of the pivot columns (Hint: Consider* Null(A).)

(b) *Prove Lemma 1.57.*

Exercise 1.8.12. *Let* $\{a_1, a_2, \ldots, a_k\} \subset \mathbb{R}^n$ *be a linearly independent set of vectors. Set* $S_j = \text{Span}\{a_1, a_2, \ldots, a_j\}$ *for* $j = 1, 2, \ldots, k$. *Show that*

(a) $\dim[S_j] = j$ *for each* $j = 1, 2, \ldots, k$ *and*

(b) $S_1 \subset S_2 \subset \cdots \subset S_k$.

Exercise 1.8.13. *Let* $S \subset \mathbb{R}^n$ *be a subspace. Show that* $S = \text{Span}(S)$, *where we define the span of a subspace to be the set of all finite linear combinations of vectors in S.*

Exercise 1.8.14. *Let* $\{a_1, a_2, \ldots, a_k\} \subset S$, *where* $S \subset \mathbb{R}^n$ *is a subspace. Show that*

$$\text{Span}\{a_1, a_2, \ldots, a_k\} \subset S.$$

Exercise 1.8.15. *Let* $S_1, S_2 \subseteq \mathbb{R}^n$ *be subspaces with* $S_1 \subseteq S_2$. *Show that* $\dim[S_1] \leq \dim[S_2]$. *(Hint: Find a basis for* S_1 *and* S_2.)

Exercise 1.8.16. *Determine if each of the following statements is true or false. Provide an explanation for your answer.*

(a) *If* $A \in \mathcal{M}_7(\mathbb{R})$ *is such that the RREF of A has two zero rows, then* rank(A) = 6.

(b) *Any set of seven linearly independent vectors is a basis for* $\mathbb{R}^7$.

(c) *If $A \in \mathcal{M}_{4\times 6}(\mathbb{R})$ is such that the RREF of A has one zero row, then* $\dim[\text{Null}(A)] = 4$.

(d) *If $A \in \mathcal{M}_9(\mathbb{R})$ is such that the RREF of A has six pivot columns,* $\dim[\text{Null}(A)] = 3$.

1.9 • Equivalence results

We now summarize the previous results. We wish to connect these results to the RREF of the appropriate matrix A. We break these results into four separate pieces.

1.9.1 • A solution exists

When we defined matrix/vector multiplication so that the linear system makes sense as $Ax = b$, we showed that the linear system is consistent if and only if for some scalars $x_1, x_2, \ldots, x_n \in \mathbb{R}$,

$$b = x_1 a_1 + x_2 a_2 + \cdots + x_n a_n, \quad A = (a_1 \, a_2 \, \cdots \, a_n).$$

Using Definition 1.30 for the span of a collection of vectors, the system is consistent if and only if

$$b \in \text{Span}\{a_1, a_2, \ldots, a_n\}.$$

On the other hand, we solve the system by using Gaussian elimination to put the augmented matrix $(A|b)$ into RREF. We know that the system is inconsistent if the RREF form of the augmented matrix has a row of the form $(0\,0\,0\,\cdots\,0|1)$; otherwise, it is consistent. These observations lead to the following equivalence result:

Theorem 1.68. *Regarding the linear system $Ax = b$, where $A = (a_1 \, a_2 \, \cdots \, a_n)$, the following are equivalent statements:*

(a) *The system is consistent.*

(b) *b is a linear combination of the columns of A.*

(c) *$b \in \text{Span}\{a_1, a_2, \ldots, a_n\}$.*

(d) *The RREF of the augmented matrix $(A|b)$ has no rows of the form $(0\,0\,0\,\cdots\,0|1)$.*

Example 1.69. Suppose that

$$A \xrightarrow{\text{RREF}} \begin{pmatrix} 1 & 0 & -5 \\ 0 & 1 & 3 \\ 0 & 0 & 0 \end{pmatrix}.$$

We have $\text{rank}(A) = 2$, and $\dim[\text{Null}(A)] = 1$. Since only the first two columns of A are pivot columns, $\{a_1, a_2\}$ is a basis for $\text{Col}(A)$. Since the RREF of A has a zero row (or $\text{rank}(A) = 2 < 3 = \dim[\mathbb{R}^3]$), the linear system will not always be consistent. Indeed, in order for the system to be consistent, it is necessary that $b \in \text{Span}(\{a_1, a_2\})$. Finally, if the linear system $Ax = b$ is consistent, solutions are not unique. ∎

Example 1.70. Suppose that $A \in \mathcal{M}_{3 \times 5}(\mathbb{R})$. Since A has more columns than rows, it is impossible for the RREF of A to have a pivot position in every column. Indeed, the linear system must have at least two free variables, and there can be no more than three pivot columns. Hence, the columns of A cannot be linearly independent. We cannot say that the columns form a spanning set for $\mathbb{R}^3$ without knowing something more about the RREF of A. If we are told that the RREF of A has two pivot positions (rank$(A) = 2$), then the RREF of A has one zero row; hence, the columns cannot form a spanning set. However, if we are told that the RREF of A has three pivot positions (the maximum number possible), then the RREF of A has no zero rows, which means that the columns do indeed form a spanning set. In any case, since dim[Null(A)] ≥ 2, there will be at least two free variables, so there will be an infinite number of solutions to any consistent linear system, $Ax = b$. ∎

1.9.2 ▪ A solution always exists

We now wish to refine Theorem 1.68 in order to determine criteria that guarantee that the linear system is consistent for *any* vector b. First, points (b) and (c) of Theorem 1.68 must be refined to say that for any $b \in \mathbb{R}^m$, $b \in \text{Span}\{a_1, \ldots, a_n\}$; in other words, Span$\{a_1, \ldots, a_n\} = \mathbb{R}^m$. Additionally, for a given b, no row of the RREF of the augmented matrix $(A|b)$ has row(s) of the form $(0\,0\,0\,\cdots\,0|0)$. Equivalently, the RREF of A must not have a zero row. This is possible if and only if there are at least as many columns as rows, $n \geq m$ (the linear system is underdetermined). The RREF of A will have no zero rows if and only if there are m pivot columns, which in turn is true if and only if rank$(A) = m$. Using Lemma 1.64, we know this condition is true if and only if dim[Null(A)] $= n - m$.

Theorem 1.71. *Consider the linear system* $Ax = b$, *where*

$$A = (a_1\, a_2\, \cdots\, a_n) \in \mathcal{M}_{m \times n}(\mathbb{R}), \quad b \in \mathbb{R}^m.$$

Assuming $n \geq m$, *the following are equivalent statements:*

(a) *The system is consistent for any* b,

(b) $\mathbb{R}^m = \text{Span}\{a_1, a_2, \ldots, a_n\}$,

(c) *the RREF of* A *has no zero rows,*

(d) rank$(A) = m$, *and*

(e) dim[Null(A)] $= n - m$.

Example 1.72. Suppose that the coefficient matrix satisfies

$$A = \begin{pmatrix} 1 & -4 & -2 \\ 3 & 1 & 7 \end{pmatrix} \xrightarrow{\text{RREF}} \begin{pmatrix} 1 & 0 & -2 \\ 0 & 1 & -1 \end{pmatrix}.$$

We have rank(A) = 2, and dim[Null(A)] = 1. By Theorem 1.71, the linear system $Ax = b$ is consistent for any $b \in \mathbb{R}^2$. ∎

1.9.3 • A solution is unique

From Theorem 1.27, we know that all solutions are given by $x = x_h + x_p$, where $x_h \in$ Null(A) is a homogeneous solution and x_p is a particular solution. Since $cx_h \in$ Null(A) for any $c \in \mathbb{R}$ (see Lemma 1.49), we know that if $x_h \neq 0$, then the linear system has an infinite number of solutions. Since Null(A) = {0}—equivalently, dim[Null(A)] = 0—if and only if the linear system has no free variables, a solution can be unique if and only if every column is a pivot column. This condition requires that the number of equations be at least as great as the number of variables, $m \geq n$ (the system is overdetermined). Now, following the discussion after Definition 1.36, we know that the columns of a matrix A are linearly independent if and only if the only solution to the homogeneous problem $Ax = 0$ is the trivial solution $x = 0$. In other words, the columns are linearly independent if and only if Null(A) = {0}. We can summarize our discussion with the following result:

> **Theorem 1.73.** *Consider the linear system $Ax = b$, where $A \in \mathcal{M}_{m \times n}(\mathbb{R})$. Assuming $n \geq m$, the following are equivalent statements:*
>
> (a) *there is at most one solution to the linear system $Ax = b$,*
>
> (b) *the linear system has no free variables,*
>
> (c) *every column of A is a pivot column,*
>
> (d) *the columns of A are linearly independent,*
>
> (e) Null(A) = {0},
>
> (f) rank(A) = n, *and*
>
> (g) dim[Null(A)] = 0.

Example 1.74. Suppose that $A \in \mathcal{M}_{6 \times 4}(\mathbb{R})$ has four pivot columns so that rank(A) = 4. Since that is the maximal number of pivot columns, by Theorem 1.73, the columns are a linearly independent set. Consequently, the solution to a consistent linear system, $Ax = b$, will be unique. The columns do not form a spanning set for $\mathbb{R}^6$, however, since the RREF of A will have two zero rows. ∎

1.9.4 • A unique solution always exists

We finally consider the problem of determining when there will always be a unique solution to the linear system. Let us first consider the size of the coefficient matrix for which this may be possible. By Theorem 1.71, it is possible for a solution to always exist if and only if the number of rows is no larger than the number of columns, $m \leq n$. On the other hand, by Theorem 1.73, consistent systems will have unique solutions if and only if the number of columns is no larger than the number of rows, $m \leq n$. Consequently, both conditions can be fulfilled if and only if $m = n$, i.e., the number of equations is the same as the number of variables.

Henceforth, assume that A is square. The RREF of A can have free variables if and only if the RREF of A has zero rows. If the RREF of A has no zero rows, then, since it is square,

(a) the RREF of A is the identity matrix, I_n, and

(b) the columns of A are linearly independent.

Going back to the definition of the determinant, we see that (a) is equivalent to $\det(A) \neq 0$. Since all the columns are pivot columns, $\text{rank}(A) = n$ and $\dim[\text{Null}(A)] = 0$. By Theorem 1.68, the lack of zero rows for the RREF of A means that the system $Ax = b$ is consistent for any b, and by Theorem 1.71, this lack of zero rows implies that the columns of A form a spanning set for $\mathbb{R}^n$.

Theorem 1.75. *Consider the linear system $Ax = b$, where $A \in \mathcal{M}_n(\mathbb{R})$. The following are equivalent statements:*

(a) *the linear system $Ax = b$ has a unique solution (given by $x = A^{-1}b$),*

(b) *the RREF of A is I_n,*

(c) *the columns of A are linearly independent,*

(d) *the columns of A form a spanning set for $\mathbb{R}^n$,*

(e) *the linear system has no free variables,*

(f) $\text{rank}(A) = n$,

(g) $\dim[\text{Null}(A)] = 0$, *and*

(h) $\det(A) \neq 0$.

On the other hand, if one of these conditions is not true, the following holds:

Theorem 1.76. *Consider the linear system $Ax = b$, where $A \in \mathcal{M}_n(\mathbb{R})$. The following are equivalent statements:*

(a) $\det(A) = 0$;

(b) *the inverse matrix A^{-1} does not exist;*

(c) $\dim[\text{Null}(A)] \geq 1$; *i.e., the linear system has free variables;*

(d) *the RREF of A has at least one zero row; and*

(e) $\text{rank}(A) \leq n - 1$.

Example 1.77. Suppose that

$$A = \begin{pmatrix} 1 & 2 & 3 \\ 4 & 5 & 9 \\ -1 & 4 & 3 \end{pmatrix} \xrightarrow{\text{RREF}} \begin{pmatrix} 1 & 0 & 1 \\ 0 & 1 & 1 \\ 0 & 0 & 0 \end{pmatrix}.$$

1.9. Equivalence results

We have rank$(A) = 2$, and dim[Null(A)] $= 1$. Since the RREF of A is not the identity, I_3, the linear system $Ax = b$ is not necessarily consistent. Moreover, if the system is consistent, the solution will not be unique. ∎

Exercises

Exercise 1.9.1. *Suppose that the RREF of $A \in \mathcal{M}_5(\mathbb{R})$ has one zero row.*

(a) *Is $Ax = b$ consistent for any $b \in \mathbb{R}^5$? Why or why not?*

(b) *If $Ax = b$ is consistent, how many solutions are there?*

Exercise 1.9.2. *Suppose that the RREF of $A \in \mathcal{M}_9(\mathbb{R})$ has seven pivot columns.*

(a) *Is $Ax = b$ consistent for any $b \in \mathbb{R}^9$? Why or why not?*

(b) *If $Ax = b$ is consistent, how many solutions are there?*

Exercise 1.9.3. *Suppose that $A \in \mathcal{M}_{5 \times 9}(\mathbb{R})$ and further suppose that the associated linear system for which A is the coefficient matrix has five free variables.*

(a) *Do the columns of A span $\mathbb{R}^5$? Explain.*

(b) *Are the columns of A linearly dependent or linearly independent, or is it not possible to say without more information? Explain.*

Exercise 1.9.4. *Suppose that $A \in \mathcal{M}_{7 \times 4}(\mathbb{R})$ and further suppose that the associated linear system for which A is the coefficient matrix has zero free variables.*

(a) *Do the columns of A span $\mathbb{R}^7$? Explain.*

(b) *Are the columns of A linearly dependent or linearly independent, or is it not possible to say without more information? Explain.*

Exercise 1.9.5. *Determine if each of the following statements is true or false. Provide an explanation for your answer.*

(a) *If $A \in \mathcal{M}_{5 \times 3}(\mathbb{R})$, then it is possible for the columns of A to span $\mathbb{R}^3$.*

(b) *If the RREF of $A \in \mathcal{M}_{9 \times 7}(\mathbb{R})$ has three zeros rows, then $Ax = b$ is consistent for any vector $b \in \mathbb{R}^9$.*

(c) *If $A \in \mathcal{M}_{5 \times 9}(\mathbb{R})$, then $Ax = b$ is consistent for any $b \in \mathbb{R}^5$.*

(d) *If the RREF of $A \in \mathcal{M}_{12 \times 16}(\mathbb{R})$ has 12 pivot columns, then $Ax = b$ is consistent for any $b \in \mathbb{R}^{12}$.*

(e) *If $Av_j = 0$ for $j = 1, 2$, then $x_1 v_1 + x_2 v_2 \in$ Null(A) for any $x_1, x_2 \in \mathbb{R}$.*

(f) *If $A \in \mathcal{M}_{5 \times 7}(\mathbb{R})$ is such that $Ax = b$ is consistent for every vector $b \in \mathbb{R}^5$, then the RREF of A has at least one zero row.*

(g) *If $A \in \mathcal{M}_{7 \times 6}(\mathbb{R})$, then $Ax = b$ is consistent for any $b \in \mathbb{R}^7$.*

1.10 • The determinant of a square matrix

We wish to derive a scalar that tells us whether a square matrix is invertible. First, suppose that $A \in \mathcal{M}_2(\mathbb{R})$ is given by

$$A = \begin{pmatrix} a & b \\ c & d \end{pmatrix}.$$

If we try to compute A^{-1}, we get

$$(A|I_2) \xrightarrow{-c\rho_1 + a\rho_2} \left(\begin{array}{cc|cc} a & b & 1 & 0 \\ 0 & ad-bc & -c & a \end{array} \right).$$

If $ad - bc \neq 0$, then we can continue with the row reduction and eventually compute A^{-1}; otherwise, A^{-1} does not exist. This fact implies that this quantity has special significance for 2×2 matrices.

Determinant

Definition 1.78. Let $A \in \mathcal{M}_2(\mathbb{R})$ be given by

$$A = \begin{pmatrix} a & b \\ c & d \end{pmatrix}.$$

The determinant of A, $\det(A)$, is given by

$$\det(A) = ad - bc.$$

We know that the matrix A has the RREF of I_2 if and only if $\det(A) \neq 0$. Continuing with the row reductions if $\det(A) \neq 0$ leads to the following:

Lemma 1.79. *Suppose that $A \in \mathcal{M}_2(\mathbb{R})$ is given by*

$$A = \begin{pmatrix} a & b \\ c & d \end{pmatrix}.$$

The matrix is invertible if and only if $\det(A) \neq 0$. Furthermore, if $\det(A) \neq 0$, then the inverse is given by

$$A^{-1} = \frac{1}{\det(A)} \begin{pmatrix} d & -b \\ -c & a \end{pmatrix}.$$

Proof. A simple calculation shows that $AA^{-1} = I_2$ if $\det(A) \neq 0$. □

Example 1.80. Suppose that

$$A = \begin{pmatrix} 4 & 7 \\ -3 & 2 \end{pmatrix}.$$

Since

$$\det(A) = (4)(2) - (7)(-3) = 29,$$

1.10. The determinant of a square matrix

the inverse of A exists, and it is given by

$$A^{-1} = \frac{1}{29}\begin{pmatrix} 2 & -7 \\ 3 & 4 \end{pmatrix}.$$

By Lemma 1.16, the unique solution to the linear system $Ax = b$ is given by $x = A^{-1}b$. ∎

Example 1.81. Suppose that

$$A = \begin{pmatrix} 4 & 1 \\ 8 & 2 \end{pmatrix}.$$

Since $\det(A) = 0$, the inverse of A does not exist. If there is a solution to $Ax = b$, it must be found by putting the augmented matrix $(A|b)$ into RREF and then solving the resultant system. ∎

We now wish to define the determinant for $A \in \mathcal{M}_n(\mathbb{R})$ for $n \geq 3$. In theory, we could derive it in a manner similar to that for the case $n = 2$: start with a matrix of a given size and then attempt to row reduce it to the identity. At some point, a scalar arises that must be nonzero in order to ensure that the RREF of the matrix is the identity. This scalar would then be denoted as the determinant. Instead of going through this derivation, we settle on the final result.

For $A \in \mathcal{M}_n(\mathbb{R})$, let $A_{ij} \in \mathcal{M}_{n-1}(\mathbb{R})$ denote the *submatrix* gotten from A after deleting the ith row and jth column. For example,

$$A = \begin{pmatrix} 1 & 4 & 7 \\ 2 & 5 & 8 \\ 3 & 6 & 9 \end{pmatrix} \rightsquigarrow A_{12} = \begin{pmatrix} 2 & 8 \\ 3 & 9 \end{pmatrix},\ A_{31} = \begin{pmatrix} 4 & 7 \\ 5 & 8 \end{pmatrix}.$$

With this notion of submatrix in mind, we note that for 2×2 matrices, the determinant can be written as

$$\det(A) = a_{11}\det(A_{11}) - a_{12}\det(A_{12}),$$

where here the determinant of a scalar is simply the scalar. The generalization to larger matrices is the following:

Determinant

Definition 1.82. *If $A \in \mathcal{M}_n(\mathbb{R})$, then the determinant of A is given by*

$$\det(A) = a_{11}\det(A_{11}) - a_{12}\det(A_{12}) + a_{13}\det(A_{13}) + \cdots + (-1)^{n+1}a_{1n}\det(A_{1n}).$$

The matrix is invertible if and only if $\det(A) \neq 0$.

Example 1.83. If

$$A = \begin{pmatrix} 1 & 4 & 7 \\ 2 & 5 & 8 \\ 3 & 6 & 9 \end{pmatrix},$$

then we have

$$A_{11} = \begin{pmatrix} 5 & 8 \\ 6 & 9 \end{pmatrix}, \quad A_{12} = \begin{pmatrix} 2 & 8 \\ 3 & 9 \end{pmatrix}, \quad A_{13} = \begin{pmatrix} 2 & 5 \\ 3 & 6 \end{pmatrix}.$$

Since

$$a_{11} = 1, \quad a_{12} = 4, \quad a_{13} = 7,$$

the determinant of the matrix is

$$\det(A) = 1 \cdot \det(A_{11}) - 4 \cdot \det(A_{12}) + 7 \cdot \det(A_{13}) = -3 + 24 - 21 = 0.$$

Thus, we know that A^{-1} does not exist; indeed, the RREF of A is

$$A \xrightarrow{\text{RREF}} \begin{pmatrix} 1 & 0 & -1 \\ 0 & 1 & 2 \\ 0 & 0 & 0 \end{pmatrix}. \quad \blacksquare$$

The determinant has many properties that are too many to detail in full here (e.g., see Eves [16, Chapter 3] and Vein and Dale [38]). We will consider only a small number that we will directly need. The first and perhaps the most important is that the expression of Definition 1.82 is not the only way to calculate the determinant. In general, the determinant can be calculated by going across any row or down any column; in particular, we have

$$\det(A) = \underbrace{\sum_{j=1}^{n} (-1)^{i+j} a_{ij} \det(A_{ij})}_{\text{across } i\text{th row}} = \underbrace{\sum_{i=1}^{n} (-1)^{i+j} a_{ij} \det(A_{ij})}_{\text{down } j\text{th column}}. \quad (1.10.1)$$

A judicious choice for the expansion of the determinant can greatly simplify the calculation. In particular, it is generally best to calculate the determinant using the row or column that has the most zeros. Note that if a matrix has a zero row or column, then by using the more general definition (1.10.1) and expanding across that zero row or column, we get $\det(A) = 0$.

Example 1.84. Suppose that

$$A = \begin{pmatrix} 4 & 3 & 6 \\ 2 & 0 & 0 \\ -1 & 7 & -5 \end{pmatrix}.$$

Going down the third column, we have

$$\det(A) = (6)\det\begin{pmatrix} 2 & 0 \\ -1 & 7 \end{pmatrix} - (0)\det\begin{pmatrix} 4 & 3 \\ -1 & 7 \end{pmatrix} + (-5)\det\begin{pmatrix} 4 & 3 \\ 2 & 0 \end{pmatrix}.$$

Going across the second row gives

$$\det(A) = -(2)\det\begin{pmatrix} 3 & 6 \\ 7 & -5 \end{pmatrix} + (0)\det\begin{pmatrix} 4 & 6 \\ -1 & -5 \end{pmatrix} - (0)\det\begin{pmatrix} 4 & 3 \\ -1 & 7 \end{pmatrix}.$$

In either case, $\det(A) = 104$. $\quad \blacksquare$

1.10. The determinant of a square matrix

A couple of other properties that may sometimes be useful are as follows. If a matrix B is formed from A by multiplying a row or column by a constant c, e.g., $A = (a_1 \, a_2 \, \cdots \, a_n)$ and $B = (c a_1 \, a_2 \, \cdots \, a_n)$, then $\det(B) = c \det(A)$. In particular, after multiplying each column by the same constant, i.e., multiplying the entire matrix by a constant,
$$\det(cA) = c^n \det(A).$$
(See Exercise 1.10.5.) Another useful property is that
$$\det(AB) = \det(A) \det(B).$$
Since $I_n A = A$, we get from this property that
$$\det(A) = \det(I_n) \det(A) = \det(I_n) \det(A) \quad \leadsto \quad \det(I_n) = 1.$$
(This could also be shown by a direct computation.) Since $AA^{-1} = I_n$, this also allows us to state that
$$1 = \det(I_n) = \det(AA^{-1}) = \det(A) \det(A^{-1}) \quad \leadsto \quad \det(A^{-1}) = \frac{1}{\det(A)}.$$
We summarize with the following:

> **Proposition 1.85.** *The determinant of matrices $A, B \in \mathcal{M}_n(\mathbb{R})$ has the properties:*
>
> (a) $\det(cA) = c^n \det(A)$,
>
> (b) $\det(AB) = \det(A) \det(B)$, *and*
>
> (c) *if A is invertible, $\det(A^{-1}) = 1/\det(A)$.*

━━━━━━━━━━━━━━━ **Exercises** ━━━━━━━━━━━━━━━

Exercise 1.10.1. *Compute by hand $\det(A)$ for each of the following matrices and then state whether or not the matrix is invertible. If the matrix is invertible, compute $\det(A^{-1})$.*

(a) $A = \begin{pmatrix} 3 & -2 \\ -2 & 3 \end{pmatrix}$

(b) $A = \begin{pmatrix} 1 & -3 & 4 \\ 1 & 2 & -1 \\ 3 & -5 & 8 \end{pmatrix}$

(c) $A = \begin{pmatrix} 1 & 2 & 3 \\ 0 & 4 & 0 \\ 2 & 8 & 5 \end{pmatrix}$

(d) $A = \begin{pmatrix} 1 & 2 & 0 & 5 \\ 2 & 4 & 0 & 6 \\ 0 & -3 & 0 & 5 \\ 6 & -1 & 2 & 4 \end{pmatrix}$

Exercise 1.10.2. *Suppose that*

$$A \xrightarrow{\text{RREF}} \begin{pmatrix} 1 & 0 & 3 \\ 0 & 1 & -2 \\ 0 & 0 & 0 \end{pmatrix}.$$

Is $\det(A) \neq 0$? *Explain.*

Exercise 1.10.3. *Suppose that* $A, B \in \mathcal{M}_n(\mathbb{R})$. *Show that the matrix product AB is invertible if and only if both A and B are invertible. (Hint: Use Proposition 1.85(b).)*

Exercise 1.10.4. *Suppose that* $D = \mathrm{diag}(\lambda_1, \lambda_2, \ldots, \lambda_n)$ *is a diagonal matrix, e.g.,*

$$\mathrm{diag}(\lambda_1, \lambda_2) = \begin{pmatrix} \lambda_1 & 0 \\ 0 & \lambda_2 \end{pmatrix}, \quad \mathrm{diag}(\lambda_1, \lambda_2, \lambda_3) = \begin{pmatrix} \lambda_1 & 0 & 0 \\ 0 & \lambda_2 & 0 \\ 0 & 0 & \lambda_3 \end{pmatrix}, \quad \text{etc.}$$

(a) *If $n = 2$, show that $\det(D) = \lambda_1 \lambda_2$.*

(b) *If $n = 3$, show that $\det(D) = \lambda_1 \lambda_2 \lambda_3$.*

(c) *Show that for any n,*

$$\det(D) = \prod_{j=1}^{n} \lambda_j.$$

Exercise 1.10.5. *Here we generalize the result of Proposition 1.85(a). For a matrix $A = (a_1\, a_2\, a_3\, \cdots\, a_n) \in \mathcal{M}_n(\mathbb{R})$, let B be defined as $B = (c_1 a_1\, c_2 a_2\, c_3 a_3\, \cdots\, c_n a_n)$.*

(a) *If $n = 2$, show that $\det(B) = c_1 c_2 \det(A)$.*

(b) *Show that $B = AC$, where $C = \mathrm{diag}(c_1, c_2, \ldots, c_n)$ is a diagonal matrix.*

(c) *Show that for $n \geq 3$,*

$$\det(B) = \left(\prod_{j=1}^{n} c_j \right) \det(A).$$

(Hint: Use Proposition 1.85(b) and Exercise 1.10.4.)

Exercise 1.10.6. *A matrix $V \in \mathcal{M}_3(\mathbb{R})$ is said to be a Vandermonde matrix if*

$$V = \begin{pmatrix} 1 & a & a^2 \\ 1 & b & b^2 \\ 1 & c & c^2 \end{pmatrix}.$$

(a) *Show that $\det(V) = (b-a)(c-a)(c-b)$.*

(b) *What conditions must the scalars a, b, c satisfy in order that V be invertible?*

Exercise 1.10.7. *Suppose that*

$$A(\lambda) = \begin{pmatrix} 3-\lambda & -2 \\ -2 & 3-\lambda \end{pmatrix}.$$

For which value(s) of λ does the system $A(\lambda)x = 0$ have a nontrivial solution? For one such value of λ, compute a corresponding nontrivial solution.

Exercise 1.10.8. *Suppose that $A \in \mathcal{M}_n(\mathbb{R})$ is such that $Ax = 0$ has infinitely many solutions. What can be said about $\det(A)$? Explain.*

Exercise 1.10.9. *Determine if each of the following statements is true or false. Provide an explanation for your answer.*

(a) *If $A \in \mathcal{M}_n(\mathbb{R})$ has a pivot position in every row, then $\det(A) = 0$.*

(b) *If $A \in \mathcal{M}_n(\mathbb{R})$ and $Ax = b$ has a unique solution for any b, then $\det(A) = 0$.*

(c) *If $A \in \mathcal{M}_n(\mathbb{R})$ is a diagonal matrix, then $\det(A)$ is the product of the diagonal entries.*

(d) *If the RREF of $A \in \mathcal{M}_n(\mathbb{R})$ has one zero row, then $\det(A) \neq 0$.*

1.11 • Linear algebra with complex-valued numbers, vectors, and matrices

Before we proceed to our last topic on matrices, we will need to understand the basics associated with complex-valued numbers and the associated algebraic manipulations. As we will see, these will be naturally encountered in future calculations with square matrices, even if the matrix in question contains only real-valued entries.

We say that $z \in \mathbb{C}$ if $z = a + ib$, where $a, b \in \mathbb{R}$, and $i^2 = -1$. The number a is the *real part* of the complex number and is sometimes denoted by $\text{Re}(z)$, i.e., $\text{Re}(z) = a$. The number b is the *imaginary part* of the complex number and is sometimes denoted by $\text{Im}(z)$, i.e., $\text{Im}(z) = b$. We say a vector $v \in \mathbb{C}^n$ if each entry is complex valued, and we will often write

$$v = p + iq, \quad p, q \in \mathbb{R}^n.$$

The vector p is the real part of v, i.e., $\text{Re}(v) = p$, and the vector q is the imaginary part, i.e., $\text{Im}(v) = q$. For example,

$$\begin{pmatrix} 1 - i5 \\ 2 + i7 \end{pmatrix} = \begin{pmatrix} 1 \\ 2 \end{pmatrix} + i \begin{pmatrix} -5 \\ 7 \end{pmatrix} \quad \leadsto \quad p = \begin{pmatrix} 1 \\ 2 \end{pmatrix}, q = \begin{pmatrix} -5 \\ 7 \end{pmatrix}.$$

We say that a matrix $A \in \mathcal{M}_{m \times n}(\mathbb{C})$ if each entry of the matrix is (possibly) complex valued.

The addition/subtraction of two complex numbers is as expected: add/subtract the real parts and imaginary parts. For example,

$$(2 - i3) + (3 + i2) = (2 + 3) + i(-3 + 2) = 5 - i.$$

As for multiplication, we multiply products of sums in the usual way and use the fact that $i^2 = -1$, e.g.,

$$(2 - i3)(3 + i2) = (2)(3) + (-i3)(3) + (2)(i2) + (-i3)(i2)$$
$$= 6 - i9 + i4 - i^2 6 = 12 - i5.$$

In particular, note that
$$c(a+ib) = ac + ibc;$$
i.e., multiplication of a complex number by a real number gives a complex number in which the both real and the imaginary parts of the original number are multiplied by the real number. For example,
$$7(-4+i9) = -28 + i63.$$

Before we can consider the problem of division, we must first think about the size of a complex number. The *complex conjugate* of a complex number z, which is denoted by $\bar{z}$, is given by taking the negative of the imaginary part, i.e.,
$$z = a + ib \rightsquigarrow \bar{z} = a - ib.$$
If the number is real valued, then $\bar{z} = z$. The complex conjugate of a vector $v \in \mathbb{C}^n$ is given by
$$\bar{v} = \overline{p + iq} = p - iq.$$
The complex conjugate of a matrix A is written as $\bar{A}$, and the definition is what is to be expected. If $A = (a_{jk})$, then $\bar{A} = (\overline{a_{jk}})$. For example,
$$A = \begin{pmatrix} 2-i5 & 3 \\ 1+i7 & -3+i5 \end{pmatrix} \rightsquigarrow \bar{A} = \begin{pmatrix} 2+i5 & 3 \\ 1-i7 & -3-i5 \end{pmatrix}.$$

Regarding the conjugate, it is not difficult to check that
$$\overline{z_1 z_2} = \bar{z}_1 \bar{z}_2;$$
i.e., the conjugate of a product is the product of the conjugates (see Exercise 1.11.1(a)). We further have
$$z\bar{z} = (a+ib)(a-ib) = a^2 + b^2 > 0,$$
and using this fact, we say that the *magnitude* (*absolute value*) of a complex number is
$$|z| = \sqrt{z\bar{z}} = \sqrt{a^2 + b^2}.$$
It is not difficult to check that
$$|z_1 z_2| = |z_1||z_2|;$$
i.e., the magnitude of a product is the product of the magnitudes (see Exercise 1.11.1(b)).

We consider the division of two complex numbers by thinking of it as a multiplication problem. We first multiply the complex number by the number 1, represented as the complex conjugate of the denominator divided by the complex conjugate of the denominator. We then write
$$\frac{z_1}{z_2} = \frac{z_1}{z_2} \frac{\bar{z}_2}{\bar{z}_2} = \frac{1}{z_2 \bar{z}_2} z_1 \bar{z}_2 = \frac{1}{|z_2|^2} z_1 \bar{z}_2,$$

1.11. Linear algebra with complex-valued numbers, vectors, and matrices

so that division is replaced by the appropriate multiplication. For example,

$$\frac{2-i3}{3+i2} = \left(\frac{2-i3}{3+i2}\right)\left(\frac{3-i2}{3-i2}\right) = \frac{1}{13}[(2-i3)(3-i2)] = \frac{1}{13}(-i13) = -i.$$

We now derive and state a very important identity—*Euler's formula*—which connects the exponential function to the sine and cosine. This will be accomplished via the use of the Maclaurin series for the exponential and trigonometric functions. Recall that

$$e^x = \sum_{j=0}^{\infty} \frac{x^j}{j!} = 1 + x + \frac{x^2}{2!} + \frac{x^3}{3!} + \cdots$$

$$\sin(x) = \sum_{j=0}^{\infty} (-1)^j \frac{x^{2j+1}}{(2j+1)!} = x - \frac{x^3}{3!} + \frac{x^5}{5!} - \frac{x^7}{7!} + \cdots$$

$$\cos(x) = \sum_{j=0}^{\infty} (-1)^j \frac{x^{2j}}{(2j)!} = 1 - \frac{x^2}{2!} + \frac{x^4}{4!} - \frac{x^6}{6!} + \cdots$$

and that each series converges for all $x \in \mathbb{R}$. Since

$$i^2 = -1, \quad i^3 = i^2 i = -i, \quad i^4 = i^2 i^2 = 1,$$

we can write for $\theta \in \mathbb{R}$

$$e^{i\theta} = \sum_{j=0}^{\infty} \frac{(i\theta)^j}{j!} = 1 + i\theta - \frac{\theta^2}{2!} - i\frac{\theta^3}{3!} + \frac{\theta^4}{4!} + i\frac{\theta^5}{5!} + \cdots$$

$$= \left(1 - \frac{\theta^2}{2!} + \frac{\theta^4}{4!} + \cdots + (-1)^j \frac{\theta^{2j}}{(2j)!} + \cdots\right)$$

$$+ i\left(\theta - \frac{\theta^3}{3!} + \frac{\theta^5}{5!} + \cdots + (-1)^j \frac{\theta^{2j+1}}{(2j+1)!} + \cdots\right).$$

Noting that the real part is the Maclaurin series for $\cos(\theta)$ and that the imaginary part is the Maclaurin series for $\sin(\theta)$, we arrive at Euler's formula:

$$e^{i\theta} = \cos(\theta) + i\sin(\theta).$$

Note that for any $\theta \in \mathbb{R}$,

$$|e^{i\theta}| = \sqrt{\cos^2(\theta) + \sin^2(\theta)} = 1.$$

Further note that Euler's formula yields the intriguing identity

$$e^{i\pi} = -1,$$

which brings into one simple formula some of the most important constants and concepts in all of mathematics.

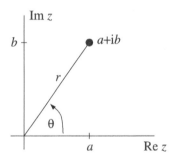

Figure 1.3. *A cartoon illustration of the polar representation of complex numbers.*

As a consequence of Euler's formula, we are able to write complex numbers using a polar representation. Let $z = a + ib$ be given. We know that if we represent the point (a, b) in the xy-plane, then the distance from the origin is $r = \sqrt{a^2 + b^2}$, and the angle from the positive x-axis satisfies $\tan(\theta) = b/a$ (see Figure 1.3). This allows us the polar coordinate representation:

$$a = r\cos(\theta), \quad b = r\sin(\theta).$$

Now, we know that the magnitude of the complex number is $|z| = \sqrt{a^2 + b^2}$, so we could write

$$a = |z|\cos(\theta), \quad b = |z|\sin(\theta).$$

On using Euler's formula, we finally see that

$$z = a + ib = |z|\cos(\theta) + i|z|\sin(\theta) = |z|[\cos(\theta) + i\sin(\theta)] = |z|e^{i\theta}, \quad (1.11.1)$$

where again

$$|z| = \sqrt{a^2 + b^2}, \quad \tan(\theta) = \frac{b}{a}.$$

As we will see in the case study of section 1.13.4, this representation of a complex number allows us to more easily understand the multiplication of complex-valued numbers.

Example 1.86. For $z = 2 + i2\sqrt{3}$, we have

$$|z| = 4, \quad \tan(\theta) = \sqrt{3} \rightsquigarrow \theta = \frac{\pi}{6},$$

so $2 + i2\sqrt{3} = 4e^{i\pi/3}$. ∎

Example 1.87. For $z = -2\sqrt{3} + i2$, we have

$$|z| = 4, \quad \tan(\theta) = -\frac{1}{\sqrt{3}} \rightsquigarrow \theta = \frac{5\pi}{6}.$$

The choice for the angle follows from the fact that the point $(-2\sqrt{3}, 2)$ is in the second quadrant of the xy-plane. In conclusion, $-2\sqrt{3} + i2 = 4e^{i5\pi/6}$. ∎

1.11. Linear algebra with complex-valued numbers, vectors, and matrices

> Does anything really change if we consider the previous linear algebra calculations and concepts under the assumption that the matrices and vectors have complex-valued entries? In summary, no. The definitions and properties of the span of a set of vectors and subspaces—in particular, the subspaces Null(A) and Col(A)—remain the same; indeed, the only difference is that the constants may now be complex valued. A basis of a subspace is still computed in the same manner, and the dimension of a subspace is still the number of basis vectors. Again, the only difference is that the vectors may have complex-valued entries. Finally, nothing changes for matrix/vector and matrix/matrix multiplication, the calculation of the inverse of a square matrix, and the calculation of the determinant for a square matrix. In conclusion, the only reason we did not start this chapter with a discussion of linear systems with complex-valued coefficients is for the sake of pedagogy, as it is easier to visualize vectors in $\mathbb{R}^n$ and subspaces that are realized as real-valued linear combinations of vectors in $\mathbb{R}^n$.

Example 1.88. Let us see how we can use our understanding of the algebra of complex numbers when doing Gaussian elimination. Consider the linear system

$$(1-i)x_1 + 4x_2 = 6$$
$$(-2+i3)x_1 + (-8+i3)x_2 = -9.$$

Performing Gaussian elimination on the augmented matrix yields

$$\begin{pmatrix} 1-i & 4 & | & 6 \\ -2+i3 & -8+i3 & | & -9 \end{pmatrix} \xrightarrow{(1/(1-i))\rho_1} \begin{pmatrix} 1 & 2+i2 & | & 3+i3 \\ -2+i3 & -8+i3 & | & -9 \end{pmatrix}$$

$$\xrightarrow{(2-i3)\rho_1+\rho_2} \begin{pmatrix} 1 & 2+i2 & | & 3+i3 \\ 0 & -2+i & | & 6-i3 \end{pmatrix} \xrightarrow{(1/(-2+i))\rho_2} \begin{pmatrix} 1 & 2+i2 & | & 3+i3 \\ 0 & 1 & | & -3 \end{pmatrix}$$

$$\xrightarrow{(-2-i2)\rho_2+\rho_1} \begin{pmatrix} 1 & 0 & | & 9+i9 \\ 0 & 1 & | & -3 \end{pmatrix}.$$

The solution is the last column:

$$x = \begin{pmatrix} 9+i9 \\ -3 \end{pmatrix} = x = \begin{pmatrix} 9 \\ -3 \end{pmatrix} + ix = \begin{pmatrix} 9 \\ 0 \end{pmatrix}. \blacksquare$$

Example 1.89. For another example, let us find Null(A) for the matrix

$$A = \begin{pmatrix} 3-i & 4 \\ 5 & 6+i2 \end{pmatrix}.$$

Since

$$A \xrightarrow{\text{RREF}} \begin{pmatrix} 1 & (6+i2)/5 \\ 0 & 0 \end{pmatrix},$$

the null space is found by solving

$$x_1 + \frac{6+i2}{5} x_2 = 0.$$

On setting $x_2 = 5t$, the solution vector is given by

$$x = \begin{pmatrix} -(6+i2)t \\ 5t \end{pmatrix} = t \begin{pmatrix} -6-i2 \\ 5 \end{pmatrix}.$$

We conclude that

$$\text{Null}(A) = \text{Span}\left\{\begin{pmatrix} -6-i2 \\ 5 \end{pmatrix}\right\}, \quad \dim[\text{Null}(A)] = 1. \quad \blacksquare$$

Example 1.90. Let us find those vectors b for which the linear system $Ax = b$ is consistent with

$$A = \begin{pmatrix} 2-i & 4 \\ 5 & 8+i4 \end{pmatrix}.$$

Gaussian elimination yields that the RREF of A is

$$A \xrightarrow{\text{RREF}} \begin{pmatrix} 1 & (8+i4)/5 \\ 0 & 0 \end{pmatrix}.$$

Since the RREF of A has a zero row, the system $Ax = b$ is not consistent for all b. Moreover, only the first column is a pivot column, so, by using Lemma 1.64, we know that a basis for $\text{Col}(A)$ is the first column of A, i.e.,

$$\text{Col}(A) = \text{Span}\left\{\begin{pmatrix} 2-i \\ 5 \end{pmatrix}\right\}, \quad \text{rank}(A) = 1.$$

The linear system is consistent if and only if $b \in \text{Col}(A)$. $\quad \blacksquare$

Exercises

Exercise 1.11.1. Let $z_1 = a_1 + ib_1$ and $z_2 = a_2 + ib_2$ be two complex numbers. Show that

(a) $\overline{z_1 z_2} = \overline{z_1}\, \overline{z_2}$, and

(b) $|z_1 z_2| = |z_1||z_2|$.

Exercise 1.11.2. Write each complex number z in the form $|z|e^{i\theta}$, where $-\pi < \theta \leq \pi$.

(a) $z = 3 - i4$

(b) $z = -2 + i5$

(c) $z = -3 - i7$

(d) $z = 6 + i$

Exercise 1.11.3. Solve each system of equations or explain why no solution exists.

(a) $(3-i)x_1 + 2x_2 = 2$, $-4x_1 + (1+i4)x_2 = -3$

(b) $x_1 + (-2+i5)x_2 = -3$, $(1-i5)x_1 + 3x_2 = 12$

Exercise 1.11.4. *For each of the problems below, compute the product Ax when it is well defined. If the product cannot be computed, explain why.*

(a) $A = \begin{pmatrix} 2+i & 3 \\ -2 & 1+i4 \\ 3 & 7 \end{pmatrix}$, $x = \begin{pmatrix} 2+i3 \\ 8 \end{pmatrix}$

(b) $A = \begin{pmatrix} 2 & -1+i3 & -4 \\ 2+i5 & 6 & 3-i7 \end{pmatrix}$, $x = \begin{pmatrix} 2 \\ 9 \\ 4+i3 \end{pmatrix}$

Exercise 1.11.5. *For each matrix A, find* Null(A) *and determine its dimension.*

(a) $A = \begin{pmatrix} 2+i3 & 26 \\ 2 & 8-i12 \end{pmatrix}$

(b) $A = \begin{pmatrix} 1-i4 & 17 \\ 2 & 2+i8 \end{pmatrix}$

Exercise 1.11.6. *Solve the following linear system and explicitly identify the homogeneous solution, x_h, and the particular solution, x_p:*

$$\begin{pmatrix} 3+i2 & -26 \\ -2 & 12-i8 \end{pmatrix} x = \begin{pmatrix} 13 \\ -6+i4 \end{pmatrix}.$$

Exercise 1.11.7. *In Example 1.62, it was shown that* $\dim[\mathbb{R}^n] = n$. *Show that* $\dim[\mathbb{C}^n] = n$.

1.12 ▪ Eigenvalues and eigenvectors

Consider a square matrix $A \in \mathcal{M}_n(\mathbb{R})$. As we will see in the two case studies in section 1.13, as well as when solving homogeneous systems of ODEs in Chapter 3, it will be especially useful to identify a set of vectors, say, $v_1, v_2, \ldots, v_n$, such that for each vector, there is a constant λ_j such that

$$Av_j = \lambda_j v_j, \quad j = 1, \ldots, n. \quad (1.12.1)$$

The vectors v_j, which are known as *eigenvectors*, and multiplicative factors λ_j, which are known as *eigenvalues*, may be complex valued (see section 1.11). If the eigenvalues are complex valued, then the corresponding eigenvector also has complex valued entries. Eigenvectors are vectors that have the property that matrix multiplication by A leads to a scalar multiple of the original vector.

1.12.1 ▪ Characterization of eigenvalues and eigenvectors

How do we find these vectors v and associated multiplicative factors λ? We can rewrite (1.12.1) as

$$Av = \lambda v \quad (A - \lambda I_n)v = 0.$$

Recalling the definition of a null space, an eigenvector v can be found if we can find an eigenvalue λ such that

$$\dim[\text{Null}(A - \lambda I_n)] \geq 1.$$

If λ is an eigenvalue, then we will call Null$(A - \lambda I_n)$ the *eigenspace*. An eigenvector is any basis vector of the eigenspace.

> Eigenvectors associated with a particular eigenvalue are not unique, as a basis is not unique (recall the discussion in section 1.8). However, the *number* of basis vectors is unique (recall Lemma 1.60), so associated with each eigenvalue, there will be a fixed number of *linearly independent* eigenvectors.

Now, if we are given an eigenvalue, then it is straightforward to compute a basis for the associated eigenspace. The problem really is in finding an eigenvalue. This requires an additional equation, for at the moment the linear system is a set of n equations with $n+1$ variables: the n components of the vector plus the associated eigenvalue. In constructing this additional equation, we can rely on the result of Theorem 1.76, in which it is stated that a square matrix has a nontrivial null space if and only if its determinant is zero. If we set

$$p_A(\lambda) = \det(A - \lambda I_n),$$

then the eigenvalues will correspond to the zeros of the *characteristic polynomial* $p_A(\lambda)$. While we will not do it here, it is not difficult to show that the characteristic polynomial is a polynomial of degree n, the size of the square matrix (see Exercise 1.12.4).

We summarize this discussion with the following result:

> **Theorem 1.91.** Let $A \in \mathcal{M}_n(\mathbb{R})$. The zeros of the nth-order characteristic polynomial $p_A(\lambda) = \det(A - \lambda I_n)$ are the eigenvalues of the matrix A. The (not unique) eigenvectors associated with an eigenvalue are a basis for Null$(A - \lambda I_n)$.

Before going any further in the discussion of the theory associated with eigenvalues and eigenvectors, let us do a relatively simple computation.

Example 1.92. Let us find the eigenvalues and associated eigenvectors for

$$A = \begin{pmatrix} 3 & 2 \\ 2 & 3 \end{pmatrix}.$$

We have

$$A - \lambda I_2 = \begin{pmatrix} 3-\lambda & 2 \\ 2 & 3-\lambda \end{pmatrix},$$

so that the characteristic polynomial is

$$p_A(\lambda) = (3-\lambda)^2 - 4.$$

The zeros of the characteristic polynomial are $\lambda = 1, 5$. As for the associated eigenvectors, we must compute a basis for Null$(A - \lambda I_2)$ for each eigenvalue. For the eigenvalue $\lambda_1 = 1$, we have

$$A - I_2 \xrightarrow{\text{RREF}} \begin{pmatrix} 1 & 1 \\ 0 & 0 \end{pmatrix},$$

which corresponds to the linear equation $v_1 + v_2 = 0$. Since

$$\text{Null}(A - I_2) = \text{Span}\left\{\begin{pmatrix} -1 \\ 1 \end{pmatrix}\right\},$$

an associated eigenvector is

$$\lambda_1 = 1; \quad v_1 = \begin{pmatrix} -1 \\ 1 \end{pmatrix}.$$

Because eigenvectors are not unique, any multiple of v_1 given above would be an eigenvector associated with the eigenvalue $\lambda_1 = 1$. For the eigenvalue $\lambda_2 = 5$, we have

$$A - 5I_2 \xrightarrow{\text{RREF}} \begin{pmatrix} 1 & -1 \\ 0 & 0 \end{pmatrix},$$

which corresponds to the linear equation $v_1 - v_2 = 0$. Since

$$\text{Null}(A - 5I_2) = \text{Span}\left\{\begin{pmatrix} 1 \\ 1 \end{pmatrix}\right\},$$

an associated eigenvector is

$$\lambda_2 = 5; \quad v_2 = \begin{pmatrix} 1 \\ 1 \end{pmatrix}. \quad \blacksquare$$

Before continuing, we need to decide how many eigenvectors are to be associated with a given eigenvalue. An eigenvalue λ_0 is such that

$$\text{m}_g(\lambda_0) := \dim[\text{Null}(A - \lambda_0 I_n)] \geq 1.$$

The integer $\text{m}_g(\lambda_0)$ is the *geometric multiplicity* of the eigenvalue λ_0. We know from section 1.8 that $\text{m}_g(\lambda_0)$ will be the number of free variables for the associated linear system. Consequently, any basis of the eigenspace will have $\text{m}_g(\lambda_0)$ vectors. Since a basis is not unique, eigenvectors are not unique; however, once a set of eigenvectors has been chosen, any other eigenvector must be a linear combination of the chosen set.

Since the characteristic polynomial $p_A(\lambda)$ is an nth-order polynomial, by the Fundamental Theorem of Algebra it can be factored as

$$p_A(\lambda) = c(\lambda - \lambda_1)(\lambda - \lambda_2) \cdots (\lambda - \lambda_n), \quad c \neq 0.$$

If $\lambda_i \neq \lambda_j$ for all $i \neq j$, then all of the eigenvalues are said to be *algebraically simple*. If an eigenvalue λ_j is algebraically simple, then the associated eigenspace will be one-dimensional, i.e., $\text{m}_g(\lambda_j) = 1$, so that all basis vectors for the eigenspace will be scalar multiples of each other. In other words, for simple eigenvalues, an associated eigenvector will be a scalar multiple of any other associated eigenvector. If an eigenvalue is not simple, then we will call it a *multiple eigenvalue*. For example, if

$$p_A(\lambda) = (\lambda + 1)(\lambda - 1)^2(\lambda - 3)^4,$$

then $\lambda = -1$ is a simple eigenvalue, and $\lambda = 1$ and $\lambda = 3$ are multiple eigenvalues. The *(algebraic) multiplicity* of a multiple eigenvalue is the order of the zero of the

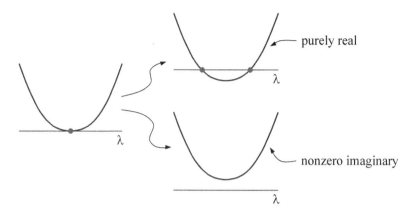

Figure 1.4. *A cartoon illustration of the change in the graph of the characteristic polynomial as a matrix is perturbed. The left figure shows the graph for the unperturbed matrix. The order of the zero is two, so the algebraic multiplicity of the corresponding eigenvalue is two. The zero is marked with a filled circle. The right two figures illustrate the two possibilities for the graph under a generic perturbation of the matrix. In the upper right figure, the double zero becomes two simple real zeros. In the lower right figure, the resulting two simple zeros have nonzero imaginary part. In both cases, the perturbed matrix has algebraically simple eigenvalues.*

characteristic polynomial and will be denoted by $m_a(\lambda_0)$. In this example, $\lambda = -1$ is such that $m_a(-1) = 2$ (a double eigenvalue), and $\lambda = 3$ is such that $m_a(3) = 4$ (a quartic eigenvalue).

It is a fundamental fact of linear algebra (indeed, a consequence of Schur's Lemma) that the two multiplicities are related via

$$1 \leq m_g(\lambda_0) \leq m_a(\lambda_0).$$

As already stated, if an eigenvalue is algebraically simple, i.e., $m_a(\lambda_0) = 1$, then it must be true that $m_g(\lambda_0) = m_a(\lambda_0) = 1$. On the other hand, if $m_a(\lambda_0) \geq 2$, it may be the case that $m_g(\lambda_0) < m_a(\lambda_0)$. This situation is nongeneric and can be rectified by a small perturbation of the matrix A (see Figure 1.4). Indeed, generically it will be the case that all of the eigenvalues for a given matrix are algebraically simple.

As a final remark, we remind the reader that while eigenvectors themselves are not unique, the number of linearly independent eigenvectors is unique. In all that follows, we will compute only a set of eigenvectors associated with a particular eigenvalue and not spend much effort discussing the associated eigenspace. The reader needs to always remember that for a given set of eigenvectors associated with a given eigenvalue, any linear combination of these given eigenvectors also counts as an eigenvector.

Example 1.93. Let us find the eigenvalues and associated eigenvectors for

$$A = \begin{pmatrix} 5 & 0 & 0 \\ 0 & 3 & 2 \\ 0 & 2 & 3 \end{pmatrix}.$$

We have

$$A - \lambda I_3 = \begin{pmatrix} 5-\lambda & 0 & 0 \\ 0 & 3-\lambda & 2 \\ 0 & 2 & 3-\lambda \end{pmatrix},$$

1.12. Eigenvalues and eigenvectors

so that the characteristic polynomial is

$$p_A(\lambda) = (5-\lambda)[(3-\lambda)^2 - 4].$$

The zeros of the characteristic polynomial are $\lambda = 1, 5$, where $\lambda = 5$ is a double root; i.e., $\lambda = 5$ is a double eigenvalue. Regarding the algebraic multiplicities, we have $m_a(1) = 1$ and $m_a(5) = 2$. As for the associated eigenvectors, we have for the eigenvalue $\lambda_1 = 1$

$$A - 1I_3 \xrightarrow{\text{RREF}} \begin{pmatrix} 1 & 0 & 0 \\ 0 & 1 & 1 \\ 0 & 0 & 0 \end{pmatrix},$$

which corresponds to the linear system $v_1 = 0$, $v_2 + v_3 = 0$. An eigenvector is then given by

$$\lambda_1 = 1; \quad v_1 = \begin{pmatrix} 0 \\ -1 \\ 1 \end{pmatrix}.$$

For the eigenvalue $\lambda_2 = \lambda_3 = 5$, we have

$$A - 5I_3 \xrightarrow{\text{RREF}} \begin{pmatrix} 0 & 1 & -1 \\ 0 & 0 & 0 \\ 0 & 0 & 0 \end{pmatrix},$$

which corresponds to the linear equation $v_2 - v_3 = 0$. There are two free variables, v_1 and v_3, so that there are two linearly independent eigenvectors:

$$\lambda_2 = \lambda_3 = 5; \quad v_2 = \begin{pmatrix} 1 \\ 0 \\ 0 \end{pmatrix}, \quad v_3 = \begin{pmatrix} 0 \\ 1 \\ 1 \end{pmatrix}. \blacksquare$$

Example 1.94. Let us find the eigenvalues and associated eigenvectors for

$$A = \begin{pmatrix} 5 & 0 & 8 \\ 0 & 3 & 2 \\ 0 & 2 & 3 \end{pmatrix}.$$

Note that only one entry in A, a_{13}, has changed from the previous example. We have

$$A - \lambda I_3 = \begin{pmatrix} 5-\lambda & 0 & 8 \\ 0 & 3-\lambda & 2 \\ 0 & 2 & 3-\lambda \end{pmatrix},$$

so that the characteristic polynomial is again

$$p_A(\lambda) = (5-\lambda)[(3-\lambda)^2 - 4].$$

As in the previous example, the eigenvalues are $\lambda = 1$ and $\lambda = 5$ with $m_a(1) = 1$ and $m_a(5) = 2$. As for the associated eigenvectors, we have for the eigenvalue $\lambda_1 = 1$,

$$A - 1I_3 \xrightarrow{\text{RREF}} \begin{pmatrix} 1 & 0 & 2 \\ 0 & 1 & 1 \\ 0 & 0 & 0 \end{pmatrix},$$

which corresponds to the linear system $v_1 + 2v_3 = 0$, $v_2 + v_3 = 0$. An associated eigenvector is then

$$\lambda_1 = 1; \quad v_1 = \begin{pmatrix} -2 \\ -1 \\ 1 \end{pmatrix}.$$

For the eigenvalue $\lambda_2 = \lambda_3 = 5$, we have

$$A - 5I_3 \xrightarrow{\text{RREF}} \begin{pmatrix} 0 & 1 & 0 \\ 0 & 0 & 1 \\ 0 & 0 & 0 \end{pmatrix},$$

which corresponds to the linear system $v_2 = v_3 = 0$. Unlike the previous example, there is now only one free variable, v_1, which means that there is only one linearly independent eigenvector associated with both of these eigenvalues:

$$\lambda_2 = \lambda_3 = 5; \quad v_2 = \begin{pmatrix} 1 \\ 0 \\ 0 \end{pmatrix}.$$

In this example, there are not as many linearly independent eigenvectors as there are eigenvalues. ∎

Example 1.95. Let us find the eigenvalues and associated eigenvectors for

$$A = \begin{pmatrix} 3 & -2 \\ 2 & 3 \end{pmatrix}.$$

We have

$$A - \lambda I_2 = \begin{pmatrix} 3-\lambda & -2 \\ 2 & 3-\lambda \end{pmatrix},$$

so that the characteristic polynomial is

$$p_A(\lambda) = (3-\lambda)^2 + 4.$$

The zeros of the characteristic polynomial are $\lambda = 3 \pm i2$. Note that this set of eigenvalues is a complex-conjugate pair. As for the associated eigenvectors, we have for the eigenvalue $\lambda_1 = 3 + i2$

$$A - (3+i2)I_2 \xrightarrow{\text{RREF}} \begin{pmatrix} 1 & -i \\ 0 & 0 \end{pmatrix},$$

which corresponds to the linear equation $v_1 - iv_2 = 0$. An eigenvector is then given by

$$\lambda_1 = 3 + i2; \quad v_1 = \begin{pmatrix} i \\ 1 \end{pmatrix} = \begin{pmatrix} 0 \\ 1 \end{pmatrix} + i\begin{pmatrix} 1 \\ 0 \end{pmatrix}.$$

For the eigenvalue $\lambda_2 = 3 - i2$, we have

$$A - (3-i2)I_2 \xrightarrow{\text{RREF}} \begin{pmatrix} 1 & i \\ 0 & 0 \end{pmatrix},$$

which corresponds to the linear equation $v_1 + iv_2 = 0$. An eigenvector is then given by

$$\lambda_2 = 3 - i2; \quad v_2 = \begin{pmatrix} -i \\ 1 \end{pmatrix} = \begin{pmatrix} 0 \\ 1 \end{pmatrix} - i \begin{pmatrix} 1 \\ 0 \end{pmatrix}.$$

As was the case for the eigenvalues, eigenvectors also come in a complex-conjugate pair. ∎

Example 1.96. Let us find the eigenvalues and associated eigenvectors for

$$A = \begin{pmatrix} 0 & 1 \\ -5 & -2 \end{pmatrix}.$$

We have

$$A - \lambda I_2 = \begin{pmatrix} -\lambda & 1 \\ -5 & -2 - \lambda \end{pmatrix},$$

so that the characteristic polynomial is

$$p_A(\lambda) = \lambda^2 + 2\lambda + 5 = (\lambda + 1)^2 + 4.$$

The zeros of the characteristic polynomial are $\lambda = -1 \pm i2$. Note that, once again the eigenvalues arise in a complex-conjugate pair. As for the associated eigenvectors, we have for the eigenvalue $\lambda_1 = -1 + i2$

$$A - (-1 + i2)I_2 \xrightarrow{\text{RREF}} \begin{pmatrix} 1 - i2 & 1 \\ 0 & 0 \end{pmatrix},$$

which corresponds to the linear equation $(1 - i2)v_1 + v_2 = 0$. An eigenvector is then given by

$$\lambda_1 = -1 + i2; \quad v_1 = \begin{pmatrix} -1 + i2 \\ 1 \end{pmatrix} = \begin{pmatrix} -1 \\ 1 \end{pmatrix} + i \begin{pmatrix} 2 \\ 0 \end{pmatrix}.$$

For the eigenvalue $\lambda_2 = -1 - i2$, we eventually see that an eigenvector is given by

$$v_2 = \begin{pmatrix} -1 - i2 \\ 1 \end{pmatrix} = \begin{pmatrix} -1 \\ 1 \end{pmatrix} - i \begin{pmatrix} 2 \\ 0 \end{pmatrix}.$$

Thus, just as in the previous example, the associated eigenvectors also come in a complex-conjugate pair. ∎

Example 1.97. We finally consider an example for which the eigenvalues and eigenvectors must be computed numerically. Here

$$A = \begin{pmatrix} 1 & 2 & 3 \\ -1 & 2 & -3 \\ 5 & 6 & 7 \end{pmatrix} \in \mathcal{M}_3(\mathbb{R}),$$

which means that $p_A(\lambda)$ is a third-order polynomial. Unless the problem is very special, it is generally the case that it is not possible to (easily) find the three roots. Using SAGE, we get

$$\lambda_1 \sim 5.43 + i2.80, \; v_1 \sim \begin{pmatrix} 1.00 \\ -1.58 \\ 2.53 \end{pmatrix} + i \begin{pmatrix} 0.00 \\ -1.13 \\ -0.18 \end{pmatrix}; \; \lambda_3 \sim -0.86, \; v_3 \sim \begin{pmatrix} 1.00 \\ -0.18 \\ -0.50 \end{pmatrix}.$$

The second eigenvalue is the complex conjugate of the second, $\lambda_2 = \overline{\lambda_1}$, and the associated eigenvector is the complex conjugate of v_1, $v_2 = \overline{v_1}$. ∎

1.12.2 ▪ Properties

The last three examples highlight a general phenomena. Suppose that $A \in \mathcal{M}_n(\mathbb{R})$ and further suppose that $\lambda = a + ib$ is an eigenvalue with associated eigenvector $v = p + iq$:

$$Av = \lambda v \quad \leadsto \quad A(p+iq) = (a+ib)(p+iq).$$

Taking the complex conjugate of both sides and using the fact that the conjugate of a product is the product of the conjugates,

$$\overline{Av} = \overline{A}\,\overline{v}, \quad \overline{\lambda v} = \overline{\lambda}\,\overline{v}$$

gives

$$\overline{A}\,\overline{v} = \overline{\lambda}\,\overline{v}.$$

Since $A \in \mathcal{M}_n(\mathbb{R})$, $\overline{A} = A$, we conclude that

$$A\overline{v} = \overline{\lambda}\,\overline{v} \quad \leadsto \quad A(p-iq) = (a-ib)(p-iq).$$

This equation is another eigenvalue/eigenvector equation for the matrix A. The eigenvalue and associated eigenvector for this equation are related to the original via complex conjugation. In conclusion, if $A \in \mathcal{M}_n(\mathbb{R})$, then the eigenvalues come in the complex-conjugate pairs $\{\lambda, \overline{\lambda}\}$, i.e., $\{a \pm ib\}$, as do the associated eigenvectors $\{v, \overline{v}\}$, i.e., $\{p \pm iq\}$.

We conclude with some additional facts about eigenvalues and eigenvectors of $A \in \mathcal{M}_n(\mathbb{R})$, each of which will be useful in applications. First, the eigenvalues tell us something about the invertibility of the matrix. We first observe that by setting $\lambda = 0$,

$$p_A(0) = \det(A);$$

thus, $\lambda = 0$ is an eigenvalue if and only if $\det(A) = 0$. Since by Theorem 1.76, A is invertible if and only if $\det(A) \neq 0$, we have that A is invertible if and only if $\lambda = 0$ is not an eigenvalue. From Exercise 1.12.4(c), we know that the characteristic polynomial is of degree n; hence, by the Fundamental Theorem of Algebra, there are precisely n eigenvalues. As we have seen in the previous examples, there may or may not be n linearly independent eigenvectors. However, if the eigenvalues are *distinct* (each one is algebraically simple), then the n eigenvectors are indeed linearly independent. Since $\dim[\mathbb{C}^n] = n$ (see Exercise 1.11.7), this means that we can use the eigenvectors as a basis for $\mathbb{C}^n$.

1.12. Eigenvalues and eigenvectors

> **Theorem 1.98.** *Consider the matrix $A \in \mathcal{M}_n(\mathbb{R})$.*
>
> (a) *If $\lambda = a + ib$ is an eigenvalue with associated eigenvector $v = p + iq$ for some vectors $p, q \in \mathbb{R}^n$, then the complex conjugate $\overline{\lambda} = a - ib$ is an eigenvalue with associated complex-conjugate eigenvector $\overline{v} = p - iq$.*
>
> (b) *$\lambda = 0$ is an eigenvalue if and only if $\det(A) = 0$.*
>
> (c) *A is invertible if and only if all of the eigenvalues are nonzero.*
>
> (d) *If the eigenvalues $\lambda_1, \lambda_2, \ldots, \lambda_n$ are distinct, i.e., all of the roots of the characteristic polynomial are simple, then a set of corresponding eigenvectors $\{v_1, v_2, \ldots, v_n\}$ forms a basis for $\mathbb{C}^n$.*

Proof. We need only to show that if the eigenvalues are distinct, then the set of corresponding eigenvectors forms a basis. Recall (1.12.1):

$$Av_j = \lambda_j v_j, \quad j = 1, \ldots, n.$$

Set $S = \text{Span}\{v_1, \ldots, v_n\}$ and suppose that $\dim[S] = r < n$. Assume without loss of generality that the set $\{v_1, \ldots, v_r\}$ is linearly independent. This means that each of the remaining eigenvectors is a linear combination of the vectors in this set; in particular, there exist constants $c_1, \ldots, c_r$ (not all zero) such that

$$v_{r+1} = c_1 v_1 + c_2 v_2 + \cdots + c_r v_r. \tag{1.12.2}$$

Multiplying both sides by A and using the linearity of matrix/vector multiplication gives

$$Av_{r+1} = c_1 Av_1 + c_2 Av_2 + \cdots + c_r Av_r,$$

which in turn implies

$$\lambda_{r+1} v_{r+1} = c_1 \lambda_1 v_1 + c_2 \lambda_2 v_2 + \cdots + c_r \lambda_r v_r.$$

Multiplying (1.12.2) by λ_{r+1} and subtracting the above yields

$$c_1(\lambda_{r+1} - \lambda_1)v_1 + c_2(\lambda_{r+1} - \lambda_2)v_2 + \cdots + c_r(\lambda_{r+1} - \lambda_r)v_r = 0.$$

Since the eigenvalues are distinct, i.e., $\lambda_{r+1} - \lambda_j \neq 0$ for $j = 1, \ldots, r$, and since not all of the constants c_j are equal to zero, we conclude that the vectors $v_1, v_2, \ldots, v_r$ are linearly dependent. This is a contradiction. Hence, we cannot write the eigenvector v_{r+1} as a linear combination of the other eigenvectors. The assumption that $r < n$ is false, and consequently $r = n$. □

1.12.3 • Eigenvectors as a basis and Fourier expansions

As we will see in the upcoming case studies as well as in our study of linear systems of ODEs, it is extremely beneficial to write a given vector in terms of a linearly independent set of eigenvectors of a given matrix. For a given matrix $A \in \mathcal{M}_n(\mathbb{R})$, suppose that a set of eigenvectors, $\{v_1, v_2, \ldots, v_n\}$, forms a basis. We know from

The moniker Fourier is used here because of the connection between expansions in terms of eigenvectors and Fourier series.

Theorem 1.98(d) that this is possible if the eigenvalues are distinct. Going back to our discussion in section 1.8, we then know that any vector $x \in \mathbb{C}^n$ can be uniquely written through the expansion

$$x = c_1 v_1 + c_2 v_2 + \cdots + c_n v_n. \tag{1.12.3}$$

Such an expansion in terms of eigenvectors is sometimes called a *Fourier expansion*, and the weights are sometimes called the *Fourier coefficients*. The Fourier coefficients are found through the solution of the linear system; i.e.,

$$c_1 v_1 + c_2 v_2 + \cdots + c_n v_n = Pc, \quad P = (v_1 \; v_2 \; \cdots \; v_n)$$

means

$$x = Pc \quad \leadsto \quad c = P^{-1} x.$$

The matrix P is invertible because the chosen eigenvectors are assumed to be linearly independent (otherwise, they would not form a basis), and the result of Theorem 1.75 states that the inverse exists if and only if the matrix has full rank.

By writing a given vector through a Fourier expansion, we develop greater insight into the geometry associated with matrix/vector multiplication. Multiplying both sides of (1.12.3) by A and using linearity gives

$$Ax = c_1 A v_1 + c_2 A v_2 + \cdots + c_n A v_n.$$

Since each vector v_j is an eigenvector with associated eigenvalue λ_j, i.e., $A v_j = \lambda_j v_j$, we can rewrite the right-hand side of the above as

$$c_1 A v_1 + c_2 A v_2 + \cdots + c_n A v_n = c_1 \lambda_1 v_1 + c_2 \lambda_2 v_2 + \cdots + c_n \lambda_n v_n.$$

Putting the pieces together yields

$$x = c_1 v_1 + c_2 v_2 + \cdots + c_n v_n \quad \leadsto \quad Ax = c_1 \lambda_1 v_1 + c_2 \lambda_2 v_2 + \cdots + c_n \lambda_n v_n.$$

Thus, the Fourier coefficients associated with the vector Ax are a scaling of those for the vector x, where the scaling is the eigenvalue associated with the associated eigenvector.

Lemma 1.99. *Suppose that for $A \in \mathcal{M}_n(\mathbb{R})$, there is a set of linearly independent eigenvectors, $\{v_1, v_2, \ldots, v_n\}$ (guaranteed if the eigenvalues are distinct). For any $x \in \mathbb{C}^n$, there is the Fourier expansion*

$$x = c_1 v_1 + c_2 v_2 + \cdots + c_n v_n,$$

where the Fourier coefficients $c_1, c_2, \ldots, c_n$ are uniquely determined. Moreover, the vector Ax has the Fourier expansion

$$Ax = c_1 \lambda_1 v_1 + c_2 \lambda_2 v_2 + \cdots + c_n \lambda_n v_n,$$

where each λ_j is the eigenvalue associated with the eigenvector v_j.

The above result is known as the *spectral decomposition* of the matrix A. While we will not go into the details here (see [31, Chapter 7.2]), it turns out to be the case that if all the eigenvalues are real, there exist unique vectors w_j such that the Fourier coefficients are given by

$$c_j = w_j^T x, \quad j = 1, \ldots, n.$$

The *transpose* of the vector w, written as w^T, takes a column vector and produces a row vector. For example,

$$w = \begin{pmatrix} 1 \\ 2 \\ 3 \end{pmatrix} \in \mathcal{M}_{3 \times 1}(\mathbb{R}) \quad \leadsto \quad w^T = (1\ 2\ 3) \in \mathcal{M}_{1 \times 3}(\mathbb{R}).$$

Consequently, we can write each term in the sum as

$$c_j v_j = \left(w_j^T x \right) v_j = v_j \left(w_j^T x \right) = \left(v_j w_j^T \right) x.$$

The middle equality follows from the fact $w_j^T x$ is a scalar. The square matrix $P_j = v_j w_j^T \in \mathcal{M}_n(\mathbb{R})$ has rank one (see Exercise 1.12.6). The matrices P_j are known as *spectral projection matrices*. We conclude with

$$Ax = \lambda_1 P_1 x + \lambda_2 P_2 x + \cdots + \lambda_n P_n x,$$

which means that original matrix can be written as a weighted sum of rank-one matrices:

$$A = \lambda_1 P_1 + \lambda_2 P_2 + \cdots + \lambda_n P_n.$$

Example 1.100. Consider the matrix

$$A = \begin{pmatrix} 0 & 1 \\ -8 & -6 \end{pmatrix}.$$

It can be checked that the eigenvalues and associated eigenvectors are

$$\lambda_1 = -2,\ v_1 = \begin{pmatrix} 1 \\ -2 \end{pmatrix}; \quad \lambda_2 = -4,\ v_2 = \begin{pmatrix} 1 \\ -4 \end{pmatrix}.$$

The eigenvectors are clearly linearly independent, so they form a basis. For a particular example, let us find the Fourier coefficients for the vector $x = (2\ -7)^T$. Using (1.12.3), we have

$$\begin{pmatrix} 2 \\ -7 \end{pmatrix} = c_1 \begin{pmatrix} 1 \\ -2 \end{pmatrix} + c_2 \begin{pmatrix} 1 \\ -4 \end{pmatrix} = \begin{pmatrix} 1 & 1 \\ -2 & -4 \end{pmatrix} \begin{pmatrix} c_1 \\ c_2 \end{pmatrix}.$$

The solution to this linear system is

$$\begin{pmatrix} c_1 \\ c_2 \end{pmatrix} = \begin{pmatrix} 1 & 1 \\ -2 & -4 \end{pmatrix}^{-1} \begin{pmatrix} 2 \\ -7 \end{pmatrix} = -\frac{1}{2} \begin{pmatrix} -4 & -1 \\ 2 & 1 \end{pmatrix} \begin{pmatrix} 2 \\ -7 \end{pmatrix} = \begin{pmatrix} 1/2 \\ 3/2 \end{pmatrix}.$$

In other words, the Fourier coefficients are

$$c_1 = \frac{1}{2}, \quad c_2 = \frac{3}{2},$$

so
$$x = \frac{1}{2}v_1 + \frac{3}{2}v_2.$$
Using the result of Lemma 1.99, we also know that
$$Ax = \frac{1}{2}(-2)v_1 + \frac{3}{2}(-4)v_2 = -v_1 - 6v_2. \blacksquare$$

Exercises

Exercise 1.12.1. *Suppose that for a given $A \in \mathcal{M}_n(\mathbb{R})$, there is a set of linearly independent eigenvectors, $\{v_1, v_2, \ldots, v_n\}$. Suppose that a given x has the Fourier expansion*
$$x = c_1 v_1 + c_2 v_2 + \cdots + c_n v_n.$$

Defining
$$A^\ell := \underbrace{A \cdot A \cdots A}_{\ell \text{ times}},$$

show that

(a) $A^2 x = c_1 \lambda_1^2 v_1 + c_2 \lambda_2^2 v_2 + \cdots + c_n \lambda_n^2 v_n$,

(b) $A^3 x = c_1 \lambda_1^3 v_1 + c_2 \lambda_2^3 v_2 + \cdots + c_n \lambda_n^3 v_n$, *and*

(c) *if $\ell \geq 4$, $A^\ell x = c_1 \lambda_1^\ell v_1 + c_2 \lambda_2^\ell v_2 + \cdots + c_n \lambda_n^\ell v_n$ (Hint: use (a), (b), and induction.)*

Exercise 1.12.2. *Compute by hand the following eigenvalues and all corresponding eigenvectors for each matrix. If the eigenvalue is complex valued, write the eigenvector in the form $v = p + iq$.*

(a) $A = \begin{pmatrix} 1 & -2 \\ -2 & 4 \end{pmatrix}$

(b) $A = \begin{pmatrix} 3 & -2 \\ 2 & 3 \end{pmatrix}$

(c) $A = \begin{pmatrix} 2 & -6 \\ 3 & -7 \end{pmatrix}$

(d) $A = \begin{pmatrix} 1 & 4 \\ -2 & -3 \end{pmatrix}$

(e) $A = \begin{pmatrix} 2 & 5 & -2 \\ 5 & 2 & 1 \\ 0 & 0 & -3 \end{pmatrix}$

(f) $A = \begin{pmatrix} 3 & 0 & 0 \\ 1 & -1 & 2 \\ -2 & -2 & -1 \end{pmatrix}$

Exercise 1.12.3. In each of the following, the characteristic polynomial of $A \in \mathcal{M}_n(\mathbb{R})$ is given. Determine n, list each eigenvalue and its algebraic multiplicity, and state whether or not the matrix is invertible.

(a) $p_A(\lambda) = (\lambda - 3)(\lambda^2 + 2\lambda + 5)(\lambda - 4)^2$

(b) $p_A(\lambda) = \lambda^2(\lambda + 3)(\lambda^2 - 4\lambda + 13)(\lambda - 1)^4$

(c) $p_A(\lambda) = (\lambda + 5)(\lambda + 2)^3(\lambda^2 + 6\lambda + 25)^2$

(d) $p_A(\lambda) = \lambda(\lambda^2 + 9)(\lambda^2 + 25)(\lambda - 8)$

Exercise 1.12.4. Suppose that $A \in \mathcal{M}_n(\mathbb{C})$, and let $p_A(\lambda) = \det(A - \lambda I_n)$ be the characteristic polynomial.

(a) If $n = 2$, show that $p_A(\lambda)$ is a polynomial of degree two.

(b) If $n = 3$, show that $p_A(\lambda)$ is a polynomial of degree three.

(c) Show that $p_A(\lambda)$ is a polynomial of degree n.

Exercise 1.12.5. Suppose that $A \in \mathcal{M}_n(\mathbb{C})$ is invertible. If for the matrix A, λ is an eigenvalue with associated eigenvector v, show that for the matrix A^{-1}, there is the eigenvalue $1/\lambda$ with associated eigenvector v.

Exercise 1.12.6. Suppose that $v, w \in \mathbb{R}^n$. Show that

(a) $vw^\mathrm{T} \in \mathcal{M}_n(\mathbb{R})$; and

(b) $\mathrm{rank}(vw^\mathrm{T}) = 1$. (Hint: Show that the matrix has only one linearly independent column.)

Exercise 1.12.7. For each of the following matrices, write the vector $x = (4 \ -3)^\mathrm{T}$ as a linear combination of eigenvectors. Explicitly give the weights (Fourier coefficients).

(a) $A = \begin{pmatrix} 1 & -2 \\ -2 & 4 \end{pmatrix}$

(b) $A = \begin{pmatrix} 3 & -2 \\ 2 & 3 \end{pmatrix}$

(c) $A = \begin{pmatrix} 2 & -6 \\ 3 & -7 \end{pmatrix}$

(d) $A = \begin{pmatrix} 1 & 4 \\ -2 & -3 \end{pmatrix}$

Exercise 1.12.8. Suppose that $A \in \mathcal{M}_2(\mathbb{R})$ has eigenvalues and associated eigenvectors given by

$$\lambda_1 = -3, \ v_1 = \begin{pmatrix} -2 \\ 5 \end{pmatrix}; \quad \lambda_2 = 7, \ v_1 = \begin{pmatrix} 3 \\ 8 \end{pmatrix}.$$

(a) Find $A(5v_1 - 3v_2)$.

(b) If $x = (6\ 4)^T$, find Ax.

Exercise 1.12.9. Let $x = (-3\ 5)^T$. For each of the following matrices, write the vector $A^{13}x$ as a linear combination of eigenvectors. Explicitly give the weights (Fourier coefficients). (Hint: Use Exercise 1.12.1.)

(a) $A = \begin{pmatrix} 1 & -2 \\ -2 & 4 \end{pmatrix}$

(b) $A = \begin{pmatrix} 3 & -2 \\ 2 & 3 \end{pmatrix}$

(c) $A = \begin{pmatrix} 2 & -6 \\ 3 & -7 \end{pmatrix}$

(d) $A = \begin{pmatrix} 1 & 4 \\ -2 & -3 \end{pmatrix}$

Exercise 1.12.10. Let $x = (5\ 2\ -7)^T$. For each of the following matrices, write the vector $A^9 x$ as a linear combination of eigenvectors. Explicitly give the weights (Fourier coefficients). (Hint: Use Exercise 1.12.1.)

(a) $A = \begin{pmatrix} 2 & 5 & -2 \\ 5 & 2 & 1 \\ 0 & 0 & -3 \end{pmatrix}$

(b) $A = \begin{pmatrix} 3 & 0 & 0 \\ 1 & -1 & 2 \\ -2 & -2 & -1 \end{pmatrix}$

Exercise 1.12.11. Let $A \in \mathcal{M}_n(\mathbb{R})$ have the properties

(a) all the entries are nonnegative, and

(b) the sum of the values in each row is one.

Show that $\lambda = 1$ is an eigenvalue of A. (Hint: Find an associated eigenvector.)

Exercise 1.12.12. Determine if each of the following statements is true or false. Provide an explanation for your answer.

(a) It is possible for $A \in \mathcal{M}_4(\mathbb{R})$ to have five eigenvalues.

(b) Every $A \in \mathcal{M}_2(\mathbb{R})$ has two real eigenvalues.

(c) If $A \in \mathcal{M}_6(\mathbb{R})$, then A has at most six linearly independent eigenvectors.

(d) If $Ax = 0$ has an infinite number of solutions, then all of the eigenvalues for A are nonzero.

(e) If $A \in \mathcal{M}_5(\mathbb{R})$, then it is possible for the characteristic polynomial to be of degree four.

1.13 ▪ Case studies

We now consider several problems in which understanding concepts like rank, eigenvalues, and eigenvectors for a given matrix is crucial.

1.13.1 ▪ Digital signal filters

A linear time-invariant (LTI) digital signal filter is important in audio engineering, as it is the only filter that preserves signal frequencies (see Smith [36]). An LTI filter that uses past output terms (known as feedback) is called a recursive digital filter. A recursive LTI filter is often represented by a transfer function:

$$G(z) = c_n + \frac{c_{n-1}z^{n-1} + c_{n-2}z^{n-2} + \cdots + c_1 z + c_0}{z^n + a_{n-1}z^{n-1} + \cdots + a_1 z + a_0}.$$

(Using transfer functions to solve ODEs is presented in Chapter 5, section 5.5). Equivalently, the filter is modeled by the difference equation:

$$y_{k+n} + a_{n-1}y_{k+n-1} + \cdots + a_1 y_{k+1} + a_0 y_k$$
$$= c_n u_{k+n} + (c_{n-1} - c_n a_{n-1})u_{k+n-1} + \cdots + (c_1 - c_n a_1)u_{k+1} + (c_0 - c_n a_0)u_k.$$

The input signal is represented by $u_0, u_1, \ldots$, and the output signal generated by the input is the sequence $y_0, y_1, \ldots$. The value of the coefficients arise from the design of the filter and are fixed.

For the sake of exposition, we will now assume $n = 3$, so the transfer function becomes

$$G(z) = c_3 + \frac{c_2 z^2 + c_1 z + c_0}{z^3 + a_2 z^2 + a_1 z + a_0}.$$

While we will not go into the details, the transfer function allows us to solve for the output signal directly from the input signal via a state-space equation coupled to an output equation:

$$x_{k+1} = \begin{pmatrix} 0 & 1 & 0 \\ 0 & 0 & 1 \\ -a_0 & -a_1 & -a_2 \end{pmatrix} x_k + u_k \begin{pmatrix} 0 \\ 0 \\ 1 \end{pmatrix}$$

$$y_k = (c_2 \ c_1 \ c_0) x_k + c_3 u_k.$$

Here $x_k \in \mathbb{R}^3$ for $k = 0, 1, \ldots$. The initial state, x_0, depends on the signal before it is input into the filter. It will be convenient to rewrite the system more abstractly as

$$x_{k+1} = A x_k + u_k b, \quad y_k = c^T x_k + d u_k,$$

where the transpose of a vector was defined just after Lemma 1.99.

The filter is said to be *controllable* if for any given output x_f, there is at least one input sequence $\{u_0, u_1, u_2\}$ such that $x_3 = x_f$. The filter is said to be *observable* if from the observations y_0, y_1, y_2, it is possible to determine the initial state, x_0. The evaluation of the controllability and observability of digital filters is a key step in finding optical structures that minimize output-noise power. We will determine when the system is controllable and leave the question as to whether the system is observable for a group project.

In order to determine if the filter is controllable, we first must have an expression for x_3. Solving the state-space equation yields

$$\begin{aligned} x_1 &= Ax_0 + u_0 b, \\ x_2 &= Ax_1 + u_1 b \\ &= A(Ax_0 + u_0 b) + u_1 b \\ &= A^2 x_0 + u_0 Ab + u_1 b, \\ x_3 &= Ax_2 + u_2 b \\ &= A(A^2 x_0 + u_0 Ab + u_1 b) + u_2 b \\ &= A^3 x_0 + u_0 A^2 b + u_1 Ab + u_2 b. \end{aligned}$$

In the above, we define A^ℓ to be the matrix A multiplied by itself ℓ times. Recalling the definition of matrix/vector multiplication, if we set

$$C_3 = \begin{pmatrix} A^2 b & Ab & b \end{pmatrix} \in \mathcal{M}_3(\mathbb{R}), \quad u = \begin{pmatrix} u_0 \\ u_1 \\ u_2 \end{pmatrix},$$

then the solution can be written as

$$x_3 = A^3 x_0 + C_3 u.$$

If the system is controllable, there is a vector u such that $x_3 = x_f$. This yields

$$x_f = A^3 x_0 + C_3 u \quad \rightsquigarrow \quad C_3 u = x_f - A^3 x_0.$$

We now have a linear system of equations in which the coefficient matrix is C_3. This system will have a unique solution for any x_f if and only if $\mathrm{rank}(C_3) = 3$, i.e., if and only if the matrix is invertible. Now, by direct computation,

$$C_3 = \begin{pmatrix} 1 & 0 & 0 \\ a_1 & 1 & 0 \\ a_2 + a_1^2 & a_1 & 1 \end{pmatrix} \quad \rightsquigarrow \quad \det(C_3) = 1.$$

Since the determinant is nonzero, the matrix is invertible. In conclusion, the system is controllable.

Now suppose it is desired that the desired output x_f be achieved for some $n \geq 4$. Following the logic leading to the representation for x_3, we find that

$$x_n = A^n x_0 + C_n u,$$

where now

$$C_n = \begin{pmatrix} A^{n-1} b & A^{n-2} b & \cdots & A^2 b & Ab & b \end{pmatrix} \in \mathcal{M}_{3 \times n}(\mathbb{R}), \quad u = \begin{pmatrix} u_0 \\ u_1 \\ \vdots \\ u_{n-1} \end{pmatrix}.$$

Since

(a) the linearly independent columns of C_3 are the rightmost three columns of C_n and

(b) C_n has three rows, so the matrix has at most three pivot columns,

we have rank(C_n) = 3. Consequently, the linear system,

$$C_n u = x_f - A^n x_0,$$

is consistent. Since $n \geq 4$, there are an infinite number of solutions; indeed, dim[Null(C_n)] = $n - 3$ means the solution will be parameterized by $n - 3$ free variables. Thus, as n increases, there is larger and larger class of input signals (as determined by the number of free variables) that lead to the desired state x_f.

1.13.2 ▪ Voter registration

Consider the following table:

	D	R	I
D	0.90	0.03	0.10
R	0.02	0.85	0.20
I	0.08	0.12	0.70

Here R represents Republicans, D Democrats, and I Independents. Let D_j, R_j, I_j represent the number of voters in each group in year j. The table provides information regarding the manner in which voters change their political affiliation from one year to the next. For example, reading down the first column, we see that from one year to the next, 90% of the Democrats remain Democrats, 2% become Republicans, and 8% become Independents. On the other hand, reading across the first row, we see that the number of Democrats in a following year is the sum of 90% of the Democrats, 3% of the Republicans, and 10% of the Independents in the preceding year.

We wish to know what is the distribution of voters amongst the three groups after many years. Using the table, we see that the number of voters in each group in year $n + 1$ given the number of voters in each group in year n follows the rule

$$D_{n+1} = 0.90 D_n + 0.03 R_n + 0.10 I_n,$$
$$R_{n+1} = 0.02 D_n + 0.85 R_n + 0.20 I_n,$$
$$I_{n+1} = 0.08 D_n + 0.12 R_n + 0.70 I_n.$$

We implicitly assume here that the total number of voters is constant from one year to the next, so $D_n + R_n + I_n = N$ for any n, where N is the total number of voters. On setting

$$x_n = \begin{pmatrix} D_n \\ R_n \\ I_n \end{pmatrix}, \quad M = \begin{pmatrix} 0.90 & 0.03 & 0.10 \\ 0.02 & 0.85 & 0.20 \\ 0.08 & 0.12 & 0.70 \end{pmatrix},$$

we can rewrite this as the discrete *dynamical system*

$$x_{n+1} = M x_n, \quad x_0 \text{ given.} \qquad (1.13.1)$$

The dynamical system (1.13.1) is known as a *Markov process*, and it is distinguished by the fact that the sum of each column of the transition (stochastic, Markov) matrix M is 1.

For a given initial distribution of voters x_0, we wish to determine the distribution of voters after many years; i.e., we wish to compute $\lim_{n \to +\infty} x_n$. First, we need to solve for x_n. Since

$$x_1 = Mx_0, \quad x_2 = Mx_1 = M(Mx_0),$$

by defining $M^k := MM \cdots M$, i.e., M^k is the matrix M multiplied by itself k times, we have

$$x_2 = M^2 x_0.$$

Continuing in this fashion gives

$$x_3 = Mx_2 = M(M^2 x_0) = M^3 x_0, \quad x_4 = Mx_3 = M(M^3 x_0) = M^4 x_0,$$

so by an induction argument, the solution to the dynamical system is

$$x_n = M^n x_0. \tag{1.13.2}$$

Thus, our question is answered by determining $\lim_{n \to +\infty} M^n$.

We now use the eigenvalues and eigenvectors of M and the Fourier expansion result of Lemma 1.99 in order to simplify the expression (1.13.2). Using SAGE,

$$\lambda_1 = 1, \; v_1 = \begin{pmatrix} 1 \\ 22/21 \\ 24/35 \end{pmatrix}; \quad \lambda_2 = \frac{43}{50}, \; v_2 = \begin{pmatrix} 1 \\ -6/7 \\ -1/7 \end{pmatrix}; \quad \lambda_3 = \frac{59}{100}, \; v_3 = \begin{pmatrix} 1 \\ 3 \\ -4 \end{pmatrix}. \tag{1.13.3}$$

Since the eigenvalues are distinct, by Theorem 1.98(d) the associated eigenvectors are linearly independent. Letting $P = (v_1 \; v_2 \; v_3)$, we know by Lemma 1.99 that the initial condition has a Fourier expansion in terms of the eigenvectors:

$$x_0 = c_1 v_1 + c_2 v_2 + c_3 v_3 = Pc \quad \leadsto \quad c = P^{-1} x_0. \tag{1.13.4}$$

Now that the initial condition has been written in terms of the eigenvectors, we can rewrite the solution in terms of the eigenvalues and eigenvectors. Via the linearity of matrix/vector multiplication and using the expansion (1.13.4), (1.13.2) can be rewritten as

$$x_n = c_1 M^n v_1 + c_2 M^n v_2 + c_3 M^n v_3. \tag{1.13.5}$$

Regarding the term $M^n v_\ell$, for each $\ell = 1, 2, 3$,

$$M v_\ell = \lambda_\ell v_\ell \quad \leadsto \quad M^2 v_\ell = M(M v_\ell) = \lambda_\ell M v_\ell = \lambda_\ell^2 v_\ell,$$

which by an induction argument leads to

$$M^n v_\ell = \lambda_\ell^n v_\ell$$

(see Exercise 1.12.1.) Substitution of the above into (1.13.5) then gives the solution in the form

$$x_n = c_1 \lambda_1^n v_1 + c_2 \lambda_2^n v_2 + c_3 \lambda_3^n v_3, \quad c = P^{-1} x_0. \tag{1.13.6}$$

We are now ready to determine the asymptotic limit of the solution. Using the eigenvalues as described in (1.13.3), we have

$$\lim_{n\to+\infty} \lambda_1^n = 1, \quad \lim_{n\to+\infty} \lambda_2^n = \lim_{n\to+\infty} \lambda_3^n = 0.$$

Consequently, for the solution formula (1.13.6), we have the asymptotic limit:

$$\lim_{n\to+\infty} x_n = c_1 v_1.$$

From this formula, we see that it is important only to determine c_1. Since the total number of people must be constant and must be the same for each n, in the limit the total number of people is the same as the beginning number of people:

$$c_1\left(1 + \frac{22}{21} + \frac{24}{35}\right) = N \rightsquigarrow c_1 = \frac{105}{287}N.$$

This observation allows us to write

$$c_1 v_1 = \frac{N}{287}\begin{pmatrix} 105 \\ 110 \\ 72 \end{pmatrix} \sim N\begin{pmatrix} 0.37 \\ 0.38 \\ 0.25 \end{pmatrix}.$$

In conclusion,

$$\lim_{n\to+\infty} x_n \sim N\begin{pmatrix} 0.37 \\ 0.38 \\ 0.25 \end{pmatrix},$$

so in the long run, 37% of the voters are Democrats, 38% are Republicans, and 25% are Independents. Note that this final distribution of voters is independent of the initial distribution of voters.

What is "long run" in this case? Since

$$\lambda_2^n, \lambda_3^n < 10^{-4} \rightsquigarrow n \geq 62,$$

the terms $c_2 \lambda_2^n v_2$ and $c_3 \lambda_3^n v_3$ in the solution Fourier expansion (1.13.6) will be negligible for $n \geq 62$. Thus, for $n \geq 62$, the solution will essentially be the asymptotic limit, which means that after 62 years, the distribution of voters will be, for all intents and purposes, that given above.

1.13.3 ▪ Discrete SIR model

The dynamics of epidemics are often based on SIR models. In a given population, there are three subgroups:

(a) susceptible (S): those who are able to get a disease but have not yet been infected;

(b) infected (I): those who are currently fighting the disease; and

(c) recovered (R): those who have had the disease or are immune.

Although it is not necessary, it is often assumed that the entire population,

$$N = S + I + R,$$

is a constant. Moreover, it is assumed that the number of people in each group does not depend on location. Consequently, the model to be given is reasonable when looking at epidemics in a school environment, but it not very good when trying to understand nationwide outbreaks of disease (which are generally more regional).

Paladini et al. [33] provide a descriptive discrete-time dynamical system of the form

$$S_{n+1} = qS_n + cR_n,$$
$$I_{n+1} = (1-q)S_n + bI_n, \quad (1.13.7)$$
$$R_{n+1} = (1-b)I_n + (1-c)R_n.$$

Here S_j is the number of susceptible people in the sampling interval j, I_j is the number of infected people in the sampling interval j, and R_j is the number of recovered people in the sampling interval j. Depending on the disease being studied, the sampling interval may be monthly, yearly, or even larger. The model assumes that

(a) susceptible people must become infected before recovering,

(b) infected people must recover before again becoming susceptible, and

(c) recovered people cannot become infected without first becoming susceptible.

As for the parameters, we have

(a) $0 \leq q \leq 1$ is the probability that a susceptible avoids infection,

(b) $0 \leq b \leq 1$ is the proportion of individuals which remain infected, and

(c) $0 \leq c \leq 1$ is the fraction of recovered individuals which lose immunity.

The probability parameter q is generally a function of both S and I, e.g.,

$$q = 1 - p\frac{I}{N},$$

where p is the probability of the infection being transmitted through a time of contact. We will assume that q is fixed; in particular, we will assume that it does not depend on the proportion of infected people. It is not difficult to check that

$$S_{n+1} + I_{n+1} + R_{n+1} = S_n + I_n + R_n,$$

so the total population remains constant for all n (see Exercise 1.13.4). We could use this fact to reduce the number of variables in (1.13.7), but we will not do so in our analysis.

We now proceed to solve (1.13.7). On setting

$$x_n = \begin{pmatrix} S_n \\ I_n \\ R_n \end{pmatrix}, \quad A = \begin{pmatrix} q & 0 & c \\ 1-q & b & 0 \\ 0 & 1-b & 1-c \end{pmatrix},$$

we can rewrite the dynamical system in the form

$$x_{n+1} = Ax_n. \quad (1.13.8)$$

Note that this dynamical system shares (at least) one feature with the Markov process associated with the voter registration problem: the sum of each column of the matrix A is 1. By following the argument leading to (1.13.2), we know that the solution is
$$x_n = A^n x_0.$$
Moreover, we know that the eigenvalues and associated eigenvectors of A can be used to simplify the form of the solution. Writing $Av_j = \lambda_j v_j$ for $j = 1,2,3$, we know that the solution can be written as
$$x_n = c_1 \lambda_1^n v_1 + c_2 \lambda_2^n v_2 + c_3 \lambda_3^n v_3. \qquad (1.13.9)$$
The underlying assumption leading to the solution formula in (1.13.9), which will be verified for specific values of b, c, q, is that the eigenvectors are linearly independent. We then know by Lemma 1.99 that the initial condition x_0 has the Fourier expansion
$$x_0 = c_1 v_1 + c_2 v_2 + c_3 v_3,$$
and the Fourier coefficients are found by solving the linear system
$$Pc = x_0 \quad \leadsto \quad c = P^{-1} x_0, \quad P = (v_1 \; v_2 \; v_3).$$

We are now ready to determine the asymptotic limit of the solution. Following [33], we will assume that
$$b = 0.5, \quad c = 0.01.$$
If we further assume that $q = 0.2$, then the matrix A becomes
$$A = \begin{pmatrix} 0.2 & 0 & 0.01 \\ 0.8 & 0.5 & 0 \\ 0 & 0.5 & 0.99 \end{pmatrix}.$$

Using SAGE, we find the eigenvalues and associated eigenvectors:
$$\lambda_1 = 1, \; v_1 = \begin{pmatrix} 1 \\ 8/5 \\ 80 \end{pmatrix}; \quad \lambda_2 \sim 0.47, \; v_2 \sim \begin{pmatrix} 1.00 \\ -28.16 \\ 27.16 \end{pmatrix};$$
$$\lambda_3 \sim 0.22, \; v_3 \sim \begin{pmatrix} 1.00 \\ -2.84 \\ 1.84 \end{pmatrix}.$$

Because the eigenvalues are distinct, we know that the associated eigenvectors are linearly independent.

Since
$$\lim_{n \to +\infty} \lambda_1^n = 1, \quad \lim_{n \to +\infty} \lambda_2^n = \lim_{n \to +\infty} \lambda_3^n = 0,$$
we have the asymptotic limit
$$\lim_{n \to +\infty} x_n = c_1 v_1.$$

We see that we now must determine c_1. Since the total number of people is constant for all n, in the limit the total number of people is the same as the beginning number of people, which leads to
$$c_1 \left(1 + \frac{8}{5} + 80\right) = N \quad \leadsto \quad c_1 = \frac{5}{413} N.$$

This observation allows us to write

$$c_1 v_1 = \frac{N}{413} \begin{pmatrix} 5 \\ 8 \\ 400 \end{pmatrix} \sim N \begin{pmatrix} 0.012 \\ 0.019 \\ 0.969 \end{pmatrix}.$$

In conclusion,

$$\lim_{n \to +\infty} x_n = N \begin{pmatrix} 0.012 \\ 0.019 \\ 0.969 \end{pmatrix},$$

so in the long run, 1.2% of the people are susceptible, 1.9% of the people are infected, and 96.9% of the people are recovered. Note that this final distribution of the population is independent of the number of people who were originally infected.

What is "long run" in this case? Since

$$\lambda_2^n, \lambda_3^n < 10^{-4} \quad \leadsto \quad n \geq 13,$$

the terms $c_2 \lambda_2^n v_2$ and $c_3 \lambda_3^n v_3$ in the solution Fourier expansion (1.13.6) will be negligible for $n \geq 13$. Thus, for $n \geq 13$, the solution will essentially be the asymptotic limit, which means that after 13 sampling intervals, the distribution of people will be, for all intents and purposes, that given above.

1.13.4 ▪ Northern spotted owl

The size of the Northern spotted owl population is closely associated with the health of the mature and old-growth coniferous forests in the Pacific Northwest. Over the past few decades, there has been loss and fragmentation of these forests, which may potentially affect the long-term survival of this species of owl. For spotted owls, there are three distinct groupings:

(a) juveniles (j) under 1 year old,

(b) subadults (s) between 1 and 2 years old, and

(c) adults (a) 2 years old and older.

The owls mate during the latter two life stages and begin breeding as adults.

In year n, let j_n be the number of juveniles, s_n be the number of subadults, and a_n be the number of adults. Mathematical ecologists have modeled a particular spotted owl population via the discrete dynamical system

$$j_{n+1} = 0.33 a_n,$$
$$s_{n+1} = 0.18 j_n,$$
$$a_{n+1} = 0.71 s_n + 0.94 a_n.$$

The juvenile population in the next year is 33% of the adult population, 18% of the juveniles in one year become subadults in the next year, 71% of the subadults in one year become adults the next year, and 94% of the adults survive from one year to the next (see Lamberson et al. [23], Lay [24, Chapter 5], and the references therein). On setting

$$x_n = \begin{pmatrix} j_n \\ s_n \\ a_n \end{pmatrix},$$

1.13. Case studies

we can rewrite this dynamical system in the form

$$x_{n+1} = Ax_n, \quad A = \begin{pmatrix} 0 & 0 & 0.33 \\ 0.18 & 0 & 0 \\ 0 & 0.71 & 0.94 \end{pmatrix}. \tag{1.13.10}$$

For a given initial distribution of owls, we wish to see what is the distribution of the owls after many years. We first solve for x_n in terms of x_0. Following the argument leading to (1.13.2), we know that the solution is

$$x_n = A^n x_0. \tag{1.13.11}$$

Following our discussion in the previous case study, we know that we next wish to use the eigenvalues and associated eigenvectors for the matrix A. Using SAGE,

$$\lambda_1 \sim 0.98, \; v_1 \sim \begin{pmatrix} 1.00 \\ 0.18 \\ 2.98 \end{pmatrix}; \quad \lambda_2 \sim -0.02 + i0.21, \; v_2 \sim \begin{pmatrix} 1.00 \\ -0.09 - i0.86 \\ -0.06 + i0.62 \end{pmatrix}.$$

The third eigenvalue and associated eigenvector satisfy $\lambda_3 = \overline{\lambda_2}$ and $v_3 = \overline{v_2}$. By following the logic leading to (1.13.6), we know that the solution is

$$x_n = c_1 \lambda_1^n v_1 + c_2 \lambda_2^n v_2 + c_3 \lambda_3^n v_3, \quad x_0 = c_1 v_1 + c_2 v_2 + c_3 v_3. \tag{1.13.12}$$

Because the eigenvectors are complex valued, the Fourier coefficients may also be.

The asymptotic behavior of the solution depends on the size of the eigenvalues. Looking back to the solution formula (1.13.12), we need to understand what happens when we take successive powers of the eigenvalues. While we understand what λ_1^n means, we do not have an intuitive understanding as to what it means when we write λ_2^n and λ_3^n. Recall that we showed in (1.11.1) that complex numbers $z = a + ib$ can be written in the polar form:

$$z = |z|e^{i\theta}; \quad e^{i\theta} = \cos\theta + i\sin\theta, \; \tan\theta = \frac{b}{a}.$$

The polar representation of complex numbers allows us to write

$$z^n = |z|^n \left(e^{i\theta}\right)^n = |z|^n e^{in\theta}.$$

In particular, we have

$$\lambda_2 = 0.21 e^{i0.53\pi} \quad \leadsto \quad \lambda_2^n = 0.21^n e^{i0.53n\pi},$$
$$\lambda_3 = 0.21 e^{-i0.53\pi} \quad \leadsto \quad \lambda_3^n = 0.21^n e^{-i0.53n\pi}.$$

Since

$$|e^{i\theta}| = 1 \quad \leadsto \quad |e^{in\theta}| = |e^{i\theta}|^n = 1,$$

the magnitude of z^n is controlled solely by the magnitude of z:

$$|z^n| = |z|^n \left|e^{in\theta}\right| = |z|^n.$$

Thus, in our example, it will be the case that $|\lambda_2^n| = |\lambda_3^n| < 10^{-4}$ for $n \geq 6$. Going back to the solution formula (1.13.12), we see that for $n \geq 6$, we can write it as

$$x_n \sim 0.98^n \, c_1 v_1.$$

In order to properly interpret this solution formula, we first want to write the eigenvector v_1 so that each entry corresponds to the percentage of the total owl population in each subgroup. This requires that the entries of the eigenvector sum to one. We know that eigenvectors are not unique and can be scaled in any desired fashion. Using SAGE, we get the mapping

$$v_1 \sim \begin{pmatrix} 1.00 \\ 0.18 \\ 2.98 \end{pmatrix} \mapsto \frac{1}{1+0.18+2.98} \begin{pmatrix} 1.00 \\ 0.18 \\ 2.98 \end{pmatrix} \sim \begin{pmatrix} 0.24 \\ 0.04 \\ 0.72 \end{pmatrix},$$

so the desired eigenvector is now

$$v_1 \sim \begin{pmatrix} 0.24 \\ 0.04 \\ 0.72 \end{pmatrix}.$$

As for the constant c_1, we first rewrite the system (1.13.12) in matrix/vector form:

$$x_0 = Pc, \quad P = (v_1 \; v_2 \; v_3).$$

We use SAGE to solve this linear system:

$$c = P^{-1} x_0 \quad \rightsquigarrow \quad c_1 \sim 0.17 j_0 + 0.93 s_0 + 1.28 a_0.$$

In conclusion, we have

$$c_1 v_1 \sim [0.17 j_0 + 0.93 s_0 + 1.28 a_0] \begin{pmatrix} 0.24 \\ 0.04 \\ 0.72 \end{pmatrix}.$$

Thus, for $n \geq 6$, we can say that

$$x_n \sim 0.98^n [0.17 j_0 + 0.93 s_0 + 1.28 a_0] \begin{pmatrix} 0.24 \\ 0.04 \\ 0.72 \end{pmatrix}.$$

Roughly 24% of the owls will be juveniles, 4% of the owls will be subadults, and 72% of the owls will be adults. The total number of owls in each group will depend on the initial distribution. The overall population will slowly decrease, and assuming no changes in the conditions leading to the original model (1.13.10), the owls will *eventually* become extinct (e.g., $0.98^n \leq 0.1$ for $n \geq 114$).

1.13. Case studies

Exercises

Exercise 1.13.1. *Consider the below table, which represents the fraction of the population in each group—City (C) and Suburban (S)—that migrates to a different group in a given year. Assume that the total population is constant. Further assume that there are initially 1500 city dwellers and 1000 suburbanites. How many people will there be in each group after many years?*

	C	S
C	0.94	0.15
S	0.06	0.85

Exercise 1.13.2. *Consider the below table, which represents the fraction of the population in each group—City (C), Suburban (S), and Rural (R)—that migrates to a different group in a given year. Assume that the total population is constant. Further assume that there are initially 1000 city dwellers, 750 suburbanites, and 250 rural dwellers. How many people will there be in each group after many years?*

	C	S	R
C	0.91	0.09	0.02
S	0.05	0.87	0.08
R	0.04	0.04	0.90

Exercise 1.13.3. *Set*

$$A = \begin{pmatrix} 0.4 & 0.2 & 0.0 \\ 0.5 & 0.7 & 0.0 \\ 0.1 & 0.3 & 0.8 \end{pmatrix}, \quad b = \begin{pmatrix} b_1 \\ b_2 \\ b_3 \end{pmatrix}.$$

For the state-space equations,

$$x_{k+1} = Ax_k + u_k b$$

determine the value(s) of $b_1, b_2, b_3 \in \mathbb{R}$ for which the pair (A, b) is NOT controllable.

Exercise 1.13.4. *Consider the SIR model given in (1.13.7).*

(a) *Show that the model supports that the total population is fixed, i.e., show that $S_n + I_n + R_n = S_0 + I_0 + R_0$ for all $n \geq 1$.*

(b) *Writing $N = S_n + I_n + R_n$, show that the system is equivalent to*

$$S_{n+1} = (q-1)S_n - cI_n + cN,$$
$$I_{n+1} = (1-q)S_n + bI_n.$$

(c) *If one solves for (S_n, I_n) in (b), how is R_n found?*

Exercise 1.13.5. *Consider the SIR case study of section 1.13.3. Suppose that $b = 0.8$ and $c = 0.01$. Further suppose that n is large.*

(a) *If $q = 0.2$, what percentage of the total population will be comprised of infected people?*

(b) *If $q = 0.7$, what percentage of the total population will be comprised of infected people?*

Exercise 1.13.6. *Consider the SIR case study of section 1.13.3. Suppose that $c = 0.1$ and $q = 0.4$. Further suppose that n is large.*

(a) *If $b = 0.05$, what percentage of the total population will be comprised of recovered people?*

(b) *If $b = 0.35$, what percentage of the total population will be comprised of recovered people?*

Exercise 1.13.7. *Consider the SIR case study of section 1.13.3. Suppose that $b = 0.4$ and $q = 0.3$. Further suppose that n is large.*

(a) *If $c = 0.02$, what percentage of the total population will be comprised of susceptible people?*

(b) *If $c = 0.25$, what percentage of the total population will be comprised of susceptible people?*

Exercise 1.13.8. *Consider the case study of the Northern spotted owl in section 1.13.4. Let the fraction of subadults who become adults be represented by the parameter r (replace 0.71 with r in the matrix of (1.13.10)). Suppose that n is large.*

(a) *If $r = 0.1$, what percentage of the total population will be comprised of subadults?*

(b) *If $r = 0.45$, what percentage of the total population will be comprised of subadults?*

Exercise 1.13.9. *Consider the case study of the Northern spotted owl in section 1.13.4. Let the fraction of adults who survive from one year to the next be represented by the parameter r (replace 0.94 with r in the matrix of (1.13.10)). Suppose that n is large.*

(a) *If $r = 0.65$, what percentage of the total population will be comprised of adults?*

(b) *If $r = 0.85$, what percentage of the total population will be comprised of adults?*

Exercise 1.13.10. *Consider the case study of the Northern spotted owl in section 1.13.4. Let the fraction of juveniles who survive from one year to the next be represented by the parameter r (replace 0.18 with r in the matrix of (1.13.10)). Suppose that n is large.*

(a) *If $r = 0.30$, what percentage of the total population will be comprised of juveniles?*

(b) *If $r = 0.50$, what percentage of the total population will be comprised of juveniles?*

Group projects

1.1. A digital filter of order n can be represented by the *state-space model*,

$$x_{k+1} = Ax_k + u_k b,$$
$$y_k = c^T x_k + d u_k,$$

for $k = 0, 1, \ldots$ (see Lu et al. [27]). Here $A \in \mathcal{M}_n(\mathbb{R})$ and $b, c \in \mathbb{R}^n$. The scalar u_k is an input signal, and the scalar y_k is the output signal generated by the filter. The filter is said to be observable if from the given sequence of signals $y_0, y_1, \ldots$, one can determine the initial state x_0 (which corresponds to the input signal before the filter was activated).

(a) For a given initial state x_0, find x_n. Write your solution only in terms of the initial state x_0 and the control scalars.

(b) Setting

$$y = \begin{pmatrix} y_0 \\ y_1 \\ \vdots \\ y_{n-1} \end{pmatrix}, \quad u = \begin{pmatrix} u_0 \\ u_1 \\ \vdots \\ u_{n-1} \end{pmatrix},$$

show that $y = Ox_0 + Cu$. Explicitly, identify $O, C \in \mathcal{M}_n(\mathbb{R})$.

(c) Show that the system is observable if and only if $\text{rank}(O) = n$.

(d) Set

$$A = \begin{pmatrix} 8/10 & 3/10 & 0 \\ 2/10 & 5/10 & 1 \\ 0 & 0 & 5/10 \end{pmatrix}, \quad c = \begin{pmatrix} c_1 \\ c_2 \\ c_3 \end{pmatrix}.$$

Determine the value(s) of $c_1, c_2, c_3 \in \mathbb{R}$ for which the system is NOT observable.

1.2. The goal of *Google's PageRank algorithm* is to determine the "importance" of a given webpage. For example, Wikipedia[7] is more important than the webpage of Todd Kapitula.[8] Importance can be quantified through a few basic rules:

(a) The importance of page A is measured by the likelihood that web surfer S will visit A.

(b) The most likely way for S to reach A from page B is to click on a link to A (versus randomly choosing A from the billions of potential pages).

(c) S is more likely to click on the link from A to B (rather than on a link to some other page C from B) if there are not many links from B to other pages.

(d) In order to click on a link from B to A, S must already be on B.

(e) S is more likely to be on B if B is also important.

---

[7] https://www.wikipedia.org/.
[8] http://www.calvin.edu/~tmk5/.

In summary, important pages are linked to by many other important pages; on the other hand, if a webpage links to many other webpages, the value of each link is watered down.

For example, consider the network of pages shown below. Webpages A, B, C, and D are least important because no other page connects to them. Webpages E and F are each linked to be three other pages (A, B, and C for E and A, D, and E for F); moreover, one of the pages linking to F is important (namely, E), but none of the pages linking to E are important. Consequently, webpage F is more important than webpage E. In addition, page A contributes less to the importance of pages E and F than do pages B, C, and D, as it links to two pages, whereas the others link to one each. These extra links, however, do not affect the ranking of A; instead, they affect only the rank of the pages linked to by A.

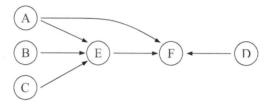

We begin in the following manner to make this discussion quantitative. Let R_A be the ranking of page A (similarly for other pages). Let ℓ_{jk} denote whether or not there is a link from webpage j to webpage k:

$$\ell_{jk} = \begin{cases} 0, & \text{no link from } j \text{ to } k, \\ 1, & \text{link from } j \text{ to } k. \end{cases}$$

For example, in the above network, $\ell_{AE} = 1$, but $\ell_{EA} = 0$. We (initially) assume that a webpage does not link to itself, so $\ell_{jj} = 0$. The total number of links from page j to other pages is given by

$$n_j = \sum_k \ell_{jk}.$$

For example, in the above network,

$$n_A = 2, \quad n_E = 1, \quad n_F = 0.$$

The ranking of a webpage will be defined to be the weighted sum of the number of number of links from other pages, with the weight being the ratio of the page rank and total number of links:

$$R_j = \sum_k \frac{\ell_{kj}}{n_k} R_k.$$

For example, in the sample network,

$$R_A = 0, \quad R_E = \frac{1}{2} R_A + R_B + R_C, \quad R_F = \frac{1}{2} R_A + R_D + R_E.$$

Finally, we assume for the network that the total rank (the sum of all the ranks) is one, so for the sample system,

$$R_A + R_B + R_C + R_D + R_E + R_F = 1.$$

The current model has the problem that a surfer who ends up on a page with no external links cannot go to a new page (e.g., page F). For those webpages, we will assume that the surfer will choose randomly from any page in the entire network. In other words, if the network has N webpages, and webpage j has no links to any other webpages (so initially $\ell_{jk} = 0$ for $k = 1,\ldots,N$), then we will set $\ell_{jk} = 1$ for each k. The total number of links then changes from 0 to N. For example, in the network shown above, we reset $\ell_{Fk} = 1$ for $k \in \{A,B,C,D,E,F\}$, so that $n_F = 6$. The modified ranking formula for pages E and F are

$$R_E = \frac{1}{2}R_A + R_B + R_C + \frac{1}{6}R_F, \quad R_F = \frac{1}{2}R_A + R_D + R_E + \frac{1}{6}R_F.$$

Note well that this modification does not change the requirement that the total rank is one.

The revised model still does not take into account that a web surfer might randomly select a new webpage from the network instead of clicking on a link from the current page. We will let $0 < d \leq 1$ (the *damping factor*) denote the probability that a web surfer uses a link on the current page to get to the next page ($1-d$ is then the probability that a new page will be randomly selected). If $d = 1$, the surfer uses only the links to go to the next webpage, and if $d = 0$, the links are never used to go to the next webpage. We will assume that the damping factor is fixed and constant for all webpages. The possibility of randomly selecting a new webpage affects the general ranking formula via

$$R_j \mapsto dR_j + \frac{1-d}{N} = dR_j + \frac{1-d}{N}\sum_k R_k.$$

The latter equality follows from the total rank being one. Going back to the specific example, the ranking formula for pages E and F are

$$R_E = d\left(\frac{1}{2}R_A + R_B + R_C + \frac{1}{6}R_F\right) + \frac{1-d}{6}(R_A + R_B + R_C + R_D + R_E + R_F),$$
$$R_F = d\left(\frac{1}{2}R_A + R_D + R_E + \frac{1}{6}R_F\right) + \frac{1-d}{6}(R_A + R_B + R_C + R_D + R_E + R_F).$$

For the project, consider the network of eight webpages as follows:

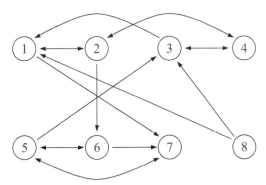

Set $r = (R_1\ R_2\ \cdots\ R_8)^T$, where R_j is the rank of page j. Suppose that the damping factor is $d = 0.9$.

(a) Find the matrix $A \in \mathcal{M}_8(\mathbb{R})$ such that
$$Ar = r \quad \leadsto \quad (A - I_8)r = 0.$$

(b) Verify that the matrix has the properties that
- all the entries are nonnegative and
- the sum of the values in each column is one.

(c) The system can be numerically solved using an iterative method (this is preferred for large networks). Set $r_0 = (1\ 0\ \cdots\ 0)^T$ and define the vectors $r_1, r_2, \ldots$ via
$$r_{n+1} = Ar_n.$$
Give an expression for r_n in terms of the initial vector r_0. Explain.

(d) Determine $r^* = \lim_{n \to \infty} r_n$.

(e) Is r^* a solution to the linear system of (a)? Why or why not?

(f) What is a value N such that if $n \geq N$, then we can expect the difference between r_n and the solution to the linear system of (a) to be $\mathcal{O}(10^{-4})$? Describe how you chose this value of N.

(g) Which webpage is most important, and what is its page rank?

(h) Which webpage is least important, and what is its page rank?

Author's note: The preceding discussion and facets of this project were provided to me by Prof. Kelly Mcquighan.

Chapter 2
Scalar first-order linear differential equations

> *One cannot understand... the universality of laws of nature, the relationship of things, without an understanding of mathematics. There is no other way to do it.*
>
> —Richard Feynman

We now have a solid foundation in linear algebra. We are ready to apply what we have learned to a study of (ODEs). A first-order ODE is a functional relationship between the rate of change of a function and the function itself. For example, it is a description of how the velocity of a particle depends on its position. For a linear ODE, the functional relationship is linear. Given that we are well versed in linearity and its implications, we will focus most of our attention in the rest of this book on *linear* ODEs.

It will turn out to be the case that if we first focus our attention on linear scalar first-order ODEs, then we will develop (almost) all of the ideas and techniques necessary to solve larger problems. Before continuing, we first consider a set of problems that are naturally modeled by scalar and linear ODEs.

2.1 • Motivating problems

Consider a reservoir that is filled with V_0 L of a brine solution and suppose that it initially contains x_0 g of salt; in other words, the initial concentration of salt in the reservoir is x_0/V_0 g/L. Suppose that the brine solution is drawn from the reservoir at a rate of $0.05 V_0$ L/day and further suppose that a new brine solution with a concentration of 0.03 g/L flows into the reservoir at the same rate as the outflow. The situation is depicted in Figure 2.1. While the total volume of the brine solution remains unchanged, the concentration changes as a function of time. It will be assumed that the reservoir is well stirred, which implies that the concentration is independent of the location in the reservoir. If this assumption is removed, so that the concentration depends on both time and location, then the resulting equation will necessarily be a partial differential equation (PDE), which is a topic of study for another course.

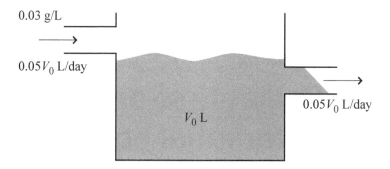

Figure 2.1. *A cartoon that depicts the mixing of a brine solution in a reservoir.*

Our initial goal is to write down a mathematical model that describes this physical situation. Let $x(t)$ represent the number of grams in the reservoir at time t, so that the concentration is given by $c(t) = x(t)/V_0$. The instantaneous rate of change of salt in the reservoir, $x'(t)$, satisfies the equation

$$x'(t) = [\text{inflow rate}] - [\text{outflow rate}].$$

The inflow rate is the rate at which salt enters the reservoir, and the outflow rate is the rate at which it leaves. Note that the units for $x'(t)$ are g/day. Now,

$$[\text{inflow rate}] = \left(0.03 \frac{\text{g}}{\text{L}}\right)\left(0.05 V_0 \frac{\text{L}}{\text{day}}\right) = 0.0015 V_0 \frac{\text{g}}{\text{day}},$$

and

$$[\text{outflow rate}] = \left(\frac{x(t)\,\text{g}}{V_0\,\text{L}}\right)\left(0.05 V_0 \frac{\text{L}}{\text{day}}\right) = 0.05 x(t) \frac{\text{g}}{\text{day}}.$$

Thus, $x(t)$ satisfies the *initial value problem* (IVP):

$$\begin{aligned} x' &= 0.0015 V_0 - 0.05x, \quad x(0) = x_0 \\ &= -0.05x + 0.0015 V_0, \quad x(0) = x_0. \end{aligned}$$

An IVP is the ODE in conjunction with an initial condition.

Now suppose that the rate for which the brine solution leaves the reservoir is lessened to $0.02 V_0$ L/day but that the inflow rate does not change. In this case, the volume in the reservoir is given by $V_0(1 + 0.03t)$, so that

$$[\text{outflow rate}] = \left(\frac{x(t)\,\text{g}}{V_0(1+0.03t)\,\text{L}}\right)\left(.02 V_0 \frac{\text{L}}{\text{day}}\right) = \frac{0.02}{1+0.03t} x(t) \frac{\text{g}}{\text{day}}.$$

The governing ODE in this scenario becomes

$$x' = -\frac{0.02}{1+0.03t} x + 0.0015 V_0. \qquad (2.1.1)$$

For a last scenario, suppose that the concentration of the incoming brine solution is given by $0.03(1 + 0.8 \sin(2\pi t)) V_0$ g/L. In this case,

$$\begin{aligned} [\text{inflow rate}] &= \left(0.03(1+0.8\sin(2\pi t)) V_0 \frac{\text{g}}{\text{L}}\right)\left(0.05 V_0 \frac{\text{L}}{\text{day}}\right) \\ &= 0.0015 V_0^2 (1+0.8\sin(2\pi t)) \frac{\text{g}}{\text{day}}, \end{aligned}$$

2.1. Motivating problems

and the outflow rate is unchanged:

$$[\text{outflow rate}] = \left(\frac{x(t)\,\text{g}}{V_0(1+0.03t)\,\text{L}}\right)\left(0.02\,V_0\,\frac{\text{L}}{\text{day}}\right) = \frac{0.02}{1+0.03t}x(t)\,\frac{\text{g}}{\text{day}}.$$

The governing ODE is then

$$x' = -\frac{0.02}{1+0.03t}x + 0.0015\,V_0^2(1+0.8\sin(2\pi t)). \tag{2.1.2}$$

All of the derived ODEs are scalar linear ODEs of the form

$$x' = a(t)x + f(t). \tag{2.1.3}$$

A nonlinear ODE is

$$x' = g(t,x),$$

where the function $g(t,x)$ has any form other than $g(t,x) = a(t)x + f(t)$. There are many other interesting physical scenarios for which we can derive linear ODEs. Some examples include the following:

(a) mathematical finance

- $x(t)$ is the amount of money in the account at time t.
- $a(t)$ is the interest rate (most often constant).
- $f(t)$ corresponds to rate of withdrawals from ($f(t) < 0$) and/or deposits into ($f(t) > 0$) the account.

(b) population growth

- $x(t)$ is the population density (of say walleye fish) at time t.
- $a(t)$ is the growth ($a(t) > 0$) or decay ($a(t) < 0$) rate (the underlying assumption is that the rate of the population growth/decay is proportional to the current population).
- $f(t)$ corresponds to some external influence (e.g., if $x(t)$ corresponds to a fish population, then $f(t) < 0$ could be a fishing rate).

(c) Newton's law of cooling

- $x(t)$ is the temperature of some object at time t.
- $a(t)$ is the rate at which the temperature changes (assuming that the rate of change of the temperature is proportional to the difference between the object temperature and the surrounding ambient temperature).
- $f(t)$ is proportional to the ambient temperature.

Exercises

Exercise 2.1.1. *Suppose that a tank initially contains V_0 L of a brine solution. Suppose that a brine solution with a concentration of $0.08(2 + 1.4\sin(2\pi t))V_0$ g/L flows into the tank at the constant rate $0.09 V_0$ L/day. Further suppose that the brine solution is drawn from that tank at the constant rate of $0.09 V_0$ L/day. If the initial concentration of the brine solution in the tank is 3.6 g/L, derive the first-order ODE for $x(t)$, which is the amount of salt in the tank at time t.*

Exercise 2.1.2. *The radioactive isotope carbon-14 emits particles and loses half its mass over a period of 5370 years. The instantaneous rate of decay is proportional to the mass present. Let $x(t)$ represent the mass of carbon-14 at time t years.*

(a) *Derive the first-order ODE for $x(t)$.*

(b) *Identify the function $a(t)$.*

(c) *Identify the forcing function $f(t)$.*

Exercise 2.1.3. *Suppose that $x(t)$ is the temperature of a body surrounded by a uniform medium of temperature x_m. Newton's law of cooling states that the temperature of the body is proportional to the difference between the temperature of the body and surrounding temperature.*

(a) *Derive the first-order ODE for $x(t)$.*

(b) *Identify the function $a(t)$.*

(c) *Identify the forcing function $f(t)$.*

2.2 • General theory

2.2.1 • Existence and uniqueness theory

Before we can study the behavior of solutions to the scalar first-order IVP

$$x' = g(t, x), \quad x(t_0) = x_0, \tag{2.2.1}$$

we first need to know that there is something to study. As we have already discussed, a linear ODE has the particular form $g(t, x) = a(t)x + f(t)$. We need to know if there is a solution and furthermore if there is only one solution. The rigorous proof of the following existence/uniqueness theorem is outside the scope of this book; however, it is not difficult to make a plausibility argument. The ODE, while it does not tell us what is the solution curve, does tell us what is the derivative of that curve. Thus, if we were to plot at every point in the tx-plane a little line whose slope is given by $g(t, x)$, then at every point we would know the slope of a given curve. Such a plot is given in Figure 2.2 for $g(t, x) = x(1 - x) - 0.2(1 + \cos(4t))$. It is reasonable to expect that as long as the *direction field* (slope field) $g(t, x)$ is continuous, then we can find a solution to the ODE. If we wish the solution to be unique, then we need an added condition.

2.2. General theory

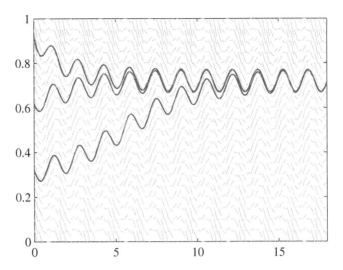

Figure 2.2. *The direction field (slope field) associated with the ODE $x' = x(1-x) - 0.2(1 + \cos(4t))$. Solution curves are calculated for three different initial conditions.*

Theorem 2.1. *Consider the IVP*

$$x' = g(t, x), \quad x(t_0) = x_0.$$

Suppose that there is a constant $C > 0$ such that in the box in the tx-plane, given by $|x - x_0|, |t - t_0| < C$, the functions $g(t, x)$ and $\partial_x g(t, x)$ are continuous. There is then a unique solution; i.e., there is a unique function $\phi(t)$ such that

$$\phi' = g(t, \phi), \quad \phi(t_0) = x_0.$$

Furthermore, the solution exists as long as the curve is contained within the box, i.e., as long as both $|t - t_0| < C$ and $|\phi(t) - x_0| < C$ are true (see Figure 2.3).

There is a simple example that shows that the continuity assumptions on $g(t, x)$ are necessary in order to guarantee that the solution is unique. Consider the IVP

$$x' = \sqrt{|x|}, \quad x(0) = 0.$$

It is clear that $g(t, x) := \sqrt{|x|}$ is continuous for all values of (t, x) and that $\partial_x g(t, x)$ is continuous except at $x = 0$. It can be checked that for any $a \geq 0$, the function

$$\phi_a(t) = \begin{cases} 0, & 0 \leq t \leq a \\ (t-a)^2/4, & a < t, \end{cases}$$

is a solution. Thus, in this case, there is an infinite number of solutions to the IVP. On the other hand, if the initial condition were $x(0) = x_0 \neq 0$, so that the existence/uniqueness theorem applies, then there would be a unique solution; in fact, it is given by $\phi(t) = (t + 2\sqrt{|x_0|})^2/4$, where it is understood that $(\sqrt{|x_0|})^2 = x_0$.

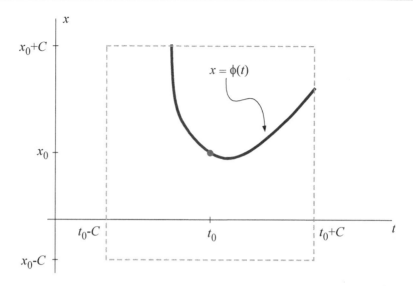

Figure 2.3. *A cartoon that illustrates the existence/uniqueness theory of Theorem 2.1. The thick (blue) curve denotes the solution curve, and the dashed (green) box is the domain on which the functions $g(t,x)$ and $\partial_x g(t,x)$ are both continuous.*

If the IVP (2.2.1) is linear, i.e., if it is of the form given in (2.1.3), then, by using the fact that

$$g(t,x) = a(t)x + f(t) \quad \Rightarrow \quad \partial_x g(t,x) = a(t),$$

we can redefine the box in Theorem 2.1 to be the strip $|t - t_0| < C$, where the constant C is chosen so that $a(t)$ is continuous on the interval $|t - t_0| < C$. For linear ODEs, we then have the more robust result:

Corollary 2.2. *Consider the IVP*

$$x' = a(t)x + f(t), \quad x(t_0) = x_0.$$

Suppose that $a(t), f(t)$ are continuous for $|t - t_0| < C$. There is then a unique solution for $|t - t_0| < C$.

It will henceforth be assumed that whenever we talk about an IVP, there is a unique solution.

2.2.2 ▪ Numerical solutions

Now that we know a unique solution exists to the IVP (2.2.1), we would like to know what it is. It turns out to be the case that if $g(t,x)$ is not linear, i.e., if it is not of the form in (2.1.3), then we may not be able to explicitly write down the solution. However, we can numerically approximate it. This topic of the numerical solution of IVPs is a topic worthy of its own chapter; we will only briefly touch on it here.

2.2. General theory

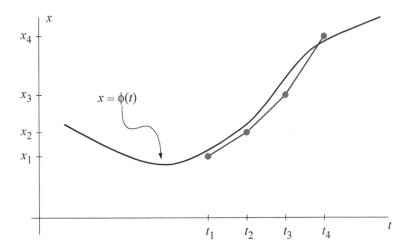

Figure 2.4. *A cartoon that depicts the solution curve $x = \phi(t)$ as well as a piecewise linear approximation.*

Euler's method

Instead of trying to find the solution curve for every single t, we will instead approximate it for discrete values of t and then, if we so desire, linearly extrapolate in order to approximate the solution for other values of t. For example, if $t_1 < t_2$ are two time values for which we have the approximations x_1 and x_2, i.e., $x_1 \sim \phi(t_1)$ and $x_2 \sim \phi(t_2)$ for the solution $\phi(t)$, then for $t_1 < t < t_2$, the approximate solution is given by the line in the tx-plane connecting these two points, which in point-slope form is

$$x(t) - x_1 = \frac{x_2 - x_1}{t_2 - t_1}(t - t_1), \quad t_1 \leq t \leq t_2.$$

(See Figure 2.4 for a pictorial description.)

In order to find the approximate values to the true solution, we will turn the ODE into a *difference equation* and then solve that equation. Recall that in Calculus I, we discussed how the derivative of a function at a point can be approximated by a difference quotient,

$$x'(t) \sim \frac{x(t+h) - x(t)}{h},$$

where $h > 0$. Here and in all that follows, the symbol "$\sim$" means "is approximately." When using this symbol, we will often not be very precise as to how good is the approximation; however, in a given context, the reader will have a reasonable idea of the quality of the approximation.

Supposing that we wish to approximate the solution on the interval $a \leq t \leq b$, for a given $N \geq 1$, set $h = (b-a)/N$ (the *step size*) and define a sequence of t values by

$$t_0 = a, \ t_1 = a + h, t_j = a + jh, \ldots, t_{N-1} = a + (N-1)h, \ t_N = b. \quad (2.2.2)$$

The approximation for the derivative of the solution when $t = t_j$ is

$$x'(t_j) \sim \frac{x(t_j + h) - x(t_j)}{h} = \frac{x(t_{j+1}) - x(t_j)}{h}.$$

For $j = 0, \ldots, N-1$, we then get

$$x'(t_j) = g(t_j, x(t_j)) \quad \leadsto \quad \frac{x(t_{j+1}) - x(t_j)}{h} \sim g(t_j, x(t_j)).$$

After setting $x_j = x(t_j)$ and replacing the approximation with an equality, we arrive at the difference equation:

$$\frac{x_{j+1} - x_j}{h} = g(t_j, x(t_j)) \quad \leadsto \quad x_{j+1} = x_j + h\, g(t_j, x_j). \tag{2.2.3}$$

Combining (2.2.2) with (2.2.3) gives us *Euler's method*:

Euler's method

Definition 2.3. *Consider the IVP*

$$x' = g(t, x), \quad x(a) = x_0.$$

Let $b > a$ be given and let $N \geq 1$ be given. Set $h = (b-a)/N$ and define a sequence of t values by

$$t_j = a + jh, \quad j = 0, \ldots, N \quad (a = t_0 < t_1 < t_2 < \cdots < t_{N-1} < t_N = b).$$

The approximate values $x_j \sim \phi(t_j)$, where $\phi(t)$ is the solution to the IVP, are given by

$$x_{j+1} = x_j + h\, g(t_j, x_j), \quad j = 0, \ldots, N-1.$$

We say $|f(h)| = \mathcal{O}(h^k)$ if there is a positive C such that $|f(h)| \leq C h^k$.

While we will not prove it here, it can be shown that Euler's method approximates the true solution up to $\mathcal{O}(h)$, i.e., $|\phi(t_j) - x_j| \leq Ch$ for some $C > 0$. In particular, this means that if the step size is halved, then the error should also be approximately halved. Moreover, as we let $h \to 0^+$, the approximate solution will converge exactly to the true solution.

For a concrete example, consider the IVP

$$x' = 5t^4, \quad x(0) = 0, \tag{2.2.4}$$

which has the solution $x(t) = t^5$. We approximate the solution at $t = 1$ using Euler's method and arrive at the first two columns of the following table:

h	Euler error	Heun error	Runge–Kutta error
0.200	0.433600	0.066400	6.6667×10^{-5}
0.100	0.233350	0.016650	0.4167×10^{-5}
0.050	0.120834	0.004166	0.0260×10^{-5}
0.025	0.061458	0.001041	0.0016×10^{-5}

As expected, we see that when the step size is halved, the error is also halved.

2.2. General theory

Other numerical methods

From a practical perspective, Euler's method is not very good. The primary problem is that Euler's method uses only the slope at $t = t_j$ before updating the approximate solution at $t = t_{j+1}$. For a refinement of Euler's method, consider the improved Euler's method (also known as Heun's method). This numerical method takes the average of the slope at $t = t_j$ and $t = t_{j+1}$ before updating the approximate solution. We first approximate the solution $\phi(t_{j+1})$ by using Euler's method, i.e.,

$$\tilde{x}_{j+1} = x_j + h\, g(t_j, x_j).$$

We now approximate the derivative at the point (t_j, x_j) not by $g(t_j, x_j)$ but instead by the average:

$$\frac{1}{2}\Big[g(t_j, x_j) + g(t_{j+1}, \tilde{x}_{j+1})\Big].$$

The term $g(t_{j+1}, \tilde{x}_{j+1})$ is the approximation to the true derivative $g(t_{j+1}, \phi(t_{j+1}))$, and this approximation is derived via Euler's method. The update takes an average of these two approximate derivatives and is

$$x_{j+1} = x_j + \frac{1}{2}h\Big[g(t_j, x_j) + g(t_{j+1}, \tilde{x}_{j+1})\Big].$$

While we will not show it here, it can be shown that the improved Euler's method approximates the solution up to $\mathcal{O}(h^2)$, i.e., $|\phi(t_j) - x_j| \leq Ch^2$ for some $C > 0$. In particular, this means that if the step size is halved, then the new error should be approximately one-fourth of the previous error. As we see using the first and third columns in the previous table associated with the IVP (2.2.4), this is precisely what happens.

An even better classical method is the fourth-order Runge–Kutta method. In this case, the update is found via

$$\begin{aligned}
k_1 &= g(t_j, x_j),\\
k_2 &= g(t_j + h/2, x_j + hk_1/2),\\
k_3 &= g(t_j + h/2, x_j + hk_2/2),\\
k_4 &= g(t_j + h, x_j + hk_3),\\
x_{j+1} &= x_j + \frac{1}{6}h(k_1 + 2k_2 + 2k_3 + k_4).
\end{aligned}$$

Each k_j is a numerical approximation of the derivative, and the update is found by using a weighted sum of these approximations. One update of the solution requires four evaluations. This method is $\mathcal{O}(h^4)$, i.e., $|\phi(t_j) - x_j| \leq Ch^4$ for some $C > 0$. In particular, this means that if the step size is halved, then the new error should be approximately one-sixteenth of the previous error. As we see using the first and last columns in the previous table associated with the IVP (2.2.4), this is precisely what happens.

The following table summarizes the above discussion regarding step size and error for one iteration:

h	Euler error	Heun error	Runge–Kutta error
10^{-1}	$\sim 10^{-1}$	$\sim 10^{-2}$	$\sim 10^{-4}$
10^{-2}	$\sim 10^{-2}$	$\sim 10^{-4}$	$\sim 10^{-8}$
10^{-3}	$\sim 10^{-3}$	$\sim 10^{-6}$	$\sim 10^{-12}$
10^{-4}	$\sim 10^{-4}$	$\sim 10^{-8}$	$\sim 10^{-16}$

We clearly see that as the number of data points used to approximate the derivative increases (one for Euler, two for Heun, and four for Runge–Kutta), the error associated with the numerical approximation to the solution decreases.

The decrease in error as a function of step size greatly affects the number of steps needed to solve the solution over one time interval. For example, suppose that you wish to solve a given ODE over one time interval ($0 \leq t \leq 1$) and you further wish that the error at $t = 1$ is no greater than 10^{-2}. If h is the step size, then $1/h$ iterations of the numerical method are required to find the solution at $t = 1$. In order to determine the step size, we must solve

$$(\text{\# steps}) \cdot (\text{error per step}) \leq 10^{-2} \quad \rightsquigarrow \quad \frac{1}{h} \cdot (\text{error per step}) \leq 10^{-2}.$$

Using Heun's method, we need to set $h \sim 10^{-2} = 0.01$ (so there will be on the order of 100 iterations) to achieve the desired accuracy. On the other hand, using the Runge–Kutta method allows us to set $h \sim 10^{-2/3} \sim 0.215$ (round down to set $h = 0.2$, which means approximately five iterations) to get the same accuracy.

As we see in the particular example of (2.2.4), this back-of-the-envelope calculation is not precise, but it does provide valuable insight into the efficiency of the various methods. From the accompanying table, we see that 20 steps ($h = 0.05$) are needed using Heun's method, while using five steps ($h = 0.2$) gives a much-better-than-desired error using the Runge–Kutta method. While we do not show it, Euler's method requires more than 200 steps ($h < 0.005$) to achieve the desired accuracy. In general, the actual step size needed for a desired accuracy depends on the ODE being solved.

The default method used by DFIELD for solving scalar ODEs is the Dormand–Prince method. This method is *adaptive* in that the step size h depends on how well the numerical solution is approximating the true solution. The user first determines what the error should be at the final time, and then the algorithm finds the appropriate step size, which ensures that the numerical solution satisfies that error bound. Unless stated otherwise, this is what is used to numerically solve ODEs in this text. The interested reader should see Polking [35, Chapter 4] and Griffiths and Higham [19] for a more thorough discussion on the different algorithms that can be used to numerically solve an ODE.

Exercises

Exercise 2.2.1. *For each of the following ODEs, use DFIELD to generate a direction field over the given box. Afterward, use DFIELD to plot the solution curve for the given initial condition.*

(a) $x' = [2 + \sin(2t)]x^2$ *with* $x(0) = 0.05$ *and* $x(0) = -2.0$ *over the box* $0 \leq t \leq 10, -3 \leq x \leq 3$

(b) $x' = e^{-t}x + \sin(5t)$ with $x(-1) = 0.0$ and $x(2) = -1.0$ over the box $-2 \leq t \leq 30, -2 \leq x \leq 2$

(c) $x' = \cos(x^2 - t^2)$ with $x(1) = -1.0$ and $x(3) = 1.0$ over the box $-3 \leq t \leq 8, -4 \leq x \leq 4$

(d) $tx' + 2x = te^t$ with $x(1) = -2$ and $x(1) = 3$ over the box $0.5 \leq t \leq 3$ and $-5 \leq x \leq 8$

Exercise 2.2.2. *Consider the IVP*

$$x' = 2t, \quad x(0) = 0,$$

which has the solution $\phi(t) = t^2$. Solve this problem numerically for $0 \leq t \leq 2$ and a step size of $h = 0.1$ using Euler's method, the improved Euler's method, and the classical fourth-order Runge–Kutta method. Plot the different numerical solutions and discuss any differences and/or similarities that you see between each solution.

Exercise 2.2.3. *Consider the IVP*

$$x' = \frac{1}{2}x, \quad x(0) = 1,$$

which has the solution $\phi(t) = e^{t/2}$. Solve this problem numerically for $0 \leq t \leq 1$ and a step size of $h = 0.1$ using Euler's method, the improved Euler's method, and the classical fourth-order Runge–Kutta method. Plot the different numerical solutions and discuss any differences and/or similarities that you see between each solution.

2.3 • The structure of the solution

The scalar linear first-order ODE is of the form

$$x' = a(t)x + f(t). \tag{2.3.1}$$

We have already seen that for a given initial condition $x(t_0) = x_0$, there will be a unique solution. For the problem without the initial condition, there will be a *general solution*. The general solution must have at least one free parameter attached to it, which is used to solve the initial value problem.

We now demonstrate that the solution structure to the linear ODE (2.3.1) has the same as that present for the linear algebraic system $Ax = b$. Regarding the algebraic system, we showed in Theorem 1.27 that all solutions are of the form

$$x = x_h + x_p,$$

where the homogeneous solution, x_h, is the general solution to $Ax = 0$ and the particular solution, x_p, is a solution to $Ax = b$. The homogeneous solution will have attached to it as many free variables as there are for the linear system in which A is a coefficient matrix. On the other hand, the particular solution does not have any free variables associated with it.

Following this idea, the homogeneous problem associated with the ODE (2.3.1) is

$$x' = a(t)x \quad \leadsto \quad x' - a(t)x = 0. \tag{2.3.2}$$

We first show that solutions to the homogeneous ODE have the same linearity property as solutions to the homogeneous linear system $Ax = 0$. Suppose that x_1 and x_2 are each a solution to the homogeneous problem

$$x_1' = a(t)x_1, \quad x_2' = a(t)x_2.$$

Since differentiation is a linear operation, we have

$$\begin{aligned}\frac{d}{dt}(c_1 x_1 + c_2 x_2) &= c_1 x_1' + x_2 x_2' \\ &= c_1 a(t) x_1 + c_2 a(t) x_2 \\ &= a(t)(c_1 x_1 + c_2 x_2).\end{aligned}$$

In other words, the linear combination of x_1 and x_2 is also a solution to the homogeneous problem. A solution to a homogeneous problem is unique only up to scalar multiplication.

Now consider the nonhomogeneous problem

$$x' = a(t)x + f(t) \quad \leadsto \quad x' - a(t)x = f(t). \tag{2.3.3}$$

Suppose that x_h is a solution to the homogeneous problem (2.3.2) and that x_p is a *particular solution* to (2.3.3). Arguing in a manner similar to that leading to Theorem 1.27, in which it is stated that all solutions to the linear system $Ax = b$ are the sum of a homogeneous solution and a particular solution, we can say that solutions to the ODE (2.3.3) are of the form $x = x_h + x_p$. This summation formulation of the solution is seen via the following sequence of calculations:

$$\begin{aligned}\frac{d}{dt}\left(x_h + x_p\right) &= x_h' + x_p' \\ &= a(t)x_h + \left(a(t)x_p + f(t)\right) \\ &= a(t)\left(x_h + x_p\right) + f(t).\end{aligned}$$

In conclusion, solutions to the linear ODE (2.3.1) have the same structure as solutions to linear algebraic systems.

Theorem 2.4. *Consider the linear nonhomogeneous ODE*

$$x' = a(t)x + f(t)$$

and its homogeneous counterpart

$$x' = a(t)x.$$

If x_h is a solution to the homogeneous problem and x_p is a particular solution, then

$$x = x_h + x_p$$

is a solution to the nonhomogeneous problem. Furthermore, if x_1, x_2 are solutions to the homogeneous problem, then so is the linear combination $c_1 x_1 + c_2 x_2$.

2.3. The structure of the solution

For linear algebraic systems, we know that the general homogeneous solution is comprised of dim[Null(A)] (the number of free variables) linearly independent vectors. What is the number of desired solutions for the scalar linear homogeneous ODE?

Consider the homogeneous IVP

$$x' = a(t)x, \quad x(t_0) = x_0.$$

Suppose that there are m solutions $x_1(t), x_2(t), \ldots, x_m(t)$, so that

$$x_h(t) = c_1 x_1(t) + c_2 x_2(t) + \cdots + c_m x_m(t)$$

is also a solution to the homogeneous problem. In order to solve for the initial condition, we must have

$$x_0 = x_h(t_0) = c_1 x_1(t_0) + c_2 x_2(t_0) + \cdots + c_m x_m(t_0).$$

This is one linear equation with m unknowns, $c_1, c_2, \ldots, c_m$. The coefficient matrix is

$$(x_1(t_0) \; x_2(t_0) \; \cdots \; x_m(t_0)) \in \mathcal{M}_{1 \times m}(\mathbb{R}).$$

From our discussion in Chapter 1, section 1.8.2, we know that if $m \geq 2$, then this equation will have an infinite number of solutions. For a unique solution, we require $m = 1$, with the additional condition that $x_1(t_0) \neq 0$. This yields

$$x_h(t) = c_1 x_1(t), \quad \text{with} \quad c_1 = \frac{x_0}{x_1(t_0)}.$$

Thus, $x_h(t)$ is the general homogeneous solution, and any other solution to the homogeneous problem must be a scalar multiple of it.

Does the general homogeneous solution always exist? In order to answer the question, consider the homogeneous IVP

$$x' = a(t)x, \quad x(t_0) = x_0 \neq 0.$$

By the existence/uniqueness Theorem 2.1, there is a unique solution to this IVP. The solution is clearly nonzero when $t = t_0$. Denoting this solution as $x_1(t)$, we conclude that the general homogeneous solution is $x_h(t) = c_1 x_1(t)$.

How do we solve the IVP for the nonhomogeneous problem (3.3.1)? Suppose that $x_1(t)$ is a solution to the homogeneous system, which is nonzero at the initial time $t = t_0$:

$$x_1' = a(t)x_1, \quad x_1(t_0) \neq 0 \quad \leadsto \quad x_h(t) = c_1 x_1(t).$$

By applying Theorem 2.4, a general solution to the nonhomogeneous problem is

$$x(t) = x_h(t) + x_p(t) = c_1 x_1(t) + x_p(t). \tag{2.3.4}$$

As for the initial condition, on evaluating the general solution at $t = t_0$,

$$x_0 = x(t_0) = c_1 x_1(t_0) + x_p(t_0) \quad \leadsto \quad c_1 x_1(t_0) = x_0 - x_p(t_0).$$

Since $x_1(t_0) \neq 0$, we can solve for the constant c_1:

$$c_1 = \frac{x_0 - x_p(t_0)}{x_1(t_0)}.$$

Thus, we again see that if a homogeneous solution that is nonzero at the initial time is chosen, then the general solution given by (2.3.4) can be used to solve the IVP for any initial condition.

We conclude with a restatement of Theorem 2.4. Using the general homogeneous solution yields a general solution to the nonhomogeneous problem.

Corollary 2.5. *The general solution to the nonhomogeneous problem*

$$x' = a(t)x + f(t)$$

is given by

$$x(t) = c_1 x_1(t) + x_p(t).$$

Here $x_p(t)$ is a particular solution, and $x_1(t)$ is a solution to the homogeneous problem, which is nonzero at the initial time $t = t_0$.

Exercises

Exercise 2.3.1. Suppose that for the scalar ODE $x' = a(t)x + f(t)$, a homogeneous solution, $x_1(t)$, and particular solution, $x_p(t)$, are given by

$$x_1(t) = e^{3t}, \quad x_p(t) = \cos^2(7t) + t^2 - 6e^{5t}.$$

(a) What is the general solution to the ODE?

(b) If $x(0) = -3$, what is the solution to the IVP?

Exercise 2.3.2. Suppose that for the scalar ODE $x' = a(t)x + f(t)$, a homogeneous solution, $x_1(t)$, and particular solution, $x_p(t)$, are given by

$$x_1(t) = \frac{1}{1+t^2}, \quad x_p(t) = \frac{t^4}{1+t^2}.$$

(a) What is the general solution to the ODE?

(b) If $x(0) = 5$, what is the solution to the IVP?

Exercise 2.3.3. Suppose that for the scalar ODE $x' = a(t)x + f(t)$, a homogeneous solution, $x_1(t)$, and particular solution, $x_p(t)$, are given by

$$x_1(t) = 3 + t^4, \quad x_p(t) = 6e^{-t}\cos(5t).$$

(a) What is the general solution to the ODE?

(b) If $x(0) = 7$, what is the solution to the IVP?

2.4 • The homogeneous solution

From Corollary 2.5, we know the general solution to the scalar linear ODE

$$x' = a(t)x + f(t) \qquad (2.4.1)$$

is given by $x(t) = x_h(t) + x_p(t)$, where $x_h(t) = c_1 x_1(t)$ is the general homogeneous solution

$$x_1' = a(t) x_1 \qquad (2.4.2)$$

and $x_p(t)$ is a particular solution to the nonhomogeneous problem (2.4.1).

Let us first focus on finding a solution to the homogeneous problem (2.4.2). Since the rate of change of the solution is proportional to the solution, this suggests that we guess the solution to be an exponential,

$$x_1(t) = e^{A(t)},$$

where $A(t)$ is an unknown function. Since

$$x_1'(t) = A'(t) e^{A(t)} = A'(t) x_1(t),$$

we have

$$x_1' = a(t) x_1 \quad \leadsto \quad A'(t) x_1(t) = a(t) x_1(t) \quad \leadsto \quad A'(t) = a(t).$$

In other words, the desired function $A(t)$ is an antiderivative of the function $a(t)$. Since the exponential function is never zero, we have the desired solution.

Lemma 2.6. *The general solution to the homogeneous problem* (2.4.2) *is given by*

$$x_h(t) = c_1 x_1(t), \quad x_1(t) = e^{\int a(t) dt}.$$

Example 2.7. For the constant coefficient problem

$$x' = \lambda x,$$

we have

$$x_1(t) = e^{\int \lambda dt} = e^{\lambda t}.$$

The general solution is

$$x_h(t) = c_1 e^{\lambda t}. \quad \blacksquare$$

Example 2.8. For the variable coefficient problem

$$x' = \frac{1}{2 + 5t} x,$$

we have

$$x_1(t) = e^{\int (2+5t)^{-1} dt} = e^{\ln(2+5t)/5} = (2+5t)^{1/5}.$$

The general solution is

$$x_h(t) = c_1 (2+5t)^{1/5}. \quad \blacksquare$$

Example 2.9. Consider the problem
$$e^{3t}x' + \sin(2t)x = 0.$$
In order to use the result of Lemma 2.6, we must first write the problem in standard form:
$$x' = -e^{-3t}\sin(2t)x.$$
We can integrate $a(t) = -e^{-3t}\sin(2t)$ via integration by parts. Alternatively, using SAGE,
$$-\int e^{-3t}\sin(2t)\,dt = \frac{1}{13}e^{-3t}(3\sin(2t) + 2\cos(2t)) + C,$$
so
$$x_1(t) = e^{\int a(t)\,dt} = e^{e^{-3t}(3\sin(2t) + 2\cos(2t))/13},$$
and the general solution is
$$x_h(t) = c_1 e^{e^{-3t}(3\sin(2t) + 2\cos(2t))/13}. \blacksquare$$

Exercises

Exercise 2.4.1. *Find the general solution for the following homogeneous linear ODEs.*

(a) $x' = (1-t)x$

(b) $(1+t^2)x' + tx = 0$

(c) $x' = -5x$

(d) $(4+t^2)x' = -2x$

(e) $(9-t^2)x' + 7x = 0$

Exercise 2.4.2. *Solve the following IVPs.*

(a) $(1-t)x' = x$, $x(0) = 3$

(b) $(4-t^2)x' + tx = 0$, $x(0) = -2$

(c) $x' = 3x$, $x(0) = 5$

(d) $(16+t^2)x' = 3x$, $x(0) = 6$

(e) $(9-t^2)x' + 7x = 0$, $x(0) = -8$

2.5 • The particular and general solution

As we saw in section 2.4, the homogeneous solution is quite easy to find (at least formally). We are now concerned with finding a particular solution. Once we have a particular solution, we will be able to write down the general solution to the full problem.

2.5. The particular and general solution

We will introduce two techniques for finding a particular solution:

(a) variation of parameters;

(b) the method of undetermined coefficients.

The first method will (at least in theory) work for any problem; however, at times it leads to the evaluation of tedious integrals. The second method reduces the search to a linear algebra problem that can be solved using the ideas and techniques of Chapter 1. Unfortunately, it does not always work; in particular, it requires that the coefficient $a(t)$ and the forcing function $f(t)$ have a special form.

2.5.1 ▪ Variation of parameters

The particular solution immediately follows from knowing the homogeneous solution. The formula we use also applies to linear systems of ODEs if we label the homogeneous solution $x_1(t)$ as

$$\Phi(t) = e^{\int a(t) dt}.$$

In the context of linear systems, the function $\Phi(t)$ will become a square matrix whose size will be determined by the number of variables in the system.

The claim is that the particular solution is given by the variation of parameters formula,

$$x_p(t) = \Phi(t) \int \Phi(t)^{-1} f(t) dt.$$

We now verify this claim. Writing

$$\int \Phi(t)^{-1} f(t) dt = \int^t \Phi(s)^{-1} f(s) ds,$$

recall that the Fundamental Theorem of Calculus states that

$$\frac{d}{dt} \int^t \Phi(s)^{-1} f(s) ds = \Phi(t)^{-1} f(t) \quad \rightsquigarrow \quad \left(\int \Phi(t)^{-1} f(t) dt \right)' = \Phi(t)^{-1} f(t).$$

Thus, on using

(a) the product rule

(b) the fact that $\Phi' = a(t)\Phi$,

we have that the derivative of the function $x_p(t)$ satisfies

$$x_p'(t) = \Phi'(t) \int \Phi(t)^{-1} f(t) dt + \Phi(t) \left(\Phi(t)^{-1} f(t) \right)$$

$$= a(t)\Phi(t) \int \Phi(t)^{-1} f(t) dt + \left(\Phi(t)\Phi(t)^{-1} \right) f(t)$$

$$= a(t) \left(\Phi(t) \int \Phi(t)^{-1} f(t) dt \right) + f(t)$$

$$= a(t) x_p(t) + f(t).$$

> **Variation of parameters**
>
> **Lemma 2.10.** *Let $x_1(t)$ be a nonzero solution to the homogeneous problem (2.4.2) and set*
> $$\Phi(t) = x_1(t).$$
> *A particular solution to (2.4.1) is given by the variation of parameters formulation*
> $$x_p(t) = \Phi(t) \int \Phi(t)^{-1} f(t) \, dt.$$

2.5.2 • The general solution

Now that we have a solution formula for the particular solution, we can write down the general solution to the full problem. As a consequence of

(a) Theorem 2.4, which states that the general solution is the sum of a particular solution and a homogeneous solution,

(b) Lemma 2.6, which states that the homogeneous solution is the exponential of the antiderivative of $a(t)$, and

(c) Lemma 2.10, which is the variation of parameters formula for the particular solution,

we have the following for the linear ODE (2.4.1):

> **Theorem 2.11.** *The general solution to the linear ODE*
> $$x' = a(t)x + f(t)$$
> *is given by*
> $$x(t) = \underbrace{c_1 x_1(t)}_{x_h(t)} + \underbrace{\Phi(t) \int \Phi(t)^{-1} f(t) \, dt}_{x_p(t)}.$$
> *Here $x_1(t)$ and $\Phi(t)$ are solutions to the homogeneous problem*
> $$x_1(t) = e^{\int a(t)\,dt}, \quad \Phi(t) = x_1(t).$$

Example 2.12. Consider the linear ODE

$$x' = 4x + 2 - 3t. \tag{2.5.1}$$

The homogeneous problem is
$$x' = 4x,$$

and using the result of Example 2.7 with $\lambda = 4$ gives the homogeneous solution:

$$x_h(t) = c_1 x_1(t), \quad \Phi(t) = x_1(t) = e^{4t}.$$

2.5. The particular and general solution

Using Lemma 2.10, the particular solution can be written as

$$x_p(t) = e^{4t} \int e^{-4t}(2-3t)\,dt = \frac{3}{4}t - \frac{5}{16}.$$

The antiderivative is found by using integration by parts (or SAGE). By Theorem 2.11, the general solution is the sum of the homogeneous and particular solutions:

$$x(t) = c_1 e^{4t} + \frac{3}{4}t - \frac{5}{16}.$$

Now suppose that in addition to the ODE, we have the initial value condition $x(0) = 2$. We have

$$2 = x(0) = c_1 - \frac{5}{16} \quad \leadsto \quad c_1 = \frac{37}{16},$$

so the solution to the IVP is

$$x(t) = \frac{37}{16}e^{4t} + \frac{3}{4}t - \frac{5}{16}. \quad \blacksquare$$

Example 2.13. Consider the variable coefficient problem

$$x' = \frac{1}{4+t}x + t^2.$$

The homogeneous problem is

$$x' = \frac{1}{4+t}x.$$

Since

$$a(t) = \frac{1}{4+t} \quad \leadsto \quad \int a(t)\,dt = \ln(4+t),$$

on taking the exponential, we see that the homogeneous solution is

$$x_h(t) = c_1 x_1(t), \quad \Phi(t) = x_1(t) = e^{\ln(4+t)} = 4+t.$$

Using Lemma 2.10, the particular solution is

$$x_p(t) = (4+t)\int \frac{1}{4+t} \cdot t^2\,dt.$$

The antiderivative can be found using a partial fraction expansion of the integrand. Instead, using SAGE, we find

$$x_p(t) = \frac{1}{2}(4+t)\left(t^2 - 8t + 32\ln(4+t)\right).$$

By Theorem 2.11, the general solution is the sum of the homogeneous and particular solutions:

$$x(t) = c_1(4+t) + \frac{1}{2}(4+t)\left(t^2 - 8t + 32\ln(4+t)\right). \quad \blacksquare$$

Example 2.14. Let us find the general solution to the ODE

$$(2+t)x' - 4x = t.$$

We first rewrite the ODE in the desired form:

$$x' = \frac{4}{2+t}x + \frac{t}{2+t}.$$

We have

$$\Phi(t) = x_1(t) = e^{4\int (2+t)^{-1} dt} = (2+t)^4,$$

so the homogeneous solution is

$$x_h(t) = c_1(2+t)^4.$$

The particular solution is given by

$$x_p(t) = (2+t)^4 \int \frac{1}{(2+t)^4} \cdot \frac{t}{2+t} dt = (2+t)^4 \int \frac{t}{(2+t)^5} dt = -\frac{1}{6}(2t+1).$$

The integration is carried out using the method of partial fractions (or SAGE). The general solution to the ODE is the sum of the homogeneous and particular solutions:

$$x(t) = c_1(2+t)^4 - \frac{1}{6}(2t+1). \quad \blacksquare$$

2.5.3 • Undetermined coefficients

The variation of parameters formulation of Lemma 2.10 always gives the particular solution; however, it may be technically arduous to carry out the integrations. We now consider a different method for finding the particular solution—the *method of undetermined coefficients*. When this method is applicable, it can be easier to use. Instead of computing several antiderivatives, we instead cleverly guess the form of the particular solution and from this guess convert the linear ODE to a linear system of algebraic equations. As we saw in Chapter 1, linear systems are easy to solve via Gaussian elimination.

When is the method applicable? The system must minimally be of the form of (2.5.1); i.e., the homogeneous problem must be constant coefficient (the coefficient $a(t) \equiv a$ is constant). Furthermore, the forcing term must be a member of a class of functions for which the derivative is also a member of the class of functions. Some careful thought reveals that this class of functions consist of linear combinations of functions of the form $p(t)e^{at}\cos(bt)$ and $p(t)e^{at}\sin(bt)$, where $p(t)$ is a polynomial. The method will be illustrated via a sequence of examples.

Example 2.15. First consider

$$x' = -x + t - 4t^3.$$

The homogeneous problem is

$$x' = -x \quad \leadsto \quad x_1(t) = e^{-t},$$

2.5. The particular and general solution

so the homogeneous solution is

$$x_h(t) = c_1 e^{-t}.$$

Since $f(t) = t - 4t^3$, which is a third-order polynomial, we will guess the particular solution to be a third-order polynomial:

$$x_p(t) = a_0 + a_1 t + a_2 t^2 + a_3 t^3.$$

Plugging this guess into the ODE yields

$$\underbrace{a_1 + 2a_2 t + 3a_3 t^2}_{x_p'} = \underbrace{-(a_0 + a_1 t + a_2 t^2 + a_3 t^3)}_{-x_p} + t - 4t^3,$$

which can be rewritten as

$$a_0 + a_1 + (a_1 + 2a_2 - 1)t + (a_2 + 3a_3)t^2 + (a_3 + 4)t^3 = 0.$$

Since a polynomial can be identically zero if and only all of its coefficients are zero, we then get the linear system of equations

$$a_0 + a_1 = 0, \quad a_1 + 2a_2 - 1 = 0, \quad a_2 + 3a_3 = 0, \quad a_3 + 4 = 0,$$

which can be rewritten in matrix form as

$$\begin{pmatrix} 1 & 1 & 0 & 0 \\ 0 & 1 & 2 & 0 \\ 0 & 0 & 1 & 3 \\ 0 & 0 & 0 & 1 \end{pmatrix} a = \begin{pmatrix} 0 \\ 1 \\ 0 \\ -4 \end{pmatrix}, \quad a = \begin{pmatrix} a_0 \\ a_1 \\ a_2 \\ a_3 \end{pmatrix}.$$

Using SAGE, we find that the solution to the linear system is

$$a = \begin{pmatrix} 23 \\ -23 \\ 12 \\ -4 \end{pmatrix}.$$

We conclude that

$$x_p(t) = 23 - 23t + 12t^2 - 4t^3,$$

so the general solution is the sum of the homogeneous and particular solutions:

$$x(t) = c_1 e^{-t} + 23 - 23t + 12t^2 - 4t^3. \quad \blacksquare$$

Example 2.16. For our next example, consider

$$x' = -x + 8e^{3t}.$$

The homogeneous solution is again $x_h(t) = c_1 e^{-t}$. Since $f(t) = 8e^{3t}$, which is an exponential function, we will guess the particular solution to be an exponential function with the same exponent:

$$x_p(t) = a_0 e^{3t}.$$

Plugging this guess into the ODE yields

$$\underbrace{3a_0 e^{3t}}_{x'_p} = \underbrace{-a_0 e^{3t}}_{-x_p} + 8e^{3t},$$

which can be rewritten as

$$(4a_0 + 8)e^{3t} = 0.$$

Since an exponential function can be identically zero if and only its coefficient is zero, we see that

$$4a_0 + 8 = 0 \quad \leadsto \quad a_0 = -2.$$

Consequently, we can conclude that

$$x_p(t) = -2e^{3t},$$

so the general solution is the sum of the homogeneous and particular solutions:

$$x(t) = c_1 e^{-t} - 2e^{3t}. \quad \blacksquare$$

Example 2.17. For our third example, we will suppose that $f(t)$ is a summation of the previous two forcing terms:

$$x' = -x + t - 4t^3 + 8e^{3t}.$$

If we break up the problem into two simpler problems,

$$x'_1 = -x_1 + t - 4t^3, \quad x'_2 = -x_2 + 8e^{3t},$$

and find the particular solution for each problem,

$$x_{1,p}(t) = 23 - 23t + 12t^2 - 4t^3, \quad x_{2,p}(t) = -2e^{3t},$$

then on using linearity, the particular solution for the full problem is the sum of these two particular solutions:

$$x_p(t) = x_{1,p}(t) + x_{2,p}(t) = 23 - 23t + 12t^2 - 4t^3 - 2e^{3t}.$$

We also see the linearity property coming into play if we instead use the variation of parameters formula in order to find the particular solution. In this case, we would use the fact that integration is a linear operation. This example illustrates an important principle regarding the method of undetermined coefficients: we can break up complicated forcing functions into smaller, more analytically tractable pieces and then form the full solution by summing the smaller pieces. $\blacksquare$

Example 2.18. For our fourth example, let us consider

$$x' = 3x + 5\cos(4t).$$

The homogeneous problem is

$$x' = 3x \quad \leadsto \quad x_1(t) = e^{3t},$$

2.5. The particular and general solution

so the homogeneous solution is

$$x_h(t) = c_1 e^{3t}.$$

Since $f(t) = 5\cos(4t)$, which is a trigonometric function, we will guess the particular solution to be a sum of trigonometric functions:

$$x_p(t) = a_0 \cos(4t) + a_1 \sin(4t).$$

Plugging this guess into the ODE yields

$$\underbrace{-4a_0 \sin(4t) + 4a_1 \cos(4t)}_{x'_p} = \underbrace{3(a_0 \cos(4t) + a_1 \sin(4t))}_{3x_p} + 5\cos(4t),$$

which can be rewritten as

$$(-3a_0 + 4a_1 - 5)\cos(4t) + (-4a_0 - 3a_1)\sin(4t) = 0.$$

In order for this equation to be hold for all t, it must again be true that each of the coefficients are zero:

$$-3a_0 + 4a_1 - 5 = 0, \quad -4a_0 - 3a_1 = 0.$$

This linear system can be rewritten in matrix form as

$$\begin{pmatrix} -3 & 4 \\ -4 & -3 \end{pmatrix} a = \begin{pmatrix} 5 \\ 0 \end{pmatrix}, \quad a = \begin{pmatrix} a_0 \\ a_1 \end{pmatrix}.$$

Since

$$\begin{pmatrix} -3 & 4 \\ -4 & -3 \end{pmatrix}^{-1} = \frac{1}{25}\begin{pmatrix} -3 & -4 \\ 4 & -3 \end{pmatrix},$$

the solution to the linear system is

$$a = \frac{1}{25}\begin{pmatrix} -3 & -4 \\ 4 & -3 \end{pmatrix}\begin{pmatrix} 5 \\ 0 \end{pmatrix} = \frac{1}{25}\begin{pmatrix} -15 \\ 20 \end{pmatrix}.$$

Consequently, on algebraic simplification we can conclude that

$$x_p(t) = -\frac{3}{5}\cos(4t) + \frac{4}{5}\sin(4t).$$

The general solution is the sum of the homogeneous and particular solutions:

$$x(t) = c_1 e^{3t} - \frac{3}{5}\cos(4t) + \frac{4}{5}\sin(4t). \quad \blacksquare$$

Example 2.19. For our final example in which we will complete the computation, consider the ODE

$$x' = -7x + 3e^{-7t}.$$

Using Example 2.16 as a guide, it appears to be the case that we should guess the particular solution to be $x_p(t) = a_0 e^{-7t}$. Plugging this guess into the ODE gives

$$-7a_0 e^{-7t} = -7a_0 e^{-7t} + 3e^{-7t} \quad \rightsquigarrow \quad 3e^{-7t} = 0.$$

Since this clearly is never true, our guess was incorrect. What went wrong? The problem is that the guess is precisely the homogeneous solution, $x_h(t) = c_1 e^{-7t}$. In the case that the forcing function $f(t)$ is also a homogeneous solution, we must modify our guess. The correct modification is to multiply our original guess by t; in other words, we will guess that

$$x_p(t) = a_0 t e^{-7t}.$$

Plugging this guess into the ODE gives

$$\underbrace{a_0 e^{-7t} - 7 a_0 t e^{-7t}}_{x_p'} = \underbrace{-7 a_0 t e^{-t}}_{-7 x_p} + 3 e^{-7t},$$

which can be rewritten as

$$(a_0 - 3) e^{-7t} = 0 \quad \rightsquigarrow \quad a_0 = 3.$$

Thus, we see that the particular solution is

$$x_p(t) = 3 t e^{-7t},$$

so the general solution is the sum of homogeneous and particular solutions:

$$x(t) = c e^{-7t} + 3 t e^{-7t}. \quad \blacksquare$$

Now let us consider a sequence of problems and simply guess the form of the particular solution without going through the work of actually computing the undetermined coefficients (the interested student can complete the calculation). We start with the problem

$$x' = 5x + f(t),$$

which has the homogeneous solution $x_h(t) = c_1 e^{5t}$. We have that for various functions $f(t)$,

(a) $f(t) = t^2 e^{-3t} \rightsquigarrow x_p(t) = (a_0 + a_1 t + a_2 t^2) e^{-3t}$,

(b) $f(t) = t e^t \cos(5t) \rightsquigarrow x_p(t) = (a_0 + a_1 t) e^t \cos(5t) + (a_2 + a_3 t) e^t \sin(5t)$,

(c) $f(t) = t \sin(2t) \rightsquigarrow x_p(t) = (a_0 + a_1 t) \cos(2t) + (a_2 + a_3 t) \sin(2t)$, and

(d) $f(t) = t^3 e^{5t} + t^2 \rightsquigarrow x_p(t) = \underbrace{t(a_0 + a_1 t + a_2 t^2 + a_3 t^3) e^{5t}}_{f_1(t) = t^3 e^{5t}} + \underbrace{a_4 + a_5 t + a_6 t^2}_{f_2(t) = t^2}.$

In each guess, we used the principle that polynomials in the forcing imply polynomials for the guess, exponentials in the forcing imply exponentials in the guess, and sine and cosine terms in the forcing imply the same for the guess. We also used the "rule of thumb" that a forcing term having a product of terms implies that the guess should be a product of guesses. For example, in (a), the forcing term is a quadratic polynomial multiplied by an exponential. The quadratic term implies the guess of a quadratic polynomial, and the exponential term implies the guess of an exponential. Consequently, the guess for the product is the quadratic polynomial multiplied

2.5. The particular and general solution

by the exponential. For the guess associated with (d), we used the principle that we can break up the forcing function into smaller pieces and then guess for each piece. Thus, in that example, we see one guess associated with $t\sin(2t)$ and another guess associated with the polynomial t^2. For the guess associated with (d), we needed to modify the expected guess in order to take into account the fact that part of our guess was also a homogeneous solution. We initially guessed a cubic multiplied by the exponential and then afterward multiplied by the smallest integer power of t, which guarantees that no part of the guess is a homogeneous solution.

Example 2.20. Let us find a particular solution for example (c) above using SAGE. Using variation of parameters, we know the particular solution is

$$x_p(t) = e^{5t} \int e^{-5t} \cdot t\sin(2t)\,dt.$$

The integral can be evaluated using a sequence of integration by parts but by using SAGE,

$$x_p(t) = -\frac{2}{841}(10+29t)\cos(2t) - \frac{1}{841}(21+145t)\sin(2t),$$

which is precisely of the expected form. The general solution to the problem is the sum of the homogeneous and particular solutions:

$$x(t) = c_1 e^{5t} - \frac{2}{841}(10+29t)\cos(2t) - \frac{1}{841}(21+145t)\sin(2t). \blacksquare$$

Exercises

Exercise 2.5.1. *Find the general solution for each of the following linear ODEs.*

(a) $x' + tx = 10t$

(b) $(100-t)x' = 0.03(100-t) - 2x$

(c) $tx' + 2x = te^t$

Exercise 2.5.2. *Solve the following IVPs.*

(a) $x' + tx = 10t$, $x(0) = 5$

(b) $(100-t)x' = 0.03(100-t) - 2x$, $x(0) = 1$

(c) $tx' + 2x = te^t$, $x(1) = 4$ *(Compare with the numerically generated plot in Exercise 2.2.1(d).)*

Exercise 2.5.3. *Use the method of undetermined coefficients to find the general solution for the following linear ODEs.*

(a) $x' = 2x + 4 - t + 3t^2$

(b) $x' = 2x + 3\cos(t) + 12e^{-4t}$

(c) $x' = 2x + 9t\sin(2t)$

Exercise 2.5.4. *Determine the functional form of the particular solution for each of the following linear ODEs. Do not actually compute the particular solution.*

(a) $x' = 2x + t^4 e^{5t} + e^{-2t} \cos(3t)$

(b) $x' = 3x + t^2 e^{3t} \cos(2t) + 7t^6$

(c) $x' = -x + 12t^3 e^{-t}$

Exercise 2.5.5. *Use SAGE to find a particular solution to the following linear ODEs.*

(a) $x' = 2x + t^4 e^{5t} + e^{-2t} \cos(3t)$

(b) $x' = 3x + t^2 e^{3t} \cos(2t) + 7t^6$

(c) $x' = -x + 12t^3 e^{-t}$

Exercise 2.5.6. *A population model under the assumptions that*

(a) *the birthrate minus the death rate is a positive constant and*

(b) *there is a constant rate of harvesting*

is

$$x' = ax - h, \quad a, h > 0.$$

Here a is the rate difference, and h is the rate of harvesting. Show that if $0 < x(0) < h/a$, then there is a finite time T—the extinction time—such that $x(t) < 0$ for $t \geq T$ (the biological interpretation is that if the initial population is not large enough, then the population dies in finite time due to overharvesting). What is T?

2.6 • Case studies

2.6.1 • One-tank mixing problem

Let us again consider the mixing problem discussed in section 2.1. Now assume that the incoming and outgoing flow rates are given by aV_0 L/day for some $a > 0$. Further assume that the incoming concentration, $c(t)$ g/L, is sinusoidal:

$$c(t) = c_0(1 - \cos(\omega t)), \quad \omega > 0.$$

Our goal here is to understand the manner in which the concentration of brine in the tank varies in the tank as a function of the frequency, ω, for large time ($t \gg 0$).

The governing equation for the amount of salt in the tank, $x(t)$, is given by

$$x' = -ax + aV_0 c(t), \quad x(0) = x_0.$$

The general solution is a sum of the homogeneous solution and the particular solution. The homogeneous problem is

$$x' = -ax \quad \rightsquigarrow \quad x_h(t) = c_1 e^{-at}.$$

Since the general solution is $x(t) = x_h(t) + x_p(t)$, for the initial condition we have

$$x(0) = x_h(0) + x_p(0) = c_1 + x_p(0) \quad \rightsquigarrow \quad c_1 = x(0) - x_p(0).$$

2.6. Case studies

The solution to the IVP is then

$$x(t) = \left(x(0) - x_p(0)\right)e^{-at} + x_p(t).$$

Since $a > 0$, $x(t) \sim x_p(t)$ for $t \gg 0$, and the value of $x(0)$ is unimportant. In other words, the original amount of salt in the tank is unimportant when trying to determine the amount of salt in the tank for large time.

We now find a particular solution. We know via variation of parameters that

$$x_p(t) = ac_0 V_0 e^{-at} \int e^{at} (1 - \cos(\omega t))\, dt$$
$$= a_0 + a_1 \cos(\omega t) + a_2 \sin(\omega t),$$

where the constants a_0, a_1, a_2 could be found using the method of undetermined coefficients. Instead, we will use the variation of parameters formula with SAGE to get

$$x_p(t) = c_0 V_0 \left(1 - \frac{a^2}{a^2 + \omega^2} \cos(\omega t) - \frac{a\omega}{a^2 + \omega^2} \sin(\omega t)\right).$$

Now that we have the approximate solution for large t, we can address our question. We first rewrite the solution to put it in a form more amenable for analysis. A standard trigonometric identity,

$$c_1 \cos\theta + c_2 \sin\theta = \sqrt{c_1^2 + c_2^2} \cos(\theta - \phi), \quad \tan\phi = \frac{c_2}{c_1},$$

yields that

$$\frac{a^2}{a^2 + \omega^2} \cos(\omega t) + \frac{a\omega}{a^2 + \omega^2} \sin(\omega t) = \frac{a}{\sqrt{a^2 + \omega^2}} \cos(\omega t - \phi), \quad \tan\phi = \frac{\omega}{a}.$$

Hence, the particular solution can be rewritten in the form

$$x_p(t) = c_0 V_0 \left(1 - \frac{a}{\sqrt{a^2 + \omega^2}} \cos(\omega t - \phi)\right), \quad \tan\phi = \frac{\omega}{a}.$$

Since the question concerns the concentration of the brine solution in the tank, we need an expression for that quantity. For $t \gg 0$, the concentration in the tank is approximately

$$c_{\text{tank}}(t) = \frac{x(t)}{V_0}$$
$$\sim \frac{x_p(t)}{V_0} = c_0 \left(1 - \frac{a}{\sqrt{a^2 + \omega^2}} \cos(\omega t - \phi)\right), \quad \tan\phi = \frac{\omega}{a}. \tag{2.6.1}$$

A plot of the concentration in the tank for $t \gg 0$ is given in Figure 2.5.

Recall that the mean, i.e., the average, of a periodic function $f(t)$ with period T is given by

$$\overline{f} = \frac{1}{T} \int_0^T f(s)\, ds.$$

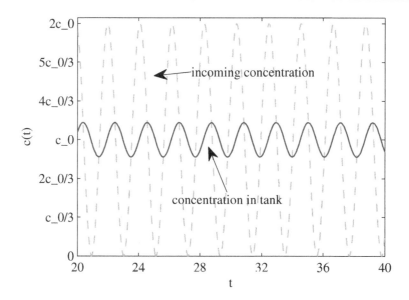

Figure 2.5. *A plot of the solution* (2.6.1) *for large t when a* = 0.45 *and* ω = 3.0. *The incoming concentration has a large variation about the mean,* c_0, *and is given by a dashed curve. The concentration in the tank has a smaller variation about the mean,* c_0, *and is denoted with a solid curve.*

Since the cosine has zero mean (see Exercise 2.6.1), the mean concentration in the tank is given by c_0, which is the same as the mean concentration of the incoming brine. On the other hand, on writing the concentration in the tank as

$$c_{\text{tank}}(t) \sim c_0 - c_0 \frac{a}{\sqrt{a^2 + \omega^2}} \cos(\omega t - \phi),$$

we see that the variation of the concentration in the tank about the mean,

$$c_{\text{var}}(t) := c_{\text{tank}}(t) - c_0,$$

is given by

$$c_{\text{var}}(t) = -c_0 \frac{a}{\sqrt{a^2 + \omega^2}} \cos(\omega t - \phi).$$

The variation about the mean depends on the frequency. The amplitude of the variation, $c_0 A^*(\omega)$, where

$$A^*(\omega) := \frac{a}{\sqrt{a^2 + \omega^2}},$$

decreases monotonically as the frequency increases. Furthermore, the frequency introduces a phase shift,

$$\phi^*(\omega) := \tan^{-1}(\omega/a),$$

that depends not only on the frequency but also on the fractional constant a. In Figure 2.6, we see plots of both of these functions for a fixed $a > 0$.

If we rewrite the variation about the mean as

$$c_0 \frac{a}{\sqrt{a^2 + \omega^2}} \cos(\omega t - \phi^*) = c_0 \frac{a}{\sqrt{a^2 + \omega^2}} \cos\left(\omega \left[t - \frac{\phi^*}{\omega}\right]\right),$$

2.6. Case studies

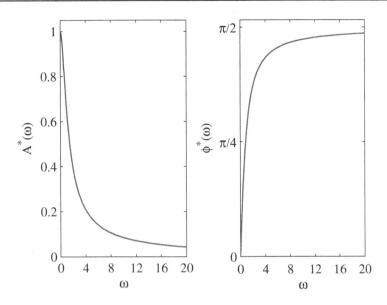

Figure 2.6. *A plot of the amplitude $A^*(\omega)$ (left figure) and the phase $\phi^*(\omega)$ (right figure) for the variation about the mean in the tank when $a = 0.45$. Note that as the frequency of the variation of the incoming concentration increases, the variation of the concentration in the tank about its mean decreases to zero. Moreover, the variation of the concentration in the tank becomes more out of phase with that of the incoming concentration.*

then we see that this variation is a shift to the right of $\cos(\omega t)$ by ϕ^*/ω. Indeed, we can think of the phase shift as a shift of the maximum of the response. Assuming that the shift is positive, the maximum of the response will be precisely ϕ^*/ω time units later than the maximum of the input. The phase shift is the cause of the delay in response of the variation of the concentration in the tank to that of the incoming concentration (again see Figure 2.5). Since $\phi^* \to \pi/2$ as $\omega \to +\infty$, the shift to the right is approximately $\pi/2\omega$ for large frequency. Since the period of forcing is $T = 2\pi/\omega$, we conclude that

(a) the variation of the concentration in the tank about the mean decreases to zero as the frequency increases, and

(b) the phase shift between the variation about the mean of the incoming brine solution and that in the tank monotonically approaches $T/4$, where $T = 2\pi/\omega$ is the period of the forcing, as the frequency increases. In particular, for large frequency, the maximal concentration in the tank occurs a quarter period after the maximum of the incoming concentration.

2.6.2 ▪ Mathematical finance

Let $x(t)$ represent the amount of money present in a bank account at time t, $r > 0$ be the fixed yearly interest rate given by the bank on the money in the account, and $f(t)$ the continuous rate of deposit ($f(t) > 0$) and/or withdrawal ($f(t) < 0$) from the account. The governing equation is given by

$$x' = rx + f(t), \quad x(0) = x_0.$$

In our previous example, the initial condition was not important. However, for this problem, the homogeneous problem being

$$x' = rx \quad \leadsto \quad x_1(t) = e^{rt}$$

means that the initial amount of money in the account, x_0, will play an important role in determining the amount of money in the account for $t \gg 0$.

We will consider the following specific problem. Suppose that the account starts with \$10,000 and that the interest rate associated with it is 3% (this is unrealistically high in the year 2014). Further suppose that money is deposited into the account according to the rule $f(t) = 300t$. We wish to know how much money will be in the account after 30 years.

What does this deposit rule mean regarding the amount of money put into the account each year? Neglecting (for the moment) the rate of change due to the interest rate, we have

$$x' = 300t \quad \leadsto \quad x(t) = 150t^2.$$

Since

$$x(j+1) - x(j) = 150\left[(j+1)^2 - j^2\right] = 150(2j+1) = 150 + 300j,$$

we have that $x(1) - x(0) = 150$, $x(2) - x(1) = 450$, $x(3) - x(2) = 750$, and so on. In other words, in year 1, \$150 is (continuously) deposited into the account; in year 2, \$450 is deposited into the account; in year 3, \$750 is deposited into the account; and so on.

Going back to the full problem, the governing ODE is

$$x' = 0.03x + 300t, \quad x(0) = 10,000.$$

The homogeneous solution is

$$x_h(t) = c_1 e^{0.03t}.$$

As for the particular solution, using the method of undetermined coefficients, we know that the solution is of the form

$$x_p(t) = a_0 + a_1 t.$$

Solving as in the previous section 2.5.3 yields that the coefficients are

$$a_0 = -\frac{1,000,000}{3}, \quad a_1 = -10,000,$$

so that

$$x_p(t) = -\frac{1,000,000}{3} - 10,000t.$$

The general solution is the sum of the homogeneous and particular solutions:

$$x(t) = c_1 e^{0.03t} - \frac{1{,}000{,}000}{3} - 10{,}000t.$$

The initial condition gives

$$10{,}000 = x(0) = c_1 - \frac{1{,}000{,}000}{3} \quad \rightsquigarrow \quad c_1 = \frac{1{,}030{,}000}{3},$$

so the solution to the IVP is

$$x(t) = \frac{1{,}030{,}000}{3} e^{0.03t} - \frac{1{,}000{,}000}{3} - 10{,}000t.$$

Going back to our question, we have

$$x(30) = \frac{1{,}030{,}000}{3} e^{0.9} - \frac{1{,}000{,}000}{3} - 300{,}000 \sim 211{,}130.40.$$

While this is a reasonable amount of saved money, it is probably not enough to fully fund a retirement (if that is the goal).

Exercises

Exercise 2.6.1. *The mean of a periodic function of period T is*

$$\overline{f} = \frac{1}{T} \int_0^T f(t)\,dt.$$

Show that

(a) $\overline{c_0} = c_0$ *for any $T > 0$,*

(b) $\overline{\cos(\omega t)} = 0$ *for $T = 2\pi/\omega$,*

(c) $\overline{\sin(\omega t)} = 0$ *for $T = 2\pi/\omega$,*

(d) $\overline{\cos(j\omega t)} = 0$ *for $T = 2\pi/\omega$ and any positive integer j, and*

(e) $\overline{\sin(j\omega t)} = 0$ *for $T = 2\pi/\omega$ and any positive integer j.*

Exercise 2.6.2. *A saltwater solution at a concentration of 6 g/m³ enters a 25 m³ tank at a flow rate of 0.25 m³/min. The uniformly mixed solution leaves the tank at the same flow rate. Assume that the tank initially holds 15 m³ of the solution at a concentration of 3 g/m³.*

(a) *State the IVP that is satisfied by $x(t)$, which is the amount of salt in grams in the tank at time t.*

(b) *What will happen to the concentration in the tank as $t \to +\infty$? Explain.*

(c) *Use DFIELD to numerically solve the IVP of (a). Plot the solution.*

(d) *At what time will there be exactly 75 g of salt in the tank?*

Exercise 2.6.3. A saltwater solution at a concentration of 6 g/m^3 enters a 25 m^3 tank at a flow rate of 0.5 m^3/min. The uniformly mixed solution leaves the tank at a flow rate of 0.25 m^3/min. Assume that the tank initially holds 5 m^3 of the solution at a concentration of 25 g/m^3.

(a) State the IVP that is satisfied by $x(t)$, which is the amount of salt in grams in the tank at time t.

(b) Solve the IVP of (a). For what t values is the solution valid? Explain.

(c) At which time will the least amount of salt be present in the tank? What will the concentration in the tank be at that time?

(d) Use DFIELD to numerically solve the IVP of (a). Plot the solution over the range of t values for which the solution is valid.

Exercise 2.6.4. A 200 million m^3 lake is polluted with mercury, which is currently present in a concentration of 5 g per million cubic meters. The water is considered to be unsafe if the concentration is above 1 g per million cubic meters. If freshwater flows into the lake at a rate of 0.5 million cubic meters per day and the uniformly mixed water flows out of the lake at the same rate, how long will it take before the water is considered to be safe?

Exercise 2.6.5. A saltwater solution at a concentration of $3 + 2\sin(t)$ g/L enters a 25 L tank at a flow rate of 0.5 L/min. The uniformly mixed solution leaves the tank at the same flow rate. Assume that the tank initially holds 20 L of the solution at a concentration of 12 g/L.

(a) State the IVP that is satisfied by $x(t)$, which is the amount of salt in grams in the tank at time t.

(b) Solve the IVP of (a).

(c) Use DFIELD to numerically solve the IVP of (a). Plot the solution.

(d) What will happen to the concentration in the tank as $t \to +\infty$? Explain.

Exercise 2.6.6. Suppose that a bank account pays a 4% interest rate. If money is deposited into the account according to the rule $f(t) = 250t$, how much money will be in the account after 25 years if there is initially $5000 in the account?

Exercise 2.6.7. Suppose that a bank account pays a 4% interest rate and that there is initially $5000 in the account. If money is deposited into the account according to the rule $f(t) = f_0 t$, what does f_0 need to be so that after 30 years there is $500,000 in the account?

Exercise 2.6.8. Suppose that a bank account pays a 4% interest rate and that there is initially $5000 in the account. If money is deposited into the account according to the rule $f(t) = f_0 t^2$, what does f_0 need to be so that after 30 years there is $500,000 in the account?

Exercise 2.6.9. *Suppose that a bank account pays a $r\%$ interest rate and that there is initially $5000 in the account. If money is deposited into the account according to the rule $f(t) = 250t$, what does r need to be so that after 30 years there is $500,000 in the account? (Hint: Use SAGE to find r.)*

Exercise 2.6.10. *Suppose that a bank account pays a $r\%$ interest rate and that there is initially $5000 in the account. If money is deposited into the account according to the rule $f(t) = 50t^2$, what does r need to be so that after 30 years there is $500,000 in the account? (Hint: Use SAGE to find r.)*

Group projects

2.1. A simple population model for cod in the North Atlantic is a nonlinear ODE:

$$x' = x(1-x) - f(1 - m\cos(2\pi t)). \tag{2.6.2}$$

Here $x = x(t) \geq 0$ is the population density of the cod (assumed to be independent of location). Further assumptions leading to this mathematical model are

(a) the fishing grounds support a finite population (the term $x(1-x)$), and

(b) the cod are harvested at a varying rate (the term $f(1 - m\cos(2\pi t))$).

Here $f > 0$ is a measure of the average amount of cod being fished, and $0 < m < 1$ determines the minimal fishing rate, which is $f(1-m)$. The time is in years, so the cosine term represents a seasonal variation in the fishing rate. The initial condition is $x(0) = x_0 > 0$. Overfishing has occurred if there is a $T > 0$ such that $x(T) = 0$. If the population density is small, it is reasonable to approximate the nonlinear term,

$$x(1-x) = x - x^2 \sim x,$$

which yields the linear approximating ODE

$$x' = x - f(1 - m\cos(2\pi t)). \tag{2.6.3}$$

The goal of this project is to compare the solution behavior between the two models:

(a) Find the general solution to the linearized problem (2.6.3).

(b) As $t \to +\infty$, does the initial condition significantly influence the solution?

(c) Find a function $x^* = x^*(f, m) > 0$ such that for the linear approximation (2.6.3),

　(1) $0 < x_0 < x^*$ means that overfishing will occur and

　(2) $x_0 > x^*$ means that the population density becomes arbitrarily large as t gets large.

(d) Now consider the nonlinear problem (2.6.2). For $f = 0.2$ and $m = 0.7$, pick an initial condition for which the linear model predicts that overfishing will occur. Numerically solve the nonlinear equation. Compare the solution of the linear equation with that for the nonlinear equation. Is there a significant difference between the solution to the linear problem and the solution to the nonlinear problem? If so, describe the difference(s).

(e) Again consider the nonlinear problem (2.6.2). For $f = 0.2$ and $m = 0.7$, pick an initial condition for which the linear model predicts that the population will become arbitrarily large (but also pick $x_0 < 1$). Numerically solve the nonlinear equation. Compare the solution of the linear equation with that for the nonlinear equation. Is there a significant difference between the solution to the linear problem and the solution to the nonlinear problem? If so, describe the difference(s).

(f) For the nonlinear problem now assume $f = 0.25$ and $m = 0.7$. What do you observe; in particular, is the threshold $x^*(f, m)$ predicted by the linear model at all valid? Choose several initial conditions on the interval $[0, 1]$ to provide supporting evidence for your answer. (*Hint*: Solve the problem for $0 \le t \le 150$.)

(g) Pick an arbitrary $0 < m < 1$ and pick three initial values x_0. What do you observe for $f \in \{0.3, 0.5, 0.8\}$?

Chapter 3

Systems of first-order linear differential equations

> *Mathematics compares the most diverse phenomena and discovers the secret analogies that unite them.*
>
> —Jean Baptiste Joseph Fourier

Now that we know how to solve scalar ODEs, it is time to begin our study of systems of linear ODEs. We will heavily use the linear algebra we learned in Chapter 1, in particular, the ideas discussed in section 1.12 (eigenvalues and eigenvectors). Before discussing how to solve these linear systems of ODEs, we will first look at some problems that we would like to solve.

3.1 ▪ Motivating problems

3.1.1 ▪ Two-tank mixing problem

Consider the two-tank mixing problem illustrated in Figure 3.1. Letting x_1 represent the pounds of salt in tank 1 and x_2 the pounds of salt in tank 2, we wish to derive an ODE which models the physical situation. Following section 2.1 we know that the rate of change of the salt in each tank is equal to the input rate minus the output rate. First, note that the amount of brine solution in each tank is constant: for tank 1, 7 gallons enter and leave per hour, and for tank 2, 8 gallons enter and leave per hour. Arguing as in section 2.1, we see that

$$x_1' = \left(3\frac{\text{lb}}{\text{gal}}\right)\left(4\frac{\text{gal}}{\text{hr}}\right) + \left(\frac{x_2}{200}\frac{\text{lb}}{\text{gal}}\right)\left(3\frac{\text{gal}}{\text{hr}}\right) - \left(\frac{x_1}{100}\frac{\text{lb}}{\text{gal}}\right)\left(7\frac{\text{gal}}{\text{hr}}\right),$$

$$x_2' = \left(5\frac{\text{lb}}{\text{gal}}\right)\left(7\frac{\text{gal}}{\text{hr}}\right) + \left(\frac{x_1}{100}\frac{\text{lb}}{\text{gal}}\right)\left(1\frac{\text{gal}}{\text{hr}}\right) - \left(\frac{x_2}{200}\frac{\text{lb}}{\text{gal}}\right)\left(8\frac{\text{gal}}{\text{hr}}\right).$$

Simplifying yields

$$x_1' = -\frac{7}{100}x_1 + \frac{3}{200}x_2 + 12,$$

$$x_2' = \frac{1}{100}x_1 - \frac{8}{200}x_2 + 35,$$

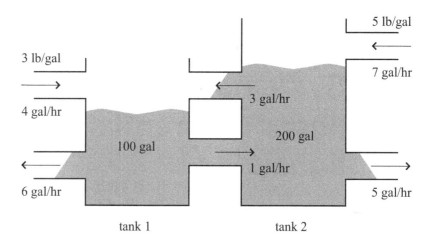

Figure 3.1. *A cartoon which depicts a two-tank mixing problem.*

which in matrix/vector form is

$$x' = \begin{pmatrix} -7/100 & 3/200 \\ 1/100 & -8/200 \end{pmatrix} x + \begin{pmatrix} 12 \\ 35 \end{pmatrix}.$$

Here we are making the (obvious) definition that the derivative of a vector is the derivative of each of its components:

$$x' = \begin{pmatrix} x'_1 \\ x'_2 \end{pmatrix}.$$

3.1.2 ▪ Glucose regulation and diabetes

Diabetes is a disease that is characterized by excessive glucose in the bloodstream. There are two types of diabetes:

(a) type 1, which is an autoimmune disease and is often called juvenile diabetes; and

(b) type 2, which is often referred to as adult-onset diabetes but which can occur in children as young as 5.

Diabetes increases the risk of heart disease, as the low insulin levels in the body can lead to atherosclerosis. Moreover, since diabetes can damage blood vessels and nerves as well as create osmotic imbalances, it can lead to

(a) kidney damage (nephropathy),

(b) nerve damage (neuropathy),

(c) blindness (retinopathy), and

(d) amputation due to decreased blood flow to the extremities.

Untreated diabetes has very severe consequences.

One way to detect diabetes is through a glucose tolerance test (GTT). A GTT is a medical test in which the patient first undertakes a 12-hour fast. Afterward,

3.1. Motivating problems

a large dose of glucose (e.g., 1.75 mg of glucose/kg of body weight) is given. The blood sugar is then monitored for the next 4–6 hours.

Ackerman et al. [2, 3] developed a mathematical model for the concentrations of glucose (G) and insulin (I) in the blood. In general, this model is of the form

$$G' = f(G,I) + J(t),$$
$$I' = h(G,I),$$

where $J(t)$ models the ingestion rate of glucose and the functions f and h model the manner in which the glucose and insulin interact with each other inside the body. These functions have parameters whose values are derived from experimental data.

Near an equilibrium, i.e., when the blood glucose and insulin concentrations are relatively constant, we can replace the nonlinear functions f and h with a linear approximation:

$$g(G,I) = aG + bI, \quad h(G,I) = cG + dI.$$

We can determine the signs of the parameters in the following manner. An increase in glucose

(a) stimulates uptake of glucose storage in the liver and

(b) results in the release of insulin.

Removing glucose from the blood implies $G' < 0$, and putting insulin into the blood means $I' > 0$. Since $G > 0$, we then have $a < 0$ and $c > 0$. On the other hand, an increase in insulin

(a) facilitates the uptake of glucose in tissues and the liver and

(b) results in the increased metabolism of excess insulin.

Since removing insulin from the blood means $I' < 0$, using $I > 0$ gives $b, d < 0$.

The equations can be written in system form. If we set

$$x = \begin{pmatrix} G \\ I \end{pmatrix},$$

then in matrix/vector form, the governing equation near equilibrium is

$$x' = \begin{pmatrix} a & b \\ c & d \end{pmatrix} x + \begin{pmatrix} J(t) \\ 0 \end{pmatrix}.$$

The constants in the matrix are experimentally determined, but we know that they satisfy the constraints

$$a, b, d < 0, \quad c > 0.$$

In a particular case, Bolie [7] found that

$$a = -2.92, \quad b = -4.34, \quad c = 0.21, \quad d = -0.78.$$

The solution to this problem will be considered in one of the group projects at the ends of Chapter 4 and 5.

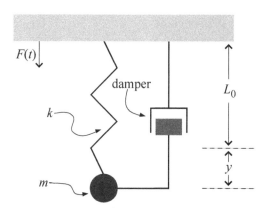

Figure 3.2. *A cartoon that depicts the mass-spring problem. The mass at the end of the spring with spring constant k is denoted by a filled circle. The attached rectangle represents a damping device.*

3.1.3 ▪ Mass-spring system

Consider a mass-spring system in which a mass is attached to the end of a spring (see Figure 3.2). This mass-spring system is then suspended from a support. The length of the spring with the attached mass is L_0: let $y(t)$ denote the distance from the equilibrium position. If $y(t) > 0$, then the spring is being extended; otherwise, it is being compressed. If m denotes the mass and g is the gravitational constant, then the force on the system due to gravity is given by $F_g = mg$. By Hooke's law, the force on the mass is given by $F_s = -k(L_0 + y)$, where $k > 0$ is the spring constant. If k is small, then the spring is weak, whereas if k is large, then the spring is stiff. When the spring is at equilibrium, then by Newton's second law, it is the case that

$$F_g + F_s = 0 \quad \leadsto \quad mg = kL_0.$$

If the spring is in motion, then by Newton's second law, we have that

$$my'' = F_g + F_s + F_d + F(t),$$

where F_d is a damping force and $F(t)$ represents a time-dependent external forcing on the system. Substituting our expressions for F_g and F_s into the above yields

$$my'' = mg - k(L_0 + y) + F_d + F(t) = -ky + F_d + F(t).$$

If we now assume that the damping force is proportional to velocity, i.e., $F_d = -cy'$ for some $c > 0$ (c small means weak damping, and c large means strong damping), the equation of motion finally becomes

$$my'' + cy' + ky = F(t).$$

On dividing through by m and setting

$$2b = \frac{c}{m}, \quad \omega_0^2 = \frac{k}{m}, \quad f(t) = \frac{1}{m}F(t),$$

we arrive at the canonical equation

$$y'' + 2by' + \omega_0^2 y = f(t). \tag{3.1.1}$$

The parameters satisfy $b \geq 0$ and $\omega_0 > 0$.

3.1. Motivating problems

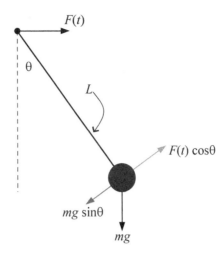

Figure 3.3. *A cartoon that depicts a horizontally forced pendulum.*

3.1.4 ▪ (Nonlinear) pendulum

Consider the pendulum depicted in Figure 3.3. A mass, m, is connected to the end of a rigid (and massless) rod of length L. Let $\theta(t)$ denote the angle from the equilibrium position in which the mass is hanging straight down; in other words, $\theta = 2k\pi$ for any $k \in \mathbb{Z}$ means that the pendulum is straight down, while $\theta = (2k+1)\pi$ for any $k \in \mathbb{Z}$ means that the pendulum is straight up. By Newton's second law, we have that

$$mL^2\theta'' = F_g + F_d + F_e,$$

where F_g is the force due to gravity, F_d represents a damping force, and F_e is an external force. We deduce from Figure 3.3 that

$$F_g = -mgL\sin\theta,$$

and, as in the previous example, we will assume that the damping is proportional to velocity, i.e.,

$$F_d = -cL\theta', \quad c > 0.$$

Regarding the external force, we will assume that the suspension point is being moved horizontally. This leads to

$$F_e = F(t)\cos\theta,$$

where $F(t)$ corresponds to the time-dependent manner in which the suspension point is being moved. For example, if the motion is periodic about a fixed point, then we would write $F(t) = F_0 \cos(\omega t)$. After some simplification, the equation of motion is

$$\theta'' + 2b\theta' + \omega_0^2 \sin\theta = f(t)\cos\theta, \qquad (3.1.2)$$

where

$$2b = \frac{c}{mL}, \quad \omega_0^2 = \frac{g}{L}, \quad f(t) = \frac{1}{mL^2}F(t).$$

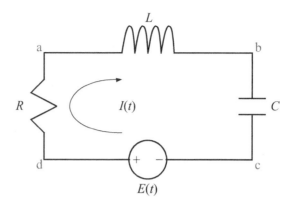

Figure 3.4. *A cartoon that depicts the closed RLC circuit with resistance R, capacitance C, inductance L, and energy source E(t). The points associated with the computation of each voltage drop are denoted by (red) letters.*

The equation of motion (3.1.2) is a *nonlinear ODE*, and the study of solutions to it is beyond the scope of this book. However, on recalling the Maclaurin series for $\sin\theta$ and $\cos\theta$,

$$\sin\theta = \theta - \frac{\theta^3}{3!} + \frac{\theta^5}{5!} + \cdots, \quad |\theta| < \infty,$$

$$\cos\theta = 1 - \frac{\theta^2}{2!} + \frac{\theta^4}{4!} + \cdots, \quad |\theta| < \infty,$$

if we assume that θ is small, i.e., the oscillations about the straight-down equilibrium position are small, then we may write $\sin\theta \sim \theta$ and $\cos\theta \sim 1$. Plugging these approximations into (3.1.2) yields the linear second-order ODE

$$\theta'' + 2b\theta' + \omega_0^2\theta = f(t),$$

which is precisely (3.1.1).

3.1.5 ▪ RLC circuit

Consider the circuit given in Figure 3.4. The current in the circuit, $I(t)$, is the rate of change of the charge, $Q(t)$. By the Fundamental Theorem of Calculus, this means that the two quantities are related via

$$Q'(t) = I(t) \quad \Leftrightarrow \quad Q(t) = \int I(t)\,dt.$$

We will henceforth assume without loss of generality that the current is flowing around the circuit in a clockwise fashion. The voltage drop between two points α and β on the circuit will be denoted as $V_{\alpha\beta}$. These various voltage drops are related by Kirchoff's law.

We begin by computing the voltage drop across each segment of the circuit. By *Faraday's law*, the voltage drop associated with an inductor is proportional to the rate of change of the current,

$$V_{ab} = LI'(t),$$

where L is the inductance produced by the inductor. *Coulomb's law* states that the voltage drop associated with a capacitor is proportional to the charge,

$$V_{bc} = \frac{1}{C}Q(t),$$

where C is the capacitance caused by the capacitor. We know from the voltage source that

$$V_{cd} = -E(t).$$

The negative sign follows from the direction of the current and the given polarity of the source. Finally, by *Ohm's law*, the voltage drop associated with a resistor is proportional to the current,

$$V_{da} = RI(t),$$

where R is the resistance produced by the resistor.

Kirchoff's law states the sum total of voltage drops around this circuit is zero, i.e.,

$$V_{ab} + V_{bc} + V_{cd} + V_{da} = 0.$$

Substitution of the above for the various voltage drops yields that the current and charge must satisfy

$$LI' + \frac{1}{C}Q - E(t) + RI = 0,$$

which can be rearranged as

$$LI' + RI + \frac{1}{C}Q = E(t). \tag{3.1.3}$$

In order to make this an ODE for the charge only, recall that the charge and current are related via $I = Q'$. If we substitute Q' for I in (3.1.3), we get

$$LQ'' + RQ' + \frac{1}{C}Q = E(t) \rightsquigarrow Q'' + \frac{R}{L}Q' + \frac{1}{LC}Q = \frac{1}{L}E(t).$$

By setting

$$2b = \frac{R}{L}, \quad \omega_0^2 = \frac{1}{LC}, \quad f(t) = \frac{1}{L}E(t),$$

we see that the equation for the charge is exactly the same as (3.1.1) for the forced-damped mass-spring system. The only difference between the two equations is the physical interpretation of the variables. If the charge Q is determined by solving the above ODE, then the current can be found by using $I = Q'$. Alternatively, the current is a solution of the ODE:

$$I'' + \frac{R}{L}I' + \frac{1}{LC}I = \frac{1}{L}E'(t).$$

Exercises

In the first three problems, set up but do not solve the system of differential equations and/or initial value problem (IVP) whose solution would give the amount of salt in each tank at time t. Write each system in matrix form.

Exercise 3.1.1. *A system of two tanks is connected such that each tank has independent inflows and outflows (drains), and each tank is connected to the other with an inflow and an outflow. The relevant information about each tank is given in the table below:*

	Tank A	Tank B
Tank volume	200 L	250 L
Rate of inflow to the tank	5 L/min	7 L/min
Rate of drain outflow	4 L/min	8 L/min
Rate of outflow to other tank	to B: 2 L/min	to A: 1 L/min
Concentration of salt in inflow	6 g/L	3 g/L

Exercise 3.1.2. *The situation is the same as in Exercise 3.1.1, except now the concentration of salt in the inflow of tank A is $5 + \cos(2t)$ g/L, and the concentration of salt in the inflow of tank B is $9 + 2e^{-2t}$ g/L. Furthermore, the concentration of the solute in tank A is initially 7 g/L, and tank B initially contains 225 g of salt.*

Exercise 3.1.3. *For a closed system of three tanks (i.e., there are no external input/output flows for any of the tanks), the following information is given:*

	Tank A	Tank B	Tank C
Tank volume	350 L	100 L	400 L
Rate of outflow	to B: 3 L/min	to C: 0 L/min	to A: 0 L/min
Rate of outflow	to C: 5 L/min	to A: 8 L/min	to B: 5 L/min

Tank A initially has 600 g of salt in the tank, the concentration of solute in tank B is initially 17 g/L, and the concentration of solute in tank C is initially 8 g/L.

Exercise 3.1.4. *We wish to find a linear second-order ODE that describes the motion of a pendulum that oscillates about the straight-up position $\theta = \pi$. The equation of motion is described by (3.1.2).*

(a) *Set $\phi = \theta - \pi$ (note that $\theta = \pi$ if and only if $\phi = 0$) and derive a nonlinear ODE for ϕ.*

(b) *Assuming that $|\phi|$ is small, find the linear second-order ODE that ϕ satisfies.*

3.2 ▪ Scalar equations and equivalent systems

The last three examples in section 3.1 led to second-order scalar linear ODEs. We now see how equations of this type can be thought of as first-order systems. We start with
$$y'' + 2by' + \omega_0^2 y = f(t).$$
Set
$$x_1 = y, \quad x_2 = y',$$
so that

$$x_1' = y' = x_2,$$
$$x_2' = y'' = -\omega_0^2 y - 2by' + f(t) = -\omega_0^2 x_1 - 2b x_2 + f(t).$$

3.2. Scalar equations and equivalent systems

The second-order ODE is then equivalent to the system

$$x' = \begin{pmatrix} 0 & 1 \\ -\omega_0^2 & -2b \end{pmatrix} x + \begin{pmatrix} 0 \\ f(t) \end{pmatrix}. \qquad (3.2.1)$$

If we are able to solve (3.2.1) for the vector x, then the solution y (which represents the position of the mass in the mass-spring system of section 3.1.3) to the original problem is given by x_1. Furthermore, the velocity of the mass (in the mass-spring context) at any given time is given by x_2. In conclusion, solving (3.2.1) allows us to *simultaneously* determine both the position and the velocity of the mass.

This type of equivalence between nth-order scalar ODEs and first-order systems holds in general. For example, suppose that we wish to convert the third-order linear ODE

$$y''' - 2ty'' + 3e^t y' + \cos(2t)y = 3t^2 - 1$$

into a system. On setting

$$x_1 = y, \quad x_2 = y', \quad x_3 = y'',$$

we get

$$x_1' = y' = x_2,$$
$$x_2' = y'' = x_3,$$
$$x_3' = y''' = -\cos(2t)y - 3e^t y' + 2ty'' + 3t^2 - 1,$$
$$= -\cos(2t)x_1 - 3e^t x_2 + 2t x_3 + 3t^2 - 1,$$

which is the system

$$x' = \begin{pmatrix} 0 & 1 & 0 \\ 0 & 0 & 1 \\ -\cos(2t) & -3e^t & 2t \end{pmatrix} x + \begin{pmatrix} 0 \\ 0 \\ 3t^2 - 1 \end{pmatrix}.$$

In general, an nth-order ODE will be converted to a system by setting

$$x_1 = y, \quad x_2 = y'', \quad , x_3 = y''', \quad \ldots \quad , x_n = y^{(n-1)},$$

noting that

$$x_1' = x_2, \quad x_2' = x_3, \quad x_3' = x_4, \quad \ldots, \quad x_{n-1}' = x_n,$$

and finding the equation for $x_n' = y^{(n)}$ from the ODE.

Exercises

Exercise 3.2.1. *Rewrite the following higher-order ODEs as an equivalent first-order system. Write each system in matrix form.*

(a) $y'' - 3y' + 6y = \sin(t)$

(b) $y'' + 4y = t^2 - e^{3t}$

(c) $y'' + t^2 y' + \sin(t)y = 0$

(d) $y''' - 8y'' + 5y' + 2y = 7t$

(e) $y''' - e^{-t}y'' + 5y' - \tan(t)y = 12\cos(4t)$

(f) $y^{(4)} - 7y''' + 13y'' + 4y = t^3 + 3t - 9\sin(2t)$

(g) $y^{(4)} + 4\ln(t)y''' - 3e^{2t}y = -6\tan(t)$

3.3 • Existence and uniqueness theory

In this section, we will state the general existence and uniqueness theory for the IVP

$$x' = A(t)x + f(t), \quad x(t_0) = x_0, \tag{3.3.1}$$

which is the system version of the scalar linear ODE discussed in Corollary 2.2. Here we will assume that the *state vector* $x = x(t) \in \mathbb{R}^n$ and that both $A(t) \in \mathcal{M}_n(\mathbb{R})$ and the *forcing vector* $f(t) \in \mathbb{R}^n$ are continuous on some interval that contains the initial time. We say that a matrix and vector are continuous if each entry of the matrix and vector is a continuous function. Systems such as (3.3.1) arise whenever the rate of change of each component of the state vector is a linear combination of all the components plus some additional term that does not depend on the state vector.

In order for there to be a unique solution (assuming that one exists at all), we must specify the *initial condition*

$$x(t_0) = x_0.$$

The ODE combined with the initial condition is the IVP. Regarding a solution to the IVP, we have the natural generalization of the existence/uniqueness result of Corollary 2.2:

Lemma 3.1. *Consider the IVP*

$$x' = A(t)x + f(t), \quad x(t_0) = x_0.$$

If $A = A(t)$ is continuous on the interval $|t - t_0| < C$, and if $f(t)$ is also continuous for $|t - t_0| < C$, then there is a unique solution for $|t - t_0| < C$. In other words, there is a smooth and unique vector-valued function $\phi(t)$: $(t_0 - C, t_0 + C) \mapsto \mathbb{R}^n$ such that

$$\phi' = A(t)\phi + f(t), \quad \phi(t_0) = x_0.$$

3.4 • The structure of the solution

By the existence and uniqueness result of Lemma 3.1, we know that there is a unique solution to the IVP

$$x' = A(t)x + f(t), \quad x(t_0) = x_0.$$

In this text, it will always be the case that $A(t)$ is continuous for all t. For the sake of exposition, it will be assumed in the rest of this chapter that $A(t)$ is continuous for all t. We now consider the structure of the solution. Here we will appropriately generalize the ideas and results of section 2.3, where we showed that the solution to the scalar ODE

$$x' = a(t)x + f(t)$$

3.4. The structure of the solution

can be written as the sum of a homogeneous solution, $x_h(t)$, and a particular solution, $x_p(t)$. The homogeneous solution solves the homogeneous ODE

$$x' = a(t)x,$$

and the particular solution is a solution to the full problem. Moreover, we showed how to construct each of these solutions.

Before we discuss finding the solution to the IVP, we first consider the solution structure to the system (3.3.1) without the initial value: this solution will again be known as the *general solution*. The following discussion is a mirror of that presented in section 2.3. Indeed, the only difference is that we now use matrices and vectors, whereas it was previously the case that everything was scalar.

First, consider the homogeneous problem:

$$x' = A(t)x \quad \leadsto \quad x' - A(t)x = 0.$$

If x_1 and x_2 are two solutions,

$$x'_1 = A(t)x_1, \quad x'_2 = A(t)x_2,$$

then the following sequence of calculations shows that a linear combination is also a homogeneous solution:

$$\begin{aligned}\frac{d}{dt}(c_1 x_1 + c_2 x_2) &= c_1 x'_1 + c_2 x'_2 \\ &= c_1 A(t)x_1 + c_2 A(t)x_2 \\ &= A(t)(cx_1 + c_2 x_2).\end{aligned}$$

The key observation here is that matrix/vector multiplication, just like differentiation, is a linear operation.

Suppose now that x_h is a solution to the homogeneous problem and that x_p is a particular solution, i.e., a solution to the full nonhomogeneous problem:

$$x' = A(t)x + f(t) \quad \leadsto \quad x' - A(t)x = f(t). \tag{3.4.1}$$

The following calculation shows that the sum of a homogeneous solution and a particular solution is also a solution to the nonhomogeneous problem:

$$\begin{aligned}\frac{d}{dt}\left(x_h + x_p\right) &= x'_h + x'_p \\ &= A(t)x_h + A(t)x_p + f(t) \\ &= A(t)\left(x_h + x_p\right) + f(t).\end{aligned}$$

Again, a key to the above calculation is the linearity of matrix/vector multiplication. In conclusion, as we saw for the scalar problem in section 2.3, solutions to the linear ODE (3.3.1) have the same structure as solutions to linear algebraic systems.

> **Theorem 3.2.** *Consider the linear nonhomogeneous ODE*
>
> $$x' = A(t)x + f(t),$$
>
> *and its homogeneous counterpart*
>
> $$x' = A(t)x.$$
>
> *If x_h is a solution to the homogeneous problem and x_p is a particular solution, then*
>
> $$x = x_h + x_p$$
>
> *is a solution to the nonhomogeneous problem. Furthermore, if x_1, x_2 are solutions to the homogeneous problem, then so is the linear combination $c_1 x_1 + c_2 x_2$.*

We must now consider the homogeneous problem in more detail. For the scalar problem, we showed that it was enough to find *one* solution to the homogeneous problem, which was nonzero at the initial time. Afterward, we constructed such a solution (see Lemma 2.6). How many solutions are needed for a linear homogeneous system?

Consider the homogeneous IVP

$$x' = A(t)x, \quad x(t_0) = x_0; \qquad A(t) \in \mathcal{M}_n(\mathbb{R}).$$

Suppose that there are m solutions $x_1(t), x_2(t), \ldots, x_m(t) \in \mathbb{R}^n$, so that

$$x_h(t) = c_1 x_1(t) + c_2 x_2(t) + \cdots + c_m x_m(t)$$

is also a solution to the homogeneous problem. In order to solve for the initial condition, we must have

$$x_0 = x_h(t_0) = c_1 x_1(t_0) + c_2 x_2(t_0) + \cdots + c_m x_m(t_0). \tag{3.4.2}$$

This is a linear system of n equations with m unknowns, $c_1, c_2, \ldots, c_m$. The coefficient matrix is

$$(x_1(t_0) \ x_2(t_0) \ \cdots \ x_m(t_0)) \in \mathcal{M}_{n \times m}(\mathbb{R}).$$

Regarding this linear system of equations, we know by Theorem 1.75 that there is a unique solution for any initial condition x_0 if and only if

(a) $m = n$, and

(b) the vectors $\{x_1(t_0), x_2(t_0), \ldots, x_m(t_0)\}$ form a linearly independent set.

Thus, in order to uniquely solve the IVP for *any* initial condition, we want to find precisely n vector-valued solutions to the homogeneous problem that, when evaluated at the initial time, form a set of linearly independent vectors. Under these conditions, we will call $x_h(t)$ the general homogeneous solution.

How do we solve the IVP for the nonhomogeneous problem (3.3.1)? Suppose that we have n solutions to the homogeneous system, $x_1, x_2, \ldots, x_n$ that are linearly independent at the initial time $t = t_0$. Define the *(fundamental) matrix-valued solution* $\Phi(t)$ by

$$\Phi(t) = (x_1(t) \ x_2(t) \ \cdots \ x_n(t))$$

3.4. The structure of the solution

and note that the moniker is well deserved because

$$\begin{aligned}\Phi' &= (x_1'\ x_2'\ \cdots\ x_n') \\ &= (A(t)x_1\ A(t)x_2\ \cdots\ A(t)x_n) \\ &= A(t)(x_1\ x_2\ \cdots\ x_n) = A(t)\Phi.\end{aligned}$$

The linear independence assumption on the solutions at $t = t_0$ implies that

$$\Phi(t_0) = (x_1(t_0)\ x_2(t_0)\ \cdots\ x_n(t_0)) \xrightarrow{\text{RREF}} I_n.$$

Since

$$\Phi(t)c = c_1 x_1(t) + c_2 x_2(t) + \cdots + c_n x_n(t)$$

is a linear combination of solutions to the homogeneous problem, we have by Theorem 3.2 that $\Phi(t)c$ is a solution to the homogeneous problem for any vector $c \in \mathbb{C}^n$. Thus, we can write the general homogeneous solution in the form

$$x_h(t) = \Phi(t)c.$$

By applying Theorem 3.2, a solution to the nonhomogeneous problem can be written as the sum of the general homogeneous solution and a particular solution:

$$\begin{aligned}x(t) &= \Phi(t)c + x_p(t) \\ &= c_1 x_1(t) + c_2 x_2(t) + \cdots + c_n x_n(t) + x_p(t).\end{aligned} \quad (3.4.3)$$

As for the initial condition, on evaluating the general solution at $t = t_0$,

$$x_0 = x(t_0) = \Phi(t_0)c + x_p(t_0) \quad \leadsto \quad \Phi(t_0)c = x_0 - x_p(t_0).$$

This is a linear system of n equations with n unknowns (the entries in the vector c). Since the initial condition is arbitrary, so is the right-hand side $x_0 - x_p(t_0)$. The RREF of $\Phi(t_0)$ is I_n, so by Theorem 1.75, there is a unique solution c to this linear algebraic system. Thus, we again see that if the n solutions to the homogeneous problem are chosen so that they are linearly independent at $t = t_0$, then the solution given by (3.4.3) can be used to solve the IVP for any initial condition. This solution will be known as the general solution.

Corollary 3.3. *The general solution to the nonhomogeneous problem*

$$x' = A(t)x + f(t)$$

is given by

$$x(t) = \underbrace{c_1 x_1(t) + c_2 x_2(t) + \cdots + c_n x_n(t)}_{\Phi(t)c} + x_p(t).$$

Here $x_p(t)$ is a particular solution, and $x_1, x_2, \ldots, x_n$ are solutions to the homogeneous problem such that the set of vectors $\{x_1(t_0), x_2(t_0), \ldots, x_n(t_0)\}$ is linearly independent.

Example 3.4. Consider the ODE

$$x' = \underbrace{\begin{pmatrix} 0 & 1 \\ -2 & -3 \end{pmatrix}}_{A}x + \underbrace{\begin{pmatrix} 0 \\ 10\cos(t) \end{pmatrix}}_{f(t)}, \quad x(0) = \underbrace{\begin{pmatrix} 2 \\ 4 \end{pmatrix}}_{x_0}.$$

Since the matrix $A(t)$ is constant coefficient, i.e., none of the coefficients are time dependent, we simply write $A(t) \equiv A$ above. It can be checked that the general homogeneous solution is

$$x_h(t) = c_1 e^{-t}\begin{pmatrix} 1 \\ -1 \end{pmatrix} + c_2 e^{-2t}\begin{pmatrix} 1 \\ -2 \end{pmatrix} = \underbrace{\begin{pmatrix} e^{-t} & e^{-2t} \\ -e^{-t} & -2e^{-2t} \end{pmatrix}}_{\Phi(t)} c,$$

and that a particular solution is

$$x_p(t) = \cos(t)\begin{pmatrix} 1 \\ 3 \end{pmatrix} + \sin(t)\begin{pmatrix} 3 \\ -1 \end{pmatrix}.$$

Thus, the general solution is the sum of the homogeneous and particular solutions:

$$x(t) = \Phi(t)c + \cos(t)\begin{pmatrix} 1 \\ 3 \end{pmatrix} + \sin(t)\begin{pmatrix} 3 \\ -1 \end{pmatrix}.$$

Regarding the initial condition, we have

$$\begin{pmatrix} 2 \\ 4 \end{pmatrix} = x(0) = \Phi(0)c + \begin{pmatrix} 1 \\ 3 \end{pmatrix},$$

which is equivalent to

$$\begin{pmatrix} 1 & 1 \\ -1 & -2 \end{pmatrix}c = \begin{pmatrix} 2 \\ 4 \end{pmatrix} - \begin{pmatrix} 1 \\ 3 \end{pmatrix} = \begin{pmatrix} 1 \\ 1 \end{pmatrix}.$$

Solving this linear system yields

$$c = \begin{pmatrix} 1 & 1 \\ -1 & -2 \end{pmatrix}^{-1}\begin{pmatrix} 1 \\ 1 \end{pmatrix} = \begin{pmatrix} 3 \\ -2 \end{pmatrix}.$$

In conclusion, the solution to the IVP is

$$x(t) = 3e^{-t}\begin{pmatrix} 1 \\ -1 \end{pmatrix} - 2e^{-2t}\begin{pmatrix} 1 \\ -2 \end{pmatrix} + \cos(t)\begin{pmatrix} 1 \\ 3 \end{pmatrix} + \sin(t)\begin{pmatrix} 3 \\ -1 \end{pmatrix}. \blacksquare$$

Exercises

Exercise 3.4.1. *Suppose that for the ODE* $x' = A(t)x + f(t)$, *two homogeneous solutions,* $x_1(t)$ *and* $x_2(t)$, *and a particular solution,* $x_p(t)$, *are given by*

$$x_1(t) = e^t\begin{pmatrix} 1 \\ 1 \end{pmatrix}, \quad x_2(t) = e^{-3t}\begin{pmatrix} 1 \\ -3 \end{pmatrix}, \quad x_p(t) = \cos(3t)\begin{pmatrix} 2 \\ 1 \end{pmatrix} + \sin(3t)\begin{pmatrix} 1 \\ -2 \end{pmatrix}.$$

(a) What is the matrix-valued solution $\Phi(t)$ to the homogeneous problem?

(b) What is the homogeneous solution to the ODE?

(c) What is the general solution to the ODE?

(d) If $x(0) = (1\ -4)^T$, what is the solution to the IVP?

Exercise 3.4.2. *Suppose that for the ODE $x' = A(t)x + f(t)$, three homogeneous solutions, $x_1(t), x_2(t), x_3(t)$, and a particular solution, $x_p(t)$, are given by*

$$x_1(t) = e^{-t}\begin{pmatrix} 1 \\ -1 \\ 1 \end{pmatrix}, \quad x_2(t) = e^{2t}\begin{pmatrix} 1 \\ 2 \\ 4 \end{pmatrix},$$

$$x_3(t) = e^{-3t}\begin{pmatrix} 1 \\ -3 \\ 9 \end{pmatrix}, \quad x_p(t) = e^{-2t}\cos(t)\begin{pmatrix} 1 \\ 6 \\ -3 \end{pmatrix}.$$

(a) What is the matrix-valued solution $\Phi(t)$ to the homogeneous problem?

(b) What is the homogeneous solution to the ODE?

(c) What is the general solution to the ODE?

(d) If $x(0) = (2\ -1\ 7)^T$, what is the solution to the IVP?

3.5 ▪ The homogeneous solution

The result of Corollary 3.3 tells us how to solve the IVP once the homogeneous and particular solutions have been found. Based on what we saw in Chapter 2, our expectation is that the key to finding a general solution to the nonhomogeneous ODE will be to first solve the homogeneous problem. In order to solve the homogeneous system

$$x' = A(t)x,$$

we know from Corollary 3.3 that we need to find n solutions, $x_1(t), x_2(t), \ldots, x_n(t)$, that have the property that the set of vectors $\{x_1(t_0), x_2(t_0), \ldots, x_n(t_0)\}$ is linearly independent. We first consider the problem of finding such a collection of solutions.

Let $\{v_1, v_2, \ldots, v_n\}$ be a basis for $\mathbb{R}^n$ and for $j = 1, \ldots, n$ consider the IVP

$$x' = A(t)x, \quad x(t_0) = v_j.$$

By Lemma 3.1, we know that there is a unique solution given by $x_j(t)$. Following the discussion after Theorem 3.2, define the matrix-valued solution $\Phi(t)$ by

$$\Phi(t) = (x_1(t)\ x_2(t)\ \cdots\ x_n(t)).$$

We have already seen that the general solution to the homogeneous problem is

$$x_h(t) = \Phi(t)c = c_1 x_1(t) + c_2 x_2(t) + \cdots + c_n x_n(t).$$

The choice of the vectors $v_1, \ldots, v_n$ ensures that $\Phi(t_0)$ is invertible. When we later discuss finding the particular solution, it will be important to know if the matrix-valued solution is invertible for *any* value of t for which it is defined. While we will not prove it here, it turns out to be the case that the determinant of the matrix solution (often called the *Wronskian*) satisfies *Abel's formula*, namely,

$$\det(\Phi(t)) = e^{\int \operatorname{tr} A(t)\,dt} \det(\Phi(t_0)).$$

Here $\operatorname{tr} A(s)$ is the *trace* of the matrix $A(s)$, and it is computed by summing the diagonal entries of the matrix. The important consequence of Abel's formula is that if $\det(\Phi(t_0)) \neq 0$, then $\det(\Phi(t)) \neq 0$ for as long as the solution exists. Since $\Phi(t_0)$ being invertible implies that $\det(\Phi(t_0)) \neq 0$ (see Theorem 1.76), it is then the case that $\det(\Phi(t)) \neq 0$. Thus, the matrix-valued solution $\Phi(t)$ is invertible as long as it is well defined.

> **Lemma 3.5.** *Consider the homogeneous system*
>
> $$x' = A(t)x, \quad A(t) \in \mathcal{M}_n(\mathbb{R}),$$
>
> *where $A(t)$ is continuous for $T_- < t < T_+$. Suppose that for some $T_- < t_0 < T_+$ the matrix-valued solution, $\Phi(t) \in \mathcal{M}_n(\mathbb{R})$, is invertible. Then $\Phi(t)$ is invertible for all $T_- < t < T_+$.*

Unfortunately, the above discussion tells us nothing about how to actually find these linearly independent homogeneous solutions. Herein we will focus primarily on the case of constant matrices, for if the matrix is time dependent, then it is generally true that no explicit solutions can be found. In other words, for the rest of this chapter, it will be assumed that $A(t) \equiv A$, so the linear homogeneous ODE to be solved is

$$x' = Ax. \tag{3.5.1}$$

In our Case Studies in section 1.13, it was especially beneficial to write a given vector in terms of the eigenvectors of A in order to solve the discrete dynamical system $x_{n+1} = Ax_n$. Since that approach was so fruitful, we will follow it here. We start with the eigenvalue/eigenvector equation:

$$Av_j = \lambda_j v_j, \quad j = 1, \ldots, n.$$

Under the assumption that the eigenvectors $\{v_1, v_2, \ldots, v_n\}$ form a basis, which by Theorem 1.98(c) is guaranteed if the eigenvalues are distinct, we can write $x(t)$ in terms of the eigenvectors with a Fourier expansion:

$$x(t) = x_1(t)v_1 + x_2(t)v_2 + \cdots + x_n(t)v_n. \tag{3.5.2}$$

Differentiation yields

$$x' = x_1' v_1 + x_2' v_2 + \cdots + x_n' v_n.$$

As for the quantity Ax, we can use Lemma 1.99 to write

$$Ax = \lambda_1 x_1 v_1 + \lambda_2 x_2 v_2 + \cdots + \lambda_n x_n v_n.$$

Thus, the ODE (3.5.1) can be rewritten as

$$x_1' v_1 + x_2' v_2 + \cdots + x_n' v_n = \lambda_1 x_1 v_1 + \lambda_2 x_2 v_2 + \cdots + \lambda_n x_n v_n,$$

3.5. The homogeneous solution

which is equivalent to

$$(x_1' - \lambda_1 x_1)v_1 + (x_2' - \lambda_2 x_2)v_2 + \cdots + (x_n' - \lambda_n x_n)v_n = 0. \quad (3.5.3)$$

The homogeneous system (3.5.1) is equivalent to the system (3.5.3). The latter system can be thought of as a linear combination of the eigenvectors, where each weight,

$$x_j' - \lambda_j x_j, \quad j = 1, \ldots, n,$$

is a scalar linear ODE of the form studied in Chapter 2. Since the eigenvectors are linearly independent, each of the weights is zero:

$$x_j' - \lambda_j x_j = 0 \rightsquigarrow x_j' = \lambda_j x_j, \quad j = 1, \ldots, n.$$

This scalar ODE is easily solved (see Lemma 2.6), and we find that

$$x_j(t) = c_j e^{\lambda_j t}, \quad j = 1, \ldots, n.$$

Going back to (3.5.2), we then see that a homogeneous solution can be written as

$$x_h(t) = c_1 e^{\lambda_1 t} v_1 + c_2 e^{\lambda_2 t} v_2 + \cdots + c_n e^{\lambda_n t} v_n.$$

This is a matrix/vector multiplication; in particular, on setting

$$\Phi(t) = \left(e^{\lambda_1 t} v_1 \; e^{\lambda_2 t} v_2 \; \cdots \; e^{\lambda_n t} v_n \right), \quad c = \begin{pmatrix} c_1 \\ c_2 \\ \vdots \\ c_n \end{pmatrix},$$

we can write $x_h(t) = \Phi(t)c$. For the constant coefficient problem, we are able to explicitly write down the matrix-valued solution, $\Phi(t)$. Since

$$\Phi(0) = (v_1 \; v_2 \; \cdots \; v_n)$$

is invertible (recall that the eigenvectors are linearly independent), by Lemma 3.5 the matrix-valued solution is invertible for all t.

Theorem 3.6. *Consider the homogeneous system with constant coefficients*

$$x' = Ax.$$

If λ is an eigenvalue with associated eigenvector v, then a solution is given by $x(t) = e^{\lambda t} v$. If the eigenvectors form a basis for $\mathbb{C}^n$ (which is ensured if the eigenvalues are distinct), then a homogeneous solution is given by

$$x_h(t) = c_1 e^{\lambda_1 t} v_1 + c_2 e^{\lambda_2 t} v_2 + \cdots + c_n e^{\lambda_n t} v_n = \Phi(t)c,$$

where the invertible matrix-valued solution is

$$\Phi(t) = \left(e^{\lambda_1 t} v_1 \; e^{\lambda_2 t} v_2 \; \cdots \; e^{\lambda_n t} v_n \right).$$

Remark 3.7. *Solutions to the homogeneous system can also be found by guessing a solution of the form*
$$x(t) = e^{\lambda t} v,$$
where v is a constant vector and λ is an unknown scalar. Noting that
$$\frac{d}{dt} e^{\lambda t} v = \lambda e^{\lambda t} v,$$
by plugging this guess into the ODE $x' = Ax$ and using the linearity of matrix/vector multiplication, we get
$$\lambda e^{\lambda t} v = e^{\lambda t} A v \quad \rightsquigarrow \quad e^{\lambda t}(Av - \lambda v) = 0.$$
Since the exponential can never be zero, this implies that
$$Av = \lambda v;$$
in other words, λ is an eigenvalue and v is a corresponding eigenvector.

Example 3.8. Let us compute the homogeneous solution for Example 3.4, for which the homogeneous ODE is
$$x' = Ax, \quad A = \begin{pmatrix} 0 & 1 \\ -2 & -3 \end{pmatrix}.$$

The solution to the ODE is found by finding the eigenvalues and corresponding eigenvectors for the matrix A. The characteristic polynomial is
$$p_A(\lambda) = \lambda^2 + 3\lambda + 2 = (\lambda+1)(\lambda+2),$$
so the eigenvalues are $\lambda_1 = -1$ and $\lambda_2 = -2$. Regarding the associated eigenvectors,
$$A - (-1)I_2 \xrightarrow{\text{RREF}} \begin{pmatrix} 1 & 1 \\ 0 & 0 \end{pmatrix} \quad \rightsquigarrow \quad v_1 = \begin{pmatrix} 1 \\ -1 \end{pmatrix},$$
and
$$A - (-2)I_2 \xrightarrow{\text{RREF}} \begin{pmatrix} 1 & 1/2 \\ 0 & 0 \end{pmatrix} \quad \rightsquigarrow \quad v_2 = \begin{pmatrix} 1 \\ -2 \end{pmatrix}.$$
From Theorem 3.6, the general solution is
$$x(t) = c_1 e^{-t} \begin{pmatrix} 1 \\ -1 \end{pmatrix} + c_2 e^{-2t} \begin{pmatrix} 1 \\ -2 \end{pmatrix},$$
which is precisely what was seen in Example 3.4. ∎

It is clear that for the homogeneous ODE $x' = Ax$, there is the trivial solution $x = 0$. We adopt the following moniker:

Equilibrium solution

Definition 3.9. *The trivial solution $x = 0$ solves the homogeneous ODE $x' = Ax$. This solution is often denoted as an equilibrium solution or a critical point.*

3.5. The homogeneous solution

As a consequence of Theorem 3.6, we know how to compute the homogeneous solution if the matrix is time independent. For the rest of this discussion on the homogeneous problem, we will concern ourselves with the graphical behavior of the solutions as a function of the eigenvalues. For the sake of exposition, we will focus on the case of $n = 2$, i.e., planar systems. This restriction allows us to easily compute the eigenvalues analytically (the characteristic polynomial is a quadratic). Moreover, it is much easier to draw parametric graphs on the $x_1 x_2$-plane than in any higher-dimensional space.

If the eigenvectors form a basis (true if the eigenvalues are distinct), the homogeneous solution is of the form

$$x_h(t) = c_1 e^{\lambda_1 t} v_1 + c_2 e^{\lambda_2 t} v_2.$$

There are several cases to consider:

(a) $\lambda_1 < 0 < \lambda_2$,

(b) $\lambda_1 < \lambda_2 < 0$,

(c) $0 < \lambda_1 < \lambda_2$, and

(d) $\lambda_{1,2}$ are complex valued, i.e., $\lambda_{1,2} = a \pm ib$ with $b \neq 0$.

The final case to be considered, which falls outside the purview of the concluding Theorem 3.15, is when there is a double eigenvalue but only one eigenvector, in other words, the case where the geometric multiplicity of the eigenvalue is one but the algebraic multiplicity is two. Here we will required to derive a second linearly independent solution to the homogeneous ODE. This task will be accomplished in Lemma 3.22.

The plotting of solutions is best considered through specific examples.

3.5.1 ▪ Real eigenvalues: Saddle point

Consider

$$x' = Ax, \quad A = \begin{pmatrix} 2 & 3 \\ 3 & 2 \end{pmatrix}. \tag{3.5.4}$$

We first find the homogeneous solution. Since the characteristic equation is

$$p_A(\lambda) = (2 - \lambda)^2 - 9,$$

the eigenvalues are $\lambda_1 = -1$, $\lambda_2 = 5$. Regarding the associated eigenvectors,

$$A - (-1)I_2 \xrightarrow{\text{RREF}} \begin{pmatrix} 1 & 1 \\ 0 & 0 \end{pmatrix} \rightsquigarrow v_1 = \begin{pmatrix} -1 \\ 1 \end{pmatrix},$$

and

$$A - 5I_2 \xrightarrow{\text{RREF}} \begin{pmatrix} 1 & -1 \\ 0 & 0 \end{pmatrix} \rightsquigarrow v_2 = \begin{pmatrix} 1 \\ 1 \end{pmatrix}.$$

Thus, from Theorem 3.6, the general solution is given by

$$x(t) = c_1 e^{-t} \begin{pmatrix} -1 \\ 1 \end{pmatrix} + c_2 e^{5t} \begin{pmatrix} 1 \\ 1 \end{pmatrix}, \tag{3.5.5}$$

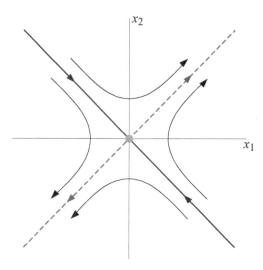

Figure 3.5. *A qualitatively correct phase plane associated with the system* (3.5.4), *for which the origin is an unstable saddle point. Solution curves are drawn for several different initial conditions.*

and the matrix-valued solution is

$$\Phi(t) = \begin{pmatrix} -e^{-t} & e^{5t} \\ e^{-t} & e^{5t} \end{pmatrix}.$$

We now consider the parametric plots of several possible solution curves on the $x_1 x_2$-plane, i.e., the *phase plane*. In components, the general solution is

$$x_1(t) = -c_1 e^{-t} + c_2 e^{5t}, \quad x_2(t) = c_1 e^{-t} + c_2 e^{5t}.$$

Based on our limited experience in calculus doing parametric plots, it is not at all clear what a curve looks like—even from a qualitative perspective—for a given c_1 and c_2. However, the plot of this solution becomes much easier to understand if we instead think of the full vector-valued solution. Indeed, this perspective is crucial in order to understand the solution behavior graphically.

First, consider the solution in (3.5.5) when $c_2 = 0$. The line through the origin and parallel to the vector $(-1, 1)^T$ is invariant; i.e., if an initial condition starts on the line, then the solution stays on the line for all $t \geq 0$. This line is represented in Figure 3.5 by the solid curve. The prefactor e^{-t} means that the length of the original vector decreases exponentially fast as $t \to +\infty$; however, the direction of the initial vector does not change. In conclusion, solutions on this line head toward the equilibrium solution $x = 0$ as $t \to +\infty$. The direction of the flow on the line is denoted by an arrow pointing toward the origin. Now suppose that $c_1 = 0$. Arguing as above, the dashed line, which is through the origin and parallel to the vector $(1, 1)^T$, is invariant. Since the exponential prefactor is now e^{5t}, solutions on this curve exponentially grow as $t \to +\infty$: the direction of the flow is again shown via an arrow but now points away from the origin.

The $c_1 = 0$ and $c_2 = 0$ lines separate the phase plane into four distinct regions. If an initial condition is in one the regions, the corresponding solution curve must stay in that region for all $t \geq 0$: it cannot cross the invariant lines because of the uniqueness of solutions. Solutions for which both c_1, and $c_2 \neq 0$ are denoted by

3.5. The homogeneous solution

thin (black) curves. Because of the nature of the exponential function, it will be the case that

$$x(t) \sim c_2 e^{5t} \begin{pmatrix} 1 \\ 1 \end{pmatrix}, \quad t \gg 0.$$

In other words, for large time, the solution for $c_2 \neq 0$ will be very close to (but not on!) the dashed line.

In conclusion, for this problem, we have that solutions other than those that start on the invariant dashed line for $c_2 = 0$ will grow exponentially fast as $t \to +\infty$. The solutions that start on the solid line $c_2 = 0$ will decay exponentially fast as $t \to +\infty$. We will label the trivial solution $x = 0$ in this case as a (unstable) *saddle point*. In general, $x = 0$ is a saddle point when the eigenvalues are real and have opposite sign.

Example 3.10. Consider the IVP

$$x' = Ax, \quad A = \begin{pmatrix} 0 & 1 \\ 3 & -2 \end{pmatrix}, \quad x(0) = \begin{pmatrix} 2 \\ -5 \end{pmatrix}.$$

Let us first consider the general solution. The characteristic equation is

$$p_A(\lambda) = \lambda^2 + 2\lambda - 3 = (\lambda - 1)(\lambda + 3).$$

The eigenvalues and associated eigenvectors are

$$\lambda_1 = -3, \, v_1 = \begin{pmatrix} 1 \\ -3 \end{pmatrix}; \quad \lambda_2 = 1, \, v_2 = \begin{pmatrix} 1 \\ 1 \end{pmatrix}.$$

The general solution is

$$x(t) = c_1 e^{-3t} \begin{pmatrix} 1 \\ -3 \end{pmatrix} + c_2 e^{t} \begin{pmatrix} 1 \\ 1 \end{pmatrix},$$

and the matrix-valued solution is

$$\Phi(t) = \begin{pmatrix} e^{-3t} & e^{t} \\ -3e^{-3t} & e^{t} \end{pmatrix}.$$

Since the eigenvalues are of opposite sign, the origin is an unstable saddle point.

In order to solve the IVP, we have

$$\begin{pmatrix} 2 \\ -5 \end{pmatrix} = x(0) = c_1 \begin{pmatrix} 1 \\ -3 \end{pmatrix} + c_2 \begin{pmatrix} 1 \\ 1 \end{pmatrix} \quad \rightsquigarrow \quad \begin{pmatrix} 1 & 1 \\ -3 & 1 \end{pmatrix} c = \begin{pmatrix} 2 \\ -5 \end{pmatrix}.$$

The solution to this linear system is

$$c = \begin{pmatrix} 1 & 1 \\ -3 & 1 \end{pmatrix}^{-1} \begin{pmatrix} 2 \\ -5 \end{pmatrix} = \frac{1}{4} \begin{pmatrix} 1 & -1 \\ 3 & 1 \end{pmatrix} \begin{pmatrix} 2 \\ -5 \end{pmatrix} = \frac{1}{4} \begin{pmatrix} 7 \\ 1 \end{pmatrix},$$

so the solution to the IVP is

$$x(t) = \frac{7}{4} e^{-3t} \begin{pmatrix} 1 \\ -3 \end{pmatrix} + \frac{1}{4} e^{t} \begin{pmatrix} 1 \\ 1 \end{pmatrix}. \quad \blacksquare$$

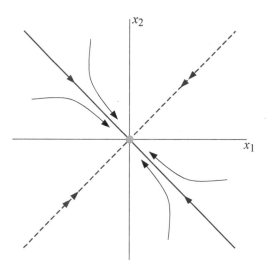

Figure 3.6. *A qualitatively correct phase plane associated with the system (3.5.6), for which the origin is a stable node. Solution curves are drawn for several different initial conditions.*

3.5.2 ▪ Real eigenvalues: Stable node

Now suppose that the eigenvalues and associated eigenvectors are given by

$$\lambda_1 = -2,\ v_1 = \begin{pmatrix} -1 \\ 1 \end{pmatrix};\quad \lambda_2 = -4,\ v_2 = \begin{pmatrix} 1 \\ 1 \end{pmatrix}. \tag{3.5.6}$$

The general solution is

$$x(t) = c_1 e^{-2t} \begin{pmatrix} -1 \\ 1 \end{pmatrix} + c_2 e^{-4t} \begin{pmatrix} 1 \\ 1 \end{pmatrix}, \tag{3.5.7}$$

and the matrix-valued solution is

$$\Phi(t) = \begin{pmatrix} -e^{-2t} & e^{-4t} \\ e^{-2t} & e^{-4t} \end{pmatrix}.$$

We plot several possible solution curves on the phase plane in Figure 3.6. First, consider the solution in (3.5.7) when $c_2 = 0$. Arguing as before, the solid line through the origin and parallel to the vector $(-1, 1)^T$ is invariant, and solutions that start on this line decay exponentially fast as $t \to +\infty$. The direction of the flow on the line is denoted by an arrow pointing toward the origin. Now suppose that $c_1 = 0$. The dashed line, which is through the origin and parallel to the vector $(1, 1)^T$, is invariant, and solutions on this curve also decay exponentially fast as $t \to +\infty$. The direction of the flow is again shown via an double arrow pointing toward the origin. The double arrow arises because e^{-4t} decays at a faster rate than e^{-2t}.

As before, the $c_1 = 0$ and $c_2 = 0$ lines separate the phase plane into four distinct regions. If an initial condition is in one the regions, the corresponding solution curve must stay in that region for all $t \geq 0$: it cannot cross the invariant lines because of the uniqueness of solutions. Solutions for which both c_1, and $c_2 \neq 0$ are

3.5. The homogeneous solution

denoted by thin curves. Because e^{-4t} decays faster than e^{-2t} as $t \to +\infty$, it will be the case that

$$x(t) \sim c_1 e^{-2t} \begin{pmatrix} -1 \\ 1 \end{pmatrix}, \quad t \gg 0.$$

In other words, for large time, the solution for $c_1 \neq 0$ will be very close to (but not on!) the solid line associated with solutions that decay at the slower rate.

In conclusion, for this problem, we have that all solutions will decay exponentially fast as $t \to +\infty$. We will label the trivial solution $x = 0$ in this case as a *stable node* (attractor). In general, $x = 0$ is a stable node when both eigenvalues are real and negative.

Example 3.11. Consider the IVP

$$x' = Ax, \quad A = \begin{pmatrix} 0 & 1 \\ -2 & -3 \end{pmatrix}, \quad x(0) = \begin{pmatrix} 5 \\ -7 \end{pmatrix}.$$

Let us first find the general solution. The characteristic equation is given by

$$p_A(\lambda) = \lambda^2 + 3\lambda + 2 = (\lambda+1)(\lambda+2).$$

The eigenvalues and associated eigenvectors are

$$\lambda_1 = -2, \; v_1 = \begin{pmatrix} 1 \\ -2 \end{pmatrix}; \quad \lambda_2 = -1, \; v_2 = \begin{pmatrix} 1 \\ -1 \end{pmatrix},$$

so that the general solution is

$$x(t) = c_1 e^{-2t} \begin{pmatrix} 1 \\ -2 \end{pmatrix} + c_2 e^{-t} \begin{pmatrix} 1 \\ -1 \end{pmatrix}.$$

The matrix-valued solution is

$$\Phi(t) = \begin{pmatrix} e^{-2t} & e^{-t} \\ -2e^{-2t} & -e^{-t} \end{pmatrix}.$$

In this case, the trivial solution is a stable node, as both of the eigenvalues are real and negative.

In order to solve the IVP, we have

$$\begin{pmatrix} 5 \\ -7 \end{pmatrix} = x(0) = c_1 \begin{pmatrix} 1 \\ -2 \end{pmatrix} + c_2 \begin{pmatrix} 1 \\ -1 \end{pmatrix} \rightsquigarrow \begin{pmatrix} 1 & 1 \\ -2 & -1 \end{pmatrix} c = \begin{pmatrix} 5 \\ -7 \end{pmatrix}.$$

The solution to this linear system is

$$c = \begin{pmatrix} 1 & 1 \\ -2 & -1 \end{pmatrix}^{-1} \begin{pmatrix} 5 \\ -7 \end{pmatrix} = \begin{pmatrix} -1 & -1 \\ 2 & 1 \end{pmatrix} \begin{pmatrix} 5 \\ -7 \end{pmatrix} = \begin{pmatrix} 2 \\ 3 \end{pmatrix}.$$

The solution to the IVP is then

$$x(t) = 2e^{-2t} \begin{pmatrix} 1 \\ -2 \end{pmatrix} + 3e^{-t} \begin{pmatrix} 1 \\ -1 \end{pmatrix}. \quad \blacksquare$$

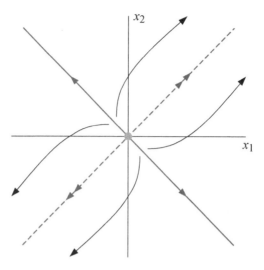

Figure 3.7. *A qualitatively correct phase plane associated with the system (3.5.8), for which the origin is an unstable node. Solution curves are drawn for several different initial conditions.*

3.5.3 ▪ Real eigenvalues: Unstable node

Next suppose that the eigenvalues and associated eigenvectors are given by

$$\lambda_1 = 2, \ v_1 = \begin{pmatrix} -1 \\ 1 \end{pmatrix}; \quad \lambda_2 = 4, \ v_2 = \begin{pmatrix} 1 \\ 1 \end{pmatrix}. \tag{3.5.8}$$

The general solution is

$$x(t) = c_1 e^{2t} \begin{pmatrix} -1 \\ 1 \end{pmatrix} + c_2 e^{4t} \begin{pmatrix} 1 \\ 1 \end{pmatrix}, \tag{3.5.9}$$

and the matrix-valued solution is

$$\Phi(t) = \begin{pmatrix} -e^{2t} & e^{4t} \\ e^{2t} & e^{4t} \end{pmatrix}.$$

We plot several possible solution curves on the phase plane in Figure 3.7. First, consider the solution in (3.5.9) when $c_2 = 0$. Arguing as before, the solid line through the origin and parallel to the vector $(-1, 1)^T$ is invariant, and solutions that start on this line grow exponentially fast as $t \to +\infty$. The direction of the flow on the line is denoted by an arrow pointing away from the origin. Now suppose that $c_1 = 0$. The dashed line, which is through the origin and parallel to the vector $(1, 1)^T$, is also invariant, and solutions on this curve also grow exponentially fast as $t \to +\infty$. The direction of the flow is again shown via a double arrow pointing away from the origin. The double arrow arises because e^{4t} grows at a faster rate than e^{2t}.

The $c_1 = 0$ and $c_2 = 0$ lines again separate the phase plane into four distinct regions. Solutions for which both c_1 and $c_2 \neq 0$ are denoted by thin curves. Because e^{4t} grows faster than e^{2t} as $t \to +\infty$, it will be the case that

$$x(t) \sim c_2 e^{4t} \begin{pmatrix} 1 \\ 1 \end{pmatrix}, \quad t \gg 0.$$

3.5. The homogeneous solution

In this case, the other term does not essentially disappear; instead, it is simply not as large as the other term. More rigorously, this is seen by noting that

$$c_1 e^{2t}\begin{pmatrix} -1 \\ 1 \end{pmatrix} + c_2 e^{4t}\begin{pmatrix} 1 \\ 1 \end{pmatrix} = e^{4t}\left[c_1 e^{-2t}\begin{pmatrix} -1 \\ 1 \end{pmatrix} + c_2 \begin{pmatrix} 1 \\ 1 \end{pmatrix}\right].$$

For large time, the solution for $c_2 \neq 0$ will be *parallel* to the line associated with solutions that grow at the faster rate.

In conclusion, for this problem, we have that all solutions will grow exponentially fast as $t \to +\infty$. We will label the trivial solution $x = 0$ in this case as an *unstable node* (repeller). In general, $x = 0$ is an unstable node when both eigenvalues are real and positive.

Example 3.12. Consider the system

$$x' = Ax, \quad A = \begin{pmatrix} 5 & 2 \\ 8 & 5 \end{pmatrix}.$$

The characteristic equation is

$$p_A(\lambda) = (5-\lambda)^2 - 16 = (\lambda-1)(\lambda-9),$$

so the eigenvalues and associated eigenvectors are

$$\lambda_1 = 1, \ v_1 = \begin{pmatrix} 1 \\ -2 \end{pmatrix}; \quad \lambda_2 = 9, \ v_2 = \begin{pmatrix} 1 \\ 2 \end{pmatrix}.$$

Since both eigenvalues are real and positive, the origin is an unstable node. The general solution is

$$x(t) = c_1 e^t \begin{pmatrix} 1 \\ -2 \end{pmatrix} + c_2 e^{9t} \begin{pmatrix} 1 \\ 2 \end{pmatrix},$$

and the matrix-valued solution is

$$\Phi(t) = \begin{pmatrix} e^t & e^{9t} \\ -2e^t & 2e^{9t} \end{pmatrix}. \quad \blacksquare$$

3.5.4 ▪ Complex-valued eigenvalues

We now consider the case that the eigenvalues are complex valued. Since $A \in \mathcal{M}_2(\mathbb{R})$, the eigenvalues and associated eigenvectors will come in complex-conjugate pairs (see Theorem 1.98(a)). Suppose that the complex-conjugate pair of eigenvalues and associated eigenvectors are

$$\lambda_1 = a + ib, \ v_1 = p + iq; \quad \lambda_2 = a - ib, \ v_2 = p - iq.$$

Formally, one of the solutions is given by

$$x(t) = e^{\lambda_1 t} v_1 = e^{(a+ib)t}(p + iq).$$

The other solution is the complex conjugate of this solution. This is a complex-valued solution, but it is not the one that we want. We want all of our components of the solution vector to be real valued. Thus, we must rewrite the solution in a different form.

Consider the exponential term. Now,

$$e^{(a+ib)t} = e^{at}e^{ibt},$$

(see Brown and Churchill [11, Section 28] for a proof of this fact), and in section 1.11, we showed Euler's formula:

$$e^{i\theta} = \cos\theta + i\sin\theta \quad \leadsto \quad e^{ibt} = \cos(bt) + i\sin(bt).$$

Since

$$e^{(a+ib)t} = e^{at}(\cos(bt) + i\sin(bt)),$$

the solution can be expressed as

$$x(t) = e^{at}(\cos(bt) + i\sin(bt))(p + iq)$$
$$= \underbrace{e^{at}(\cos(bt)p - \sin(bt)q)}_{x_1(t)} + i\underbrace{e^{at}(\sin(bt)p + \cos(bt)q)}_{x_2(t)}.$$

We now extract the desired solutions with real-valued components. By the linearity of differentiation and matrix/vector multiplication,

$$x' = Ax \quad \leadsto \quad x_1' + ix_2' = A(x_1 + ix_2) = Ax_1 + iAx_2,$$

so that

$$(x_1' - Ax_1) + i(x_2' - Ax_2) = 0.$$

The left-hand side is a complex-valued vector, so the real and imaginary parts are each the zero vector:

$$x_1' - Ax_1 = 0, \quad x_2' - Ax_2 = 0.$$

Thus, since

$$x_1' = Ax_1, \quad x_2' = Ax_2,$$

the single complex-valued solution yields *two* vector-valued solutions for which each component is real valued. All that is left to check is to make sure that these solutions are linearly independent.

Lemma 3.13. *Consider the linear ODE $x' = Ax$ for $A \in \mathcal{M}_n(\mathbb{R})$. Suppose that there is a complex-conjugate pair of eigenvalues and associated eigenvectors given by*

$$\lambda_1 = a + ib, \; v_1 = p + iq; \quad \lambda_2 = a - ib, \; v_2 = p - iq.$$

A pair of linearly independent solutions with real-valued components is given by

$$x_1(t) = e^{at}(\cos(bt)p - \sin(bt)q), \quad x_2(t) = e^{at}(\sin(bt)p + \cos(bt)q).$$

3.5. The homogeneous solution

Proof. While we know that $x_1(t)$ and $x_2(t)$ are solutions, we need to verify that they are linearly independent at $t = t_0$ for some t_0. We choose $t_0 = 0$. We have

$$x_1(0) = p, \quad x_2(0) = q,$$

so we need to show that $\{p, q\}$ is a linearly independent set of vectors. Suppose otherwise; in other words, suppose that there is a nonzero real constant c such that $q = cp$. It is then the case that the complex-valued eigenvalue λ has associated with it the eigenvector

$$v = p + iq = (1 + ic)p.$$

Since eigenvalues and eigenvectors come in complex-conjugate pairs, an eigenvector associated with the eigenvalue $\overline{\lambda}$ is $\overline{(1+ic)p} = (1-ic)p$. Now, the eigenvectors associated with distinct eigenvalues are necessarily linearly independent, which means that there can be no constant a such that

$$(1 + ic)p = a(1 - ic)p.$$

But this is obviously false, for $a = (1+ic)/(1-ic)$ does the trick. In conclusion, there is no real constant c such that $q = cp$, so the solutions are linearly independent. □

Now that we have an explicit expression for the general solution in the case that the eigenvalues have nonzero imaginary parts, we can investigate the behavior of the solutions. Consider the system

$$x' = Ax, \quad A = \begin{pmatrix} a & b \\ -b & a \end{pmatrix}, \tag{3.5.10}$$

where $a \in \mathbb{R}$ and $b > 0$. The characteristic equation is $p_A(\lambda) = (\lambda - a)^2 + b^2$, so that the complex-valued eigenvalues are $\lambda = a \pm ib$. Regarding the complex-valued eigenvector associated with the eigenvalue $a + ib$, we have

$$A - (a+ib)I_2 \xrightarrow{\text{RREF}} \begin{pmatrix} -i & 1 \\ 0 & 0 \end{pmatrix} \rightsquigarrow v = \begin{pmatrix} -i \\ 1 \end{pmatrix} = \begin{pmatrix} 0 \\ 1 \end{pmatrix} + i \begin{pmatrix} -1 \\ 0 \end{pmatrix}.$$

On using Lemma 3.13, the two linearly independent solutions are

$$x_1(t) = e^{at} \left(\cos(bt) \begin{pmatrix} 0 \\ 1 \end{pmatrix} - \sin(bt) \begin{pmatrix} -1 \\ 0 \end{pmatrix} \right) = e^{at} \begin{pmatrix} \sin(bt) \\ \cos(bt) \end{pmatrix}$$

and

$$x_2(t) = e^{at} \left(\sin(bt) \begin{pmatrix} 0 \\ 1 \end{pmatrix} + \cos(bt) \begin{pmatrix} -1 \\ 0 \end{pmatrix} \right) = e^{at} \begin{pmatrix} -\cos(bt) \\ \sin(bt) \end{pmatrix}.$$

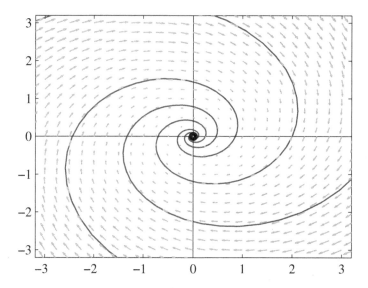

Figure 3.8. *The numerically generated phase plane for the system* (3.5.10) *when* $a = -1$ *and* $b = 3$. *The origin is a stable spiral.*

The general solution is then

$$x(t) = c_1 e^{at} \begin{pmatrix} \sin(bt) \\ \cos(bt) \end{pmatrix} + c_2 e^{at} \begin{pmatrix} -\cos(bt) \\ \sin(bt) \end{pmatrix},$$

and the matrix-valued solution is

$$\Phi(t) = \begin{pmatrix} e^{at} \sin(bt) & -e^{at} \cos(bt) \\ e^{at} \cos(bt) & e^{at} \sin(bt) \end{pmatrix}.$$

Now let us consider the phase plane associated with (3.5.10). Consider the solution $x_1(t)$. The parametric plot of the vector-valued function

$$c_1 \begin{pmatrix} \sin(bt) \\ \cos(bt) \end{pmatrix}$$

is a circle of radius $|c_1|$; furthermore, the circle is traversed in a counterclockwise direction for increasing t. The value of b, which is the frequency for the motion, determines the rate at which the circle is traversed: if b is small, the circle is slowly traversed, and if b is large, the rate of motion is large. Multiplying this vector by the function e^{at} means that the radius of the circle is decreasing ($a < 0$; see Figure 3.8), increasing ($a > 0$; see Figure 3.9), or constant ($a = 0$; see Figure 3.10). Thus, for $a \neq 0$, the curve becomes a spiral. The case $a < 0$ is called a *stable spiral* (sink), the case $a > 0$ is known as an *unstable spiral* (source), and when $a = 0$, the solution $x = 0$ is called a (linear) *center*.

3.5. The homogeneous solution

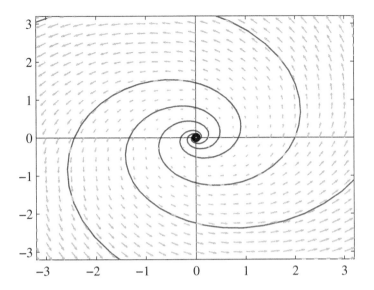

Figure 3.9. *The numerically generated phase plane for the system (3.5.10) when $a = 1$ and $b = 3$. The origin is an unstable spiral.*

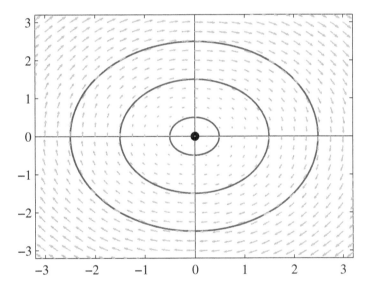

Figure 3.10. *The numerically generated phase plane for the system (3.5.10) when $a = 0$ and $b = 3$. The origin is a (linear) center.*

Is this characterization of the equilibrium solution still valid for more general matrices? The answer is yes. While we will not prove it here, it turns out to be the case that all that changes in our analysis is the *shape* of the solution curve. For our example, the form of the real and imaginary parts of the eigenvectors guaranteed that the solution curves were influenced by circles. In general, the solution curves are influenced by rotated ellipses, where the angle of rotation and the length of the major and minor axes are determined by the real and imaginary parts of the eigenvectors.

Example 3.14. Consider the IVP

$$x' = Ax, \quad A = \begin{pmatrix} 0 & 1 \\ -13 & 4 \end{pmatrix}, \quad x(0) = \begin{pmatrix} 3 \\ 0 \end{pmatrix}.$$

Let us first consider the general solution. The characteristic equation is

$$p_A(\lambda) = \lambda^2 - 4\lambda + 13 = (\lambda - 2)^2 + 9,$$

so that the complex-valued eigenvalues are $\lambda = 2 \pm i3$. The origin is an unstable spiral. Regarding the complex-valued eigenvector associated with the eigenvalue $2+i3$, we have

$$A - (2+i3)I_2 \xrightarrow{\text{RREF}} \begin{pmatrix} -2-i3 & 1 \\ 0 & 0 \end{pmatrix} \rightsquigarrow v = \begin{pmatrix} 1 \\ 2+i3 \end{pmatrix} = \begin{pmatrix} 1 \\ 2 \end{pmatrix} + i \begin{pmatrix} 0 \\ 3 \end{pmatrix}.$$

On using Lemma 3.13, the two linearly independent solutions are

$$x_1(t) = e^{2t}\left(\cos(3t)\begin{pmatrix} 1 \\ 2 \end{pmatrix} - \sin(3t)\begin{pmatrix} 0 \\ 3 \end{pmatrix}\right) = e^{2t}\begin{pmatrix} \cos(3t) \\ 2\cos(3t) - 3\sin(3t) \end{pmatrix},$$

and

$$x_2(t) = e^{2t}\left(\sin(3t)\begin{pmatrix} 1 \\ 2 \end{pmatrix} + \cos(3t)\begin{pmatrix} 0 \\ 3 \end{pmatrix}\right) = e^{2t}\begin{pmatrix} \sin(3t) \\ 2\sin(3t) + 3\cos(3t) \end{pmatrix}.$$

The general solution to the ODE system is

$$x(t) = c_1 e^{2t}\begin{pmatrix} \cos(3t) \\ 2\cos(3t) - 3\sin(3t) \end{pmatrix} + c_2 e^{2t}\begin{pmatrix} \sin(3t) \\ 2\sin(3t) + 3\cos(3t) \end{pmatrix}.$$

In order to solve the IVP, we have

$$\begin{pmatrix} 3 \\ 0 \end{pmatrix} = x(0) = c_1 \begin{pmatrix} 1 \\ 2 \end{pmatrix} + c_2 \begin{pmatrix} 0 \\ 3 \end{pmatrix} \rightsquigarrow \begin{pmatrix} 1 & 0 \\ 2 & 3 \end{pmatrix} c = \begin{pmatrix} 3 \\ 0 \end{pmatrix}.$$

The solution to this linear system is

$$c = \begin{pmatrix} 1 & 0 \\ 2 & 3 \end{pmatrix}^{-1} \begin{pmatrix} 3 \\ 0 \end{pmatrix} = \frac{1}{3}\begin{pmatrix} 3 & 0 \\ -2 & 1 \end{pmatrix} \begin{pmatrix} 3 \\ 0 \end{pmatrix} = \begin{pmatrix} 3 \\ -2 \end{pmatrix},$$

so the solution to the IVP is

$$x(t) = 3e^{2t}\begin{pmatrix} \cos(3t) \\ 2\cos(3t) - 3\sin(3t) \end{pmatrix} - 2e^{2t}\begin{pmatrix} \sin(3t) \\ 2\sin(3t) + 3\cos(3t) \end{pmatrix}. \blacksquare$$

3.5.5 ▪ Classification result: $n = 2$

We summarize the preceding discussion with the following classification of the trivial solution $x = 0$, which is based on our understanding of the solution behavior:

> **Theorem 3.15.** *Consider the homogeneous system with constant coefficients*
>
> $$x' = Ax, \quad A \in \mathcal{M}_2(\mathbb{R}).$$
>
> *Let λ_1, λ_2 denote the eigenvalues of A. The trivial solution $x = 0$ is classified for real eigenvalues as follows:*
>
> (a) $\lambda_1 < 0 < \lambda_2$: *unstable saddle point (see Figure 3.5);*
>
> (b) $\lambda_1 < \lambda_2 < 0$: *stable node (see Figure 3.6); and*
>
> (c) $0 < \lambda_1 < \lambda_2$: *unstable node (see Figure 3.7).*
>
> *If the eigenvalues are complex valued with $\lambda_1 = a + ib$ and $\lambda_2 = a - ib$, then $x = 0$ is said to be a*
>
> (a) $a < 0$: *stable spiral (see Figure 3.8);*
>
> (b) $a > 0$: *unstable spiral (see Figure 3.9); or*
>
> (c) $a = 0$: *(linear) center (see Figure 3.10).*

For pencil-and-paper calculations, the direction of spiraling (clockwise or counterclockwise) is typically not important. If it is necessary to determine the direction for a particular problem, then it is probably easiest to let PPLANE do the work for you. In the following set of examples, we use the characteristic polynomial in order to classify the equilibrium solution to the system $x' = Ax$.

Example 3.16. Suppose that for a given $A \in \mathcal{M}_2(\mathbb{R})$, the characteristic polynomial is

$$p_A(\lambda) = \lambda^2 + 5\lambda + 4 = (\lambda + 1)(\lambda + 4).$$

Since the roots are

$$\lambda_1 = -1, \quad \lambda_2 = -4,$$

the equilibrium solution is a stable node. ∎

Example 3.17. Suppose that for a given $A \in \mathcal{M}_2(\mathbb{R})$, the characteristic polynomial is

$$p_A(\lambda) = \lambda^2 - 4\lambda - 21 = (\lambda + 3)(\lambda - 7).$$

Since the roots are

$$\lambda_1 = -3, \quad \lambda_2 = 7,$$

the equilibrium solution is an unstable saddle point. ∎

Example 3.18. Suppose that for a given $A \in \mathcal{M}_2(\mathbb{R})$, the characteristic polynomial is

$$p_A(\lambda) = \lambda^2 - 4\lambda - 21 = (\lambda - 2)^2 + 36.$$

Since the roots are
$$\lambda_1 = 2+i6, \quad \lambda_2 = 2-i6,$$
the equilibrium solution is an unstable spiral point. ∎

3.5.6 • Degenerate real eigenvalue

In the previous cases, the eigenvalues were distinct, so that the eigenvectors formed a basis. We now consider what happens not only when the eigenvalues are not distinct but also when there are not enough eigenvectors to form a basis. Here we will focus solely on the case that $n = 2$: the general case requires that we consider the *Jordan canonical form* of a matrix, which is beyond the scope of this text. We begin by taking a close look at the eigenvalue problem.

Starting with the general matrix

$$A = \begin{pmatrix} a & b \\ c & d \end{pmatrix},$$

the characteristic polynomial is

$$p_A(\lambda) = \lambda^2 - (a+d)\lambda + ad - bc.$$

The roots of the characteristic polynomial are

$$\lambda = \lambda_\pm := \frac{1}{2}\left(a+d \pm \sqrt{(a+d)^2 - 4(ad-bc)}\right),$$

which, on using the identity

$$(a+d)^2 - 4(ad-bc) = (a-d)^2 + 4bc,$$

can be rewritten as

$$\lambda_\pm = \frac{1}{2}\left(a+d \pm \sqrt{(a-d)^2 + 4bc}\right).$$

This situation is not generic, as a small perturbation to any one of the matrix coefficients leads to two simple eigenvalues. The perturbed eigenvalues will be real, or a complex-conjugate pair with nonzero imaginary part.

In order to have a double eigenvalue, i.e., $\lambda_- = \lambda_+$, we need

$$bc = -\frac{1}{4}(a-d)^2 < 0.$$

We will henceforth assume that this is the case.

Now consider the eigenvalue problem with the double eigenvalue

$$\lambda_1 = \frac{1}{2}(a+d), \quad bc = -\frac{1}{4}(a-d)^2.$$

After some algebraic simplification, we have

$$A - \lambda_1 I_2 = \begin{pmatrix} (a-d)/2 & b \\ c & (d-a)/2 \end{pmatrix}.$$

In order for there to be two linearly independent eigenvectors, it must be the case that the RREF of $A - \lambda_1 I_2$ is the zero matrix. We will henceforth assume that this

3.5. The homogeneous solution

is not true. The RREF will then have only one zero row and rank$(A-\lambda_1 I_2) = 1$; in other words, the eigenvalue has geometric multiplicity one and algebraic multiplicity two. Thus, one of the columns will be a scalar multiple of the other. Without loss of generality, we will use the second column and write

$$\text{Col}(A-\lambda_1 I_2) = \text{Span}\left\{\begin{pmatrix} b \\ (d-a)/2 \end{pmatrix}\right\}.$$

The null space will be one-dimensional, and since

$$A-\lambda_1 I_2 \xrightarrow{\text{RREF}} \begin{pmatrix} (a-d)/2 & b \\ 0 & 0 \end{pmatrix},$$

a straightforward computation gives

$$\text{Null}(A-\lambda_1 I_2) = \text{Span}\left\{\begin{pmatrix} -b \\ (a-d)/2 \end{pmatrix}\right\}.$$

Noting that the two basis vectors are a scalar multiple of each other, we conclude as follows:

> **Proposition 3.19.** *Suppose that*
>
> $$A = \begin{pmatrix} a & b \\ c & d \end{pmatrix}$$
>
> *has a double eigenvalue:*
>
> $$\lambda = \lambda_1 = \frac{1}{2}(a+d), \quad bc = -\frac{1}{4}(a-d)^2.$$
>
> *Further suppose that* rank$(A-\lambda_1 I_2) = 1$, *i.e., the RREF of* $A-\lambda_1 I_2$, *has only one zero row. The column space and null space are the same,*
>
> $$\text{Col}(A-\lambda_1 I_2) = \text{Null}(A-\lambda_1 I_2).$$

Remark 3.20. *The argument leading to Proposition 3.19 implicitly assumes* $b \neq 0$. *If* $b = 0$, *then the double eigenvalue condition implies* $a = d$. *In this case,*

$$A-\lambda_1 I_2 = \begin{pmatrix} 0 & 0 \\ c & 0 \end{pmatrix}, \quad c \neq 0.$$

We would then write

$$\text{Col}(A-\lambda_1 I_2) = \text{Span}\left\{\begin{pmatrix} 0 \\ c \end{pmatrix}\right\}, \quad \text{Null}(A-\lambda_1 I_2) = \text{Span}\left\{\begin{pmatrix} 0 \\ 1 \end{pmatrix}\right\},$$

from which we still recover the desired result.

The importance of this result is that it implies the linear system

$$(A-\lambda_1 I_2)w = v_1$$

is consistent, as an eigenvector $v_1 \in \text{Col}(A - \lambda_1 I_2)$. Before solving the ODE, we first show that the vectors w and v_1 associated with this linear system are linearly independent.

The vector w is known as a generalized eigenvector.

Proposition 3.21. *Suppose that $A \in \mathcal{M}_2(\mathbb{R})$ satisfies the assumptions of Proposition 3.19. If the vectors v_1 and w are defined by the coupled linear systems,*

$$(A - \lambda_1 I_2) v_1 = 0,$$
$$(A - \lambda_1 I_2) w = v_1,$$

then the vectors are linearly independent.

Proof. Suppose not, so that $w = c v_1$ for some nonzero constant c. The left-hand side of the second linear system then becomes

$$(A - \lambda_1 I_2)(c v_1) = c(A - \lambda_1 I_2) v_1 = 0.$$

The last equality follows from the fact that v_1 is an eigenvector. Thus, if w is a scalar multiple of v_1, we require $v_1 = 0$, which contradicts the fact that eigenvectors are nonzero. In conclusion, w is not parallel to v_1, so the two vectors are linearly independent. □

Under the assumptions on matrix A presented in Proposition 3.19 and Proposition 3.21, we now consider the ODE

$$x' = Ax. \qquad (3.5.11)$$

From Remark 3.7 following Theorem 3.6, we know that one solution to the homogeneous problem is $x_1(t) = e^{\lambda_1 t} v_1$. By assumption, there is no second linearly independent eigenvector, so at the moment, we have no second linearly independent solution. But from the existence Corollary 3.3, we know such a solution exists. What is it?

The second linearly independent solution can be more easily derived using the matrix exponential and Jordan canonical form of a matrix. The interested reader should consult Perko [34, Chapter 1] for the details.

Lemma 3.22. *Consider the linear ODE $x' = Ax$ for $A \in \mathcal{M}_2(\mathbb{R})$. Suppose that there is a double eigenvalue, λ_1, but that there is only one linearly independent associated eigenvector, v_1 (see Proposition 3.19). Two linearly independent solutions are given by*

$$x_1(t) = e^{\lambda_1 t} v_1, \quad x_2(t) = t e^{\lambda_1 t} v_1 + e^{\lambda_1 t} w.$$

The vectors v_1 and w are found by solving the coupled linear systems

$$(A - \lambda_1 I_2) v_1 = 0$$
$$(A - \lambda_1 I_2) w = v_1.$$

The general solution is

$$x(t) = c_1 e^{\lambda_1 t} v_1 + c_2 \left(t e^{\lambda_1 t} v_1 + e^{\lambda_1 t} w \right).$$

3.5. The homogeneous solution

Proof. We construct the second solution. Since $\{v_1, w\}$ are linearly independent (see Proposition 3.21), the vectors form a basis for $\mathbb{R}^2$. The second solution can be written in the form

$$x_2(t) = f(t)v_1 + g(t)w, \qquad (3.5.12)$$

where the functions $f(t)$ and $g(t)$ have to be determined. Plugging this solution into the ODE (3.5.11) and using the linearity of matrix/vector multiplication gives

$$f'(t)v_1 + g'(t)w = f(t)Av_1 + g(t)Aw.$$

On using $Av_1 = \lambda_1 v_1$ and

$$(A - \lambda_1)w = v_1 \quad \leadsto \quad Aw = \lambda_1 w + v_1,$$

we see that

$$f(t)Av_1 + g(t)Aw = \lambda_1 f(t)v_1 + g(t)(\lambda_1 w + v_1)$$
$$= (\lambda_1 f(t) + g(t))v_1 + \lambda_1 g(t)w.$$

Collecting like terms gives

$$\bigl(f'(t) - \lambda_1 f(t) - g(t)\bigr)v_1 + \bigl(g'(t) - \lambda_1 g(t)\bigr)w = 0.$$

Since the vectors are linearly independent, the weights must be zero. This observation leads to the coupled set of ODEs:

$$\begin{aligned} f'(t) &= \lambda_1 f(t) + g(t), \\ g'(t) &= \lambda_1 g(t). \end{aligned} \qquad (3.5.13)$$

A solution to the second equation in (3.5.13) is

$$g(t) = e^{\lambda_1 t}, \qquad (3.5.14)$$

so the first equation becomes

$$f'(t) = \lambda_1 f(t) + e^{\lambda_1 t}. \qquad (3.5.15)$$

Using the method of undetermined coefficients as outlined in section 2.5.3, a solution to (3.5.15) is

$$f(t) = e^{\lambda_1 t} + t e^{\lambda_1 t}.$$

A second solution to the ODE (3.5.11) is now

$$\begin{aligned} x_2(t) &= \bigl(e^{\lambda_1 t} + t e^{\lambda_1 t}\bigr)v_1 + e^{\lambda_1 t}w \\ &= e^{\lambda_1 t}v_1 + \bigl(t e^{\lambda_1 t}v_1 + e^{\lambda_1 t}w\bigr). \end{aligned}$$

The first term in the sum is $x_1(t)$, so by linearity, we can ignore this term and choose the second solution to be

$$x_2(t) = t e^{\lambda_1 t}v_1 + e^{\lambda_1 t}w.$$

Now that we have two solutions, we need to check that they are linearly independent. As we saw in our discussion leading to Corollary 3.3, all that is required is that they be linearly independent at one point in time. Since

$$x_1(0) = v_1, \quad x_2(0) = w,$$

it is then enough to have $\{v_1, w\}$ be a linearly independent set. But we started the proof by showing that these two vectors are linearly independent. In conclusion, we have the two desired solutions to the ODE. □

Example 3.23. Consider

$$x' = Ax, \quad A = \begin{pmatrix} 1 & -2 \\ 2 & 5 \end{pmatrix}.$$

The characteristic equation is $p_A(\lambda) = (\lambda - 3)^2$, so that $\lambda = 3$ is a double eigenvalue. Regarding an associated eigenvector, we have

$$A - 3I_2 = \begin{pmatrix} -2 & -2 \\ 2 & 2 \end{pmatrix} \xrightarrow{\text{RREF}} \begin{pmatrix} 1 & 1 \\ 0 & 0 \end{pmatrix} \rightsquigarrow v_1 = \begin{pmatrix} -1 \\ 1 \end{pmatrix}.$$

Since the RREF is not the zero matrix, the theory leading to Lemma 3.22 applies. One solution is given by

$$x_1(t) = e^{3t} \begin{pmatrix} -1 \\ 1 \end{pmatrix}.$$

Since there is only one linearly independent eigenvector associated with the double eigenvalue, we must use Lemma 3.22 in order to find the second linearly independent solution. The RREF of the augmented matrix associated with the linear system $(A - 3I_2)w = v_1$ is

$$(A - 3I_2 | v_1) \xrightarrow{\text{RREF}} \begin{pmatrix} 1 & 1 & | & 1/2 \\ 0 & 0 & | & 0 \end{pmatrix}.$$

A solution is the last column:

$$w = \begin{pmatrix} 1/2 \\ 0 \end{pmatrix}.$$

The second solution is then

$$x_2(t) = te^{3t} \begin{pmatrix} -1 \\ 1 \end{pmatrix} + e^{3t} \begin{pmatrix} 1/2 \\ 0 \end{pmatrix} = e^{3t} \begin{pmatrix} 1/2 - t \\ t \end{pmatrix}.$$

The general solution to the ODE is then

$$x(t) = c_1 e^{3t} \begin{pmatrix} -1 \\ 1 \end{pmatrix} + c_2 e^{3t} \begin{pmatrix} 1/2 - t \\ t \end{pmatrix},$$

and the matrix-valued solution is

$$\Phi(t) = \begin{pmatrix} -e^{3t} & (1/2 - t)e^{3t} \\ e^{3t} & te^{3t} \end{pmatrix}. \quad \blacksquare$$

Example 3.24. For a second example, consider the IVP

$$x' = Ax, \quad A = \begin{pmatrix} 0 & 1 \\ -1 & -2 \end{pmatrix}, \quad x(0) = \begin{pmatrix} 4 \\ 1 \end{pmatrix}.$$

3.5. The homogeneous solution

First, consider the general solution. The characteristic equation is
$$p_A(\lambda) = \lambda^2 + 2\lambda + 1 = (\lambda+1)^2,$$
so that $\lambda = -1$ is a double eigenvalue. Regarding the associated eigenvector(s), we have
$$A - (-1)I_2 = \begin{pmatrix} 1 & 1 \\ -1 & -1 \end{pmatrix} \xrightarrow{\text{RREF}} \begin{pmatrix} 1 & 1 \\ 0 & 0 \end{pmatrix} \rightsquigarrow v_1 = \begin{pmatrix} 1 \\ -1 \end{pmatrix}.$$

One solution is then
$$x_1(t) = e^{-t}\begin{pmatrix} 1 \\ -1 \end{pmatrix}.$$

Since there is only one linearly independent eigenvector associated with the double eigenvalue, we must use Lemma 3.22 in order to find the second linearly independent solution. The RREF of the augmented matrix associated with the linear system $(A + I_2)w = v_1$ is
$$(A + I_2 | v_1) \xrightarrow{\text{RREF}} \left(\begin{array}{cc|c} 1 & 1 & 1 \\ 0 & 0 & 0 \end{array} \right).$$

A solution is the last column,
$$w = \begin{pmatrix} 1 \\ 0 \end{pmatrix},$$
so the second solution is
$$x_2(t) = te^{-t}\begin{pmatrix} 1 \\ -1 \end{pmatrix} + e^{-t}\begin{pmatrix} 1 \\ 0 \end{pmatrix} = e^{-t}\begin{pmatrix} 1+t \\ -t \end{pmatrix}.$$

The general solution to the ODE system is
$$x(t) = c_1 e^{-t}\begin{pmatrix} 1 \\ -1 \end{pmatrix} + c_2 e^{-t}\begin{pmatrix} 1+t \\ -t \end{pmatrix}.$$

In order to solve the IVP, we have
$$\begin{pmatrix} 4 \\ 1 \end{pmatrix} = x(0) = c_1\begin{pmatrix} 1 \\ -1 \end{pmatrix} + c_2\begin{pmatrix} 1 \\ 0 \end{pmatrix} \rightsquigarrow \begin{pmatrix} 1 & 1 \\ -1 & 0 \end{pmatrix} c = \begin{pmatrix} 4 \\ 1 \end{pmatrix}.$$

The solution to this linear system is
$$c = \begin{pmatrix} 1 & 1 \\ -1 & 0 \end{pmatrix}^{-1}\begin{pmatrix} 4 \\ 1 \end{pmatrix} = \begin{pmatrix} 0 & -1 \\ 1 & 1 \end{pmatrix}\begin{pmatrix} 4 \\ 1 \end{pmatrix} = \begin{pmatrix} -1 \\ 5 \end{pmatrix}.$$

The solution to the IVP is
$$x(t) = -e^{-t}\begin{pmatrix} 1 \\ -1 \end{pmatrix} + 5e^{-t}\begin{pmatrix} 1+t \\ -t \end{pmatrix}. \blacksquare$$

Exercises

Exercise 3.5.1. *The general solution of $x' = Ax$ is given by*
$$x(t) = c_1 e^{3t}\begin{pmatrix} 2 \\ -3 \end{pmatrix} + c_2 e^{-5t}\begin{pmatrix} 3 \\ 1 \end{pmatrix}.$$

(a) Classify the stability of the equilibrium solution $x = 0$.

(b) Sketch several representative solutions in the phase plane (do not use a computer).

Exercise 3.5.2. The general solution of $x' = Ax$ is given by

$$x(t) = c_1 e^{-5t} \begin{pmatrix} \cos(2t) - 3\sin(2t) \\ -4\cos(2t) + \sin(2t) \end{pmatrix} + c_2 e^{-5t} \begin{pmatrix} \sin(2t) + 2\cos(2t) \\ -4\sin(2t) - \cos(2t) \end{pmatrix}.$$

(a) Classify the stability of the equilibrium solution $x = 0$.

(b) Sketch several representative solutions in the phase plane (do not use a computer).

Exercise 3.5.3. The general solution of $x' = Ax$ is given by

$$x(t) = c_1 e^{-2t} \begin{pmatrix} 2 \\ 7 \end{pmatrix} + c_2 e^{-8t} \begin{pmatrix} 1 \\ -2 \end{pmatrix}.$$

(a) Classify the stability of the equilibrium solution $x = 0$.

(b) Sketch several representative solutions in the phase plane (do not use a computer).

Exercise 3.5.4. The general solution of $x' = Ax$ is given by

$$x(t) = c_1 e^{4t} \begin{pmatrix} \cos(3t) \\ -2\cos(3t) + 4\sin(3t) \end{pmatrix} + c_2 e^{4t} \begin{pmatrix} \sin(3t) \\ -2\sin(3t) - 4\cos(3t) \end{pmatrix}.$$

(a) Classify the stability of the equilibrium solution $x = 0$.

(b) Sketch several representative solutions in the phase plane (do not use a computer).

Exercise 3.5.5. The general solution of $x' = Ax$ is given by

$$x(t) = c_1 \begin{pmatrix} 3\cos(6t) + 4\sin(6t) \\ -5\cos(6t) + 7\sin(6t) \end{pmatrix} + c_2 \begin{pmatrix} 3\sin(6t) - 4\cos(6t) \\ -5\sin(6t) - 7\cos(6t) \end{pmatrix}.$$

(a) Classify the stability of the equilibrium solution $x = 0$.

(b) Sketch several representative solutions in the phase plane (do not use a computer).

Exercise 3.5.6. For the following, the characteristic polynomial $p_A(\lambda)$ for a matrix $A \in \mathcal{M}_2(\mathbb{R})$ is given. For each characteristic polynomial below, classify the stability of the equilibrium solution for the system $x' = Ax$.

(a) $p_A(\lambda) = \lambda^2 - 8\lambda + 12$

(b) $p_A(\lambda) = \lambda^2 + 64$

(c) $p_A(\lambda) = \lambda^2 - 5\lambda - 14$

(d) $p_A(\lambda) = \lambda^2 - 6\lambda + 25$

3.5. The homogeneous solution

(e) $p_A(\lambda) = \lambda^2 + 8\lambda + 25$

(f) $p_A(\lambda) = \lambda^2 + 3\lambda + 2$

Exercise 3.5.7. *For the following, the eigenvalues and corresponding eigenvectors for a matrix A are given. Find the general solution to the ODE $x' = Ax$.*

(a) $\lambda_1 = -2,\ v_1 = \begin{pmatrix} 2 \\ -3 \end{pmatrix};\ \lambda_2 = 4,\ v_2 = \begin{pmatrix} 3 \\ 7 \end{pmatrix}$

(b) $\lambda_1 = 5,\ v_1 = \begin{pmatrix} 1 \\ 7 \end{pmatrix};\ \lambda_2 = -3,\ v_2 = \begin{pmatrix} 4 \\ -9 \end{pmatrix}$

(c) $\lambda_1 = -2+i3,\ v_1 = \begin{pmatrix} 2-i \\ i6 \end{pmatrix};\ \lambda_2 = -2-i3,\ v_2 = \begin{pmatrix} 2+i \\ -i6 \end{pmatrix}$

(d) $\lambda_1 = 5+i2,\ v_1 = \begin{pmatrix} 3+i4 \\ -2-i7 \end{pmatrix};\ \lambda_2 = 5-i2,\ v_2 = \begin{pmatrix} 3-i4 \\ -2+i7 \end{pmatrix}$

(e) $\lambda_1 = i4,\ v_1 = \begin{pmatrix} 1+i4 \\ -2+i9 \end{pmatrix};\ \lambda_2 = -i4,\ v_2 = \begin{pmatrix} 1-i4 \\ -2-i9 \end{pmatrix}$

Exercise 3.5.8. *Consider the system of ODEs given by*

$$x_1' = x_2,\quad x_2' = 2x_1 - x_2.$$

(a) *Determine the general solution to the system.*

(b) *Graph the phase plane for the system.*

(c) *Find the solution if $x_1(0) = 4$, $x_2(0) = -7$.*

Exercise 3.5.9. *Consider the system of ODEs given by*

$$x_1' = -10x_1 - 9x_2,\quad x_2' = x_1.$$

(a) *Determine the general solution to the system.*

(b) *Graph the phase plane for the system.*

(c) *Find the solution if $x_1(0) = 2$, $x_2(0) = 7$.*

Exercise 3.5.10. *Consider the system of ODEs given by*

$$x_1' = 3x_1 - x_2,\quad x_2' = 5x_1 - 3x_2.$$

(a) *Determine the general solution to the system.*

(b) *Graph the phase plane for the system.*

(c) *Find the solution if $x_1(0) = 2$, $x_2(0) = -6$.*

Exercise 3.5.11. *Find the general solution to the system*

$$x' = \begin{pmatrix} -2 & 1 \\ -1 & -4 \end{pmatrix} x.$$

Exercise 3.5.12. *Find the general solution for the system*

$$x' = \begin{pmatrix} 0 & 1 \\ -16 & -8 \end{pmatrix} x.$$

Exercise 3.5.13. *Consider the system*

$$x' = \begin{pmatrix} -2 & 1 \\ 1 & -2 \end{pmatrix} x.$$

(a) *Classify the stability of the equilibrium solution $x = 0$.*

(b) *Sketch several representative trajectories in the phase plane.*

Exercise 3.5.14. *Consider the system*

$$x' = \begin{pmatrix} -1 & -2 \\ -2 & -1 \end{pmatrix} x.$$

(a) *Classify the stability of the equilibrium solution $x = 0$.*

(b) *Sketch several representative trajectories in the phase plane.*

Exercise 3.5.15. *Consider the system*

$$x' = \begin{pmatrix} -1 & -2 \\ 2 & -1 \end{pmatrix} x.$$

(a) *Classify the stability of the equilibrium solution $x = 0$.*

(b) *Sketch several representative trajectories in the phase plane.*

Exercise 3.5.16. *Consider the system*

$$x' = \begin{pmatrix} 4 & 1 \\ 1 & 4 \end{pmatrix} x.$$

(a) *Classify the stability of the equilibrium solution $x = 0$.*

(b) *Sketch several representative trajectories in the phase plane.*

Exercise 3.5.17. *Consider the system*

$$x' = \begin{pmatrix} 5 & 3 \\ -3 & 5 \end{pmatrix} x.$$

3.6. The particular solution

(a) *Classify the stability of the equilibrium solution $x = 0$.*

(b) *Sketch several representative trajectories in the phase plane.*

Exercise 3.5.18. *Sketch several representative trajectories in the phase plane for the system*

$$x' = \begin{pmatrix} 1 & 2 \\ 2 & 4 \end{pmatrix} x.$$

Exercise 3.5.19. *Sketch several representative trajectories in the phase plane for the system*

$$x' = \begin{pmatrix} 3 & -6 \\ 1 & -2 \end{pmatrix} x.$$

Exercise 3.5.20. *Solve the IVP*

$$x' = \begin{pmatrix} 4 & 3 \\ -3 & -2 \end{pmatrix} x, \quad x(0) = \begin{pmatrix} 7 \\ -1 \end{pmatrix}.$$

Exercise 3.5.21. *Solve the IVP*

$$x' = \begin{pmatrix} 0 & 1 \\ -25 & -6 \end{pmatrix} x, \quad x(0) = \begin{pmatrix} -3 \\ 8 \end{pmatrix}.$$

Exercise 3.5.22. *Let $A \in \mathcal{M}_2(\mathbb{R})$ be such that Proposition 3.19 holds. Let λ_1 be the eigenvalue with associated eigenvector v_1. Suppose that $\alpha \neq \lambda_1$. Show that*

(a) *the matrix $A - \alpha I_2$ is invertible, and*

(b) *the unique solution to $(A - \alpha I_2)x = v_1$ is*

$$x = \frac{1}{\lambda_1 - \alpha} v_1.$$

3.6 ▪ The particular solution

As was the case when finding the particular solution for the scalar ODE, we will consider two solution techniques for finding the particular solution for linear systems of ODEs. The first technique is variation of parameters, and the formulation will be quite similar to what is seen in Lemma 2.10. The second technique is the method of undetermined coefficients, but, as we will see, it may be better to label it as the method of undetermined *vectors*.

3.6.1 ▪ Variation of parameters

We know from our discussion in the beginning of section 3.5 that a matrix-valued solution, $\Phi(t)$, to the homogeneous problem

$$x' = A(t)x$$

satisfies the property that if $\Phi(t_0)$ is invertible, then $\Phi(t)$ is invertible for all values of t for which it is defined. Each column of the matrix-valued solution, say, $x_j(t)$ for $j = 1, \ldots, n$, is a solution to the homogeneous problem. Since $\Phi(t)$ is invertible, the collection of solutions $\{x_1(t), x_2(t), \ldots, x_n(t)\}$ is linearly independent, and the general solution to the homogeneous problem is

$$x_h(t) = \Phi(t)c = c_1 x_1(t) + c_2 x_2(t) + \cdots + c_n x_n(t).$$

As was the case for the scalar problem, we can use this matrix-valued solution to construct a particular solution. Namely, set

$$x_p(t) = \Phi(t) \int \Phi(t)^{-1} f(t) \, dt.$$

Following the line of reasoning that the scalar problem leads to Lemma 2.10—in particular, using the Fundamental Theorem of Calculus, the product rule, and the fact that $\Phi' = A(t)\Phi$—we see that

$$x_p'(t) = A(t) x_p(t) + f(t).$$

The analogue to Theorem 2.11 is then the following:

Theorem 3.25. *If $\Phi(t)$ is a matrix-valued solution to the homogeneous problem $x' = A(t)x$ that is invertible at $t = t_0$, then the general solution to the nonhomogeneous problem*

$$x' = A(t)x + f(t)$$

is the sum of the homogeneous and particular solutions:

$$x(t) = \underbrace{\Phi(t)c}_{x_h(t)} + \underbrace{\Phi(t) \int \Phi(t)^{-1} f(t) \, dt}_{x_p(t)}.$$

If $A(t)$ is a constant matrix, we know how to find the homogeneous solution. In particular, it is found by using the eigenvalues and associated eigenvectors of the matrix A. Thus, all that is left to do is a couple of examples that illustrate the theory.

Example 3.26. Consider the IVP

$$x' = \begin{pmatrix} 3 & 1 \\ 4 & 3 \end{pmatrix} x + \begin{pmatrix} 0 \\ 4e^t \end{pmatrix}, \quad x(0) = \begin{pmatrix} -2 \\ 5 \end{pmatrix}.$$

We first find the homogeneous solution. The eigenvalues and associated eigenvectors for the matrix A are

$$\lambda_1 = 1, \, v_1 = \begin{pmatrix} 1 \\ -2 \end{pmatrix}; \quad \lambda_2 = 5, \, v_2 = \begin{pmatrix} 1 \\ 2 \end{pmatrix},$$

so the general solution to the homogeneous problem is

$$x_h(t) = c_1 e^t \begin{pmatrix} 1 \\ -2 \end{pmatrix} + c_2 e^{5t} \begin{pmatrix} 1 \\ 2 \end{pmatrix} \quad \leadsto \quad \Phi(t) = \begin{pmatrix} e^t & e^{5t} \\ -2e^t & 2e^{5t} \end{pmatrix}.$$

3.6. The particular solution

Let us now construct the particular solution. Since

$$\Phi(t)^{-1} = \frac{1}{4}\begin{pmatrix} 2e^{-t} & -e^{-t} \\ -2e^{-5t} & e^{-5t} \end{pmatrix},$$

the integrand associated with the particular solution becomes

$$\Phi(t)^{-1}f(t) = \frac{1}{4}\begin{pmatrix} 2e^{-t} & -e^{-t} \\ -2e^{-5t} & e^{-5t} \end{pmatrix}\begin{pmatrix} 0 \\ 4e^{t} \end{pmatrix} = \begin{pmatrix} -1 \\ e^{-4t} \end{pmatrix}.$$

Consequently,

$$\int \Phi(t)^{-1}f(t)\,dt = \begin{pmatrix} -t \\ -e^{-4t}/4 \end{pmatrix},$$

so the particular solution is

$$x_p(t) = \Phi(t)\int \Phi(t)^{-1}f(t)\,dt = \begin{pmatrix} -te^{t} - e^{t}/4 \\ 2te^{t} - e^{t}/2 \end{pmatrix} = -te^{t}\begin{pmatrix} 1 \\ -2 \end{pmatrix} - \frac{1}{4}e^{t}\begin{pmatrix} 1 \\ 2 \end{pmatrix}.$$

From Theorem 3.25, the general solution is the sum of the homogeneous and particular solutions:

$$x(t) = c_1 e^{t}\begin{pmatrix} 1 \\ -2 \end{pmatrix} + c_2 e^{5t}\begin{pmatrix} 1 \\ 2 \end{pmatrix} - te^{t}\begin{pmatrix} 1 \\ -2 \end{pmatrix} - \frac{1}{4}e^{t}\begin{pmatrix} 1 \\ 2 \end{pmatrix}.$$

Now consider the initial condition. We have

$$\begin{pmatrix} -2 \\ 5 \end{pmatrix} = x(0) = c_1\begin{pmatrix} 1 \\ -2 \end{pmatrix} + c_2\begin{pmatrix} 1 \\ 2 \end{pmatrix} - \frac{1}{4}\begin{pmatrix} 1 \\ 2 \end{pmatrix},$$

which can be rewritten as

$$\begin{pmatrix} 1 & 1 \\ -2 & 2 \end{pmatrix}\begin{pmatrix} c_1 \\ c_2 \end{pmatrix} = \begin{pmatrix} -2 \\ 5 \end{pmatrix} + \frac{1}{4}\begin{pmatrix} 1 \\ 2 \end{pmatrix} = \frac{1}{4}\begin{pmatrix} -7 \\ 22 \end{pmatrix}.$$

The solution to the linear system is

$$\begin{pmatrix} c_1 \\ c_2 \end{pmatrix} = \frac{1}{4}\begin{pmatrix} 1 & 1 \\ -2 & 2 \end{pmatrix}^{-1}\begin{pmatrix} -7 \\ 22 \end{pmatrix} = \frac{1}{4}\begin{pmatrix} -9 \\ 2 \end{pmatrix}.$$

In conclusion, the solution to the IVP is

$$x(t) = -\frac{9}{4}e^{t}\begin{pmatrix} 1 \\ -2 \end{pmatrix} + \frac{1}{2}e^{5t}\begin{pmatrix} 1 \\ 2 \end{pmatrix} - te^{t}\begin{pmatrix} 1 \\ -2 \end{pmatrix} - \frac{1}{4}e^{t}\begin{pmatrix} 1 \\ 2 \end{pmatrix}. \blacksquare$$

Example 3.27. Let us find the general solution for the ODE

$$x' = \begin{pmatrix} 2 & -1 \\ 3 & -2 \end{pmatrix}x + \begin{pmatrix} 2/(1+e^{t}) \\ 0 \end{pmatrix}.$$

We first find the homogeneous solution. The eigenvalues and associated eigenvectors for the matrix A are given by

$$\lambda_1 = -1,\ v_1 = \begin{pmatrix} 1 \\ 3 \end{pmatrix};\quad \lambda_2 = 1,\ v_2 = \begin{pmatrix} 1 \\ 1 \end{pmatrix},$$

so that the general solution to the homogeneous problem is

$$x_h(t) = c_1 e^{-t}\begin{pmatrix} 1 \\ 3 \end{pmatrix} + c_2 e^{t}\begin{pmatrix} 1 \\ 1 \end{pmatrix} \quad\rightsquigarrow\quad \Phi(t) = \begin{pmatrix} e^{-t} & e^{t} \\ 3e^{-t} & e^{t} \end{pmatrix}.$$

Let us now construct the particular solution. Since

$$\Phi(t)^{-1} = \frac{1}{2}\begin{pmatrix} -e^t & e^t \\ 3e^{-t} & -e^{-t} \end{pmatrix},$$

the integrand associated with the particular solution becomes

$$\Phi(t)^{-1}f(t) = \frac{1}{2}\begin{pmatrix} -e^t & e^t \\ 3e^{-t} & -e^{-t} \end{pmatrix}\begin{pmatrix} 2/(1+e^t) \\ 0 \end{pmatrix} = \begin{pmatrix} -e^t/(1+e^t) \\ 3e^{-t}/(1+e^t) \end{pmatrix}.$$

It can be checked using WolframAlpha that on integrating,

$$\int \Phi(t)^{-1}f(t)\,dt = \begin{pmatrix} -\ln(1+e^t) \\ 3\ln(1+e^{-t}) - 3e^{-t} \end{pmatrix}.$$

The particular solution is

$$x_p(t) = \Phi(t)\int \Phi(t)^{-1}f(t)\,dt = \begin{pmatrix} -e^{-t}\ln(1+e^t) + 3e^t\ln(1+e^{-t}) - 3 \\ -3e^{-t}\ln(1+e^t) + 3e^t\ln(1+e^{-t}) - 3 \end{pmatrix}.$$

Theorem 3.25, the general solution, is the sum of the homogeneous and particular solutions:

$$x(t) = c_1 e^{-t}\begin{pmatrix} 1 \\ 3 \end{pmatrix} + c_2 e^t \begin{pmatrix} 1 \\ 1 \end{pmatrix} + \begin{pmatrix} -e^{-t}\ln(1+e^t) + 3e^t\ln(1+e^{-t}) - 3 \\ -3e^{-t}\ln(1+e^t) + 3e^t\ln(1+e^{-t}) - 3 \end{pmatrix}. \blacksquare$$

3.6.2 ▪ Undetermined coefficients (vectors)

Recall that in section 2.5.3, we discussed how to find the particular solution for scalar problems of the form

$$x' = ax + f(t),$$

where $f(t)$ was composed of products and sums of exponential functions, polynomials, and sines and cosines. We can do the same thing for linear systems of the form

$$x' = Ax + f(t),$$

where each component of $f(t)$ is composed of products and sums of exponential functions, polynomials, and sines and cosines. The only difference will be that instead of using undetermined coefficients as part of our guess, we will use undetermined *vectors*. We will illustrate the method for only a few simple examples for which the forcing function is of a type that often occurs in applications. For anything relatively complicated, it is generally best to use the variation or parameters formula and use some type of CAS to negotiate all of the indefinite integrals that arise.

The homogeneous problem for each example is

$$x' = Ax, \quad A = \begin{pmatrix} 2 & -1 \\ 3 & -2 \end{pmatrix}.$$

As we have already seen in Example 3.27, the general solution is

$$x_h(t) = c_1 e^{-t}\begin{pmatrix} 1 \\ 3 \end{pmatrix} + c_2 e^t \begin{pmatrix} 1 \\ 1 \end{pmatrix}.$$

3.6. The particular solution

Example 3.28. For our first example, suppose that

$$f(t) = \begin{pmatrix} 4 \\ 5 \end{pmatrix}.$$

A natural guess for the particular solution is

$$x_p(t) = a_0,$$

where a_0 is an unknown vector. Plugging this guess into the ODE yields

$$0 = Aa_0 + \begin{pmatrix} 4 \\ 5 \end{pmatrix} \quad \rightsquigarrow \quad Aa_0 = -\begin{pmatrix} 4 \\ 5 \end{pmatrix}.$$

The solution to this linear system is

$$a_0 = -A^{-1}\begin{pmatrix} 4 \\ 5 \end{pmatrix} = -\begin{pmatrix} 3 \\ 2 \end{pmatrix},$$

so a particular solution is

$$x_p(t) = -\begin{pmatrix} 3 \\ 2 \end{pmatrix}.$$

In conclusion, the general solution is the sum of the homogeneous and particular solutions:

$$x(t) = c_1 e^{-t}\begin{pmatrix} 1 \\ 3 \end{pmatrix} + c_2 e^{t}\begin{pmatrix} 1 \\ 1 \end{pmatrix} - \begin{pmatrix} 3 \\ 2 \end{pmatrix}. \blacksquare$$

Example 3.29. For the second example, suppose that

$$f(t) = e^{3t}\begin{pmatrix} 2 \\ 1 \end{pmatrix}.$$

A natural guess for the particular solution is

$$x_p(t) = e^{3t} a_0,$$

where a_0 is an unknown vector. Plugging this guess into the ODE yields

$$3e^{3t} a_0 = e^{3t} A a_0 + e^{3t}\begin{pmatrix} 2 \\ 1 \end{pmatrix} \quad \rightsquigarrow \quad (A - 3I_2)a_0 = -\begin{pmatrix} 2 \\ 1 \end{pmatrix}.$$

Since $\lambda = 3$ is not an eigenvalue of A, the matrix $A - 3I_2$ is invertible. The solution to this linear system is

$$a_0 = -(A - 3I_2)^{-1}\begin{pmatrix} 2 \\ 1 \end{pmatrix} = \frac{1}{8}\begin{pmatrix} 9 \\ 7 \end{pmatrix},$$

so the particular solution is

$$x_p(t) = \frac{1}{8}e^{3t}\begin{pmatrix} 9 \\ 7 \end{pmatrix}.$$

In conclusion, the general solution is the sum of the homogeneous and particular solutions:

$$x(t) = c_1 e^{-t}\begin{pmatrix} 1 \\ 3 \end{pmatrix} + c_2 e^{t}\begin{pmatrix} 1 \\ 1 \end{pmatrix} + \frac{1}{8}e^{3t}\begin{pmatrix} 9 \\ 7 \end{pmatrix}. \blacksquare$$

Example 3.30. For our final example, suppose that

$$f(t) = \cos(2t)\begin{pmatrix} 1 \\ 4 \end{pmatrix}.$$

A natural guess for the particular solution is

$$x_p(t) = \cos(2t)a_0 + \sin(2t)a_1,$$

where a_0, a_1 are unknown vectors. Plugging this guess into the ODE yields

$$-2\sin(2t)a_0 + 2\cos(2t)a_1 = \cos(2t)Aa_0 + \sin(2t)Aa_1 + \cos(2t)\begin{pmatrix} 1 \\ 4 \end{pmatrix},$$

which after equating the cosine and sine terms is equivalent to the pair of linear systems

$$Aa_0 - 2a_1 = -\begin{pmatrix} 1 \\ 4 \end{pmatrix}, \quad 2a_0 + Aa_1 = 0.$$

Using the identity matrix, we can rewrite the above linear system in the form

$$Aa_0 - 2I_2 a_1 = -\begin{pmatrix} 1 \\ 4 \end{pmatrix}, \quad 2I_2 a_0 + Aa_1 = 0.$$

In *block-matrix form*, this linear system is equivalent to

$$\left(\begin{array}{c|c} A & -2I_2 \\ \hline 2I_2 & A \end{array}\right)\begin{pmatrix} a_0 \\ a_1 \end{pmatrix} = \begin{pmatrix} -1 \\ -4 \\ 0 \\ 0 \end{pmatrix} \rightsquigarrow \left(\begin{array}{cc|cc} 2 & -1 & -2 & 0 \\ 3 & -2 & 0 & -2 \\ \hline 2 & 0 & 2 & -1 \\ 0 & 2 & 3 & -2 \end{array}\right) x = \begin{pmatrix} -1 \\ -4 \\ 0 \\ 0 \end{pmatrix},$$

$$x = \begin{pmatrix} a_0 \\ a_1 \end{pmatrix}.$$

We can easily solve this system using SAGE and find that

$$x = \begin{pmatrix} 2/5 \\ 1 \\ 2/5 \\ 8/5 \end{pmatrix} \rightsquigarrow a_0 = \begin{pmatrix} 2/5 \\ 1 \end{pmatrix}, \quad a_1 = \begin{pmatrix} 2/5 \\ 8/5 \end{pmatrix}.$$

A particular solution is then

$$x_p(t) = \frac{1}{5}\cos(2t)\begin{pmatrix} 2 \\ 5 \end{pmatrix} + \frac{1}{5}\sin(2t)\begin{pmatrix} 2 \\ 8 \end{pmatrix}.$$

In conclusion, the general solution is the sum of the homogeneous and particular solutions:

$$x(t) = c_1 e^{-t}\begin{pmatrix} 1 \\ 3 \end{pmatrix} + c_2 e^{t}\begin{pmatrix} 1 \\ 1 \end{pmatrix} + \frac{1}{5}\cos(2t)\begin{pmatrix} 2 \\ 5 \end{pmatrix} + \frac{1}{5}\sin(2t)\begin{pmatrix} 2 \\ 8 \end{pmatrix}. \blacksquare$$

The last example, while relatively simple, clearly illustrates the limitations of using the method of undetermined coefficients for systems of ODEs with only pencil

3.6. The particular solution

and paper. While the guess is straightforward, it is often the case that the resulting set of linear equations to be solved gets large fairly quickly. For example, if

$$f(t) = v_0 + tv_1 + t^2 v_2,$$

then a natural guess for the particular solution is

$$x_p(t) = a_0 + ta_1 + t^2 a_2.$$

Plugging this guess into the ODE and equating like terms yields the system of equations

$$-Aa_0 + a_1 = v_0, \quad -Aa_1 + 2a_2 = v_1, \quad -Aa_2 = v_2.$$

In block-matrix form, the system is

$$\begin{pmatrix} -A & I_2 & 0_2 \\ 0_2 & -A & 2I_2 \\ 0_2 & 0_2 & -A \end{pmatrix} \begin{pmatrix} a_0 \\ a_1 \\ a_2 \end{pmatrix} = \begin{pmatrix} v_0 \\ v_1 \\ v_2 \end{pmatrix},$$

which is a linear system of six equations in six unknowns. Here $0_2 \in \mathcal{M}_2(\mathbb{R})$ is the zero matrix. You will most likely use SAGE, WolframAlpha, or some other CAS to solve this system.

Exercises

Exercise 3.6.1. *Find the general solution for the system*

$$x' = \begin{pmatrix} -3 & 2 \\ 2 & -3 \end{pmatrix} x + \cos(2t) \begin{pmatrix} 4 \\ -3 \end{pmatrix} + \begin{pmatrix} 3 \\ 7 \end{pmatrix}.$$

Exercise 3.6.2. *Find the general solution for the system*

$$x' = \begin{pmatrix} 0 & 2 \\ 1 & -1 \end{pmatrix} x + \sin(t) \begin{pmatrix} 4 \\ 2 \end{pmatrix}.$$

Exercise 3.6.3. *Find the general solution for the system*

$$x' = \begin{pmatrix} 1 & 0 \\ -1 & 3 \end{pmatrix} x + e^{2t} \begin{pmatrix} 1 \\ -1 \end{pmatrix}.$$

Exercise 3.6.4. *Find the general solution for the system*

$$x' = \begin{pmatrix} 2 & 1 \\ 3 & 0 \end{pmatrix} x + t \begin{pmatrix} 1 \\ 1 \end{pmatrix}.$$

Exercise 3.6.5. *Find the general solution for the system*

$$x' = \begin{pmatrix} 0 & 1 \\ -1 & 0 \end{pmatrix} x + \sec(t) \begin{pmatrix} 2 \\ 3 \end{pmatrix}.$$

Exercise 3.6.6. *Solve the IVP (compare with Exercise 3.6.1)*

$$x' = \begin{pmatrix} -3 & 2 \\ 2 & -3 \end{pmatrix} x + \cos(2t) \begin{pmatrix} 4 \\ -3 \end{pmatrix} + \begin{pmatrix} 3 \\ 7 \end{pmatrix}, \quad x(0) = \begin{pmatrix} 2 \\ -3 \end{pmatrix}.$$

Exercise 3.6.7. *Solve the IVP (compare with Exercise 3.6.2)*

$$x' = \begin{pmatrix} 0 & 2 \\ 1 & -1 \end{pmatrix} x + \sin(t) \begin{pmatrix} 4 \\ 2 \end{pmatrix}, \quad x(0) = \begin{pmatrix} -4 \\ 1 \end{pmatrix}.$$

Exercise 3.6.8. *Solve the IVP (compare with Exercise 3.6.3)*

$$x' = \begin{pmatrix} 1 & 0 \\ -1 & 3 \end{pmatrix} x + e^{2t} \begin{pmatrix} 1 \\ -1 \end{pmatrix}, \quad x(0) = \begin{pmatrix} 2 \\ 5 \end{pmatrix}.$$

Exercise 3.6.9. *Solve the IVP (compare with Exercise 3.6.4)*

$$x' = \begin{pmatrix} 2 & 1 \\ 3 & 0 \end{pmatrix} x + t \begin{pmatrix} 10 \\ 10 \end{pmatrix}, \quad x(0) = \begin{pmatrix} 6 \\ -1 \end{pmatrix}.$$

Exercise 3.6.10. *Solve the IVP (compare with Exercise 3.6.5)*

$$x' = \begin{pmatrix} 0 & 1 \\ -1 & 0 \end{pmatrix} x + \sec(t) \begin{pmatrix} 2 \\ 3 \end{pmatrix}, \quad x(0) = \begin{pmatrix} 2 \\ 5 \end{pmatrix}.$$

Exercise 3.6.11. *Consider the ODE*

$$x' = Ax + e^{2t} \begin{pmatrix} 4 \\ -7 \end{pmatrix},$$

where the homogeneous solution is given by

$$x_h(t) = c_1 e^{2t} \begin{pmatrix} 1 \\ 3 \end{pmatrix} + c_2 e^{4t} \begin{pmatrix} 1 \\ 4 \end{pmatrix}.$$

Find a particular solution $x_p(t)$.

Exercise 3.6.12. *Consider the ODE*

$$x' = Ax + \sin(3t) \begin{pmatrix} 2 \\ 3 \end{pmatrix},$$

where the homogeneous solution is given by

$$x_h(t) = c_1 \begin{pmatrix} \cos(3t) \\ 2\cos(3t) - 4\sin(3t) \end{pmatrix} + c_2 \begin{pmatrix} \sin(3t) \\ 2\sin(3t) + 4\cos(3t) \end{pmatrix}.$$

Find a particular solution $x_p(t)$.

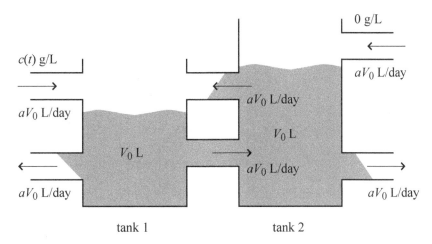

Figure 3.11. *A cartoon that depicts a two-tank extension of the one-tank problem considered in section 2.6.1. Here two identical tanks are connected, except that a solution flows into one tank but freshwater flows into the other. As for the one-tank problem, the incoming concentration $c(t)$ is not necessarily constant.*

3.7 ▪ Case studies

We now consider three problems. In the first, we look at a two-tank mixing problem in which we connect two identical tanks. In the second, we consider how the body reacts to the ingestion of lead. In particular, we want to see if the lead concentrates in any particular component (here meaning blood, tissue, and bone) of the body. In the final problem, we consider an SIR model for zoonosis—the transmission of disease from animals to humans.

3.7.1 ▪ Two-tank mixing problem

Consider the two-tank mixing problem illustrated in Figure 3.11. This system is a natural extension of the one-tank problem considered in section 2.6.1. Following the same argument as that presented in section 3.1.1, if we let x_1 denote the grams of salt in tank 1 and x_2 the grams of salt in tank 2, then the governing equation for the system is the IVP

$$x' = Ax + \begin{pmatrix} aV_0 c(t) \\ 0 \end{pmatrix}, \quad x(0) = \begin{pmatrix} x_1(0) \\ x_2(0) \end{pmatrix}; \quad A = \begin{pmatrix} -2a & a \\ a & -2a \end{pmatrix}. \quad (3.7.1)$$

We will be interested in the concentrations in each tank for large time.

We first solve the ODE. Consider the homogeneous system. The eigenvalues and associated eigenvectors for the matrix A are

$$\lambda_1 = -a, \; v_1 = \begin{pmatrix} 1 \\ 1 \end{pmatrix}; \quad \lambda_2 = -3a, \; v_2 = \begin{pmatrix} -1 \\ 1 \end{pmatrix}.$$

The homogeneous solution is then

$$x_h(t) = c_1 e^{-at} \begin{pmatrix} 1 \\ 1 \end{pmatrix} + c_2 e^{-3at} \begin{pmatrix} -1 \\ 1 \end{pmatrix}.$$

Since both eigenvalues are negative, the homogeneous solution $x_h(t)$ satisfies $x_h(t) \to 0$ as $t \to +\infty$.

The matrix-valued solution is

$$\Phi(t) = \begin{pmatrix} e^{-at} & e^{-3at} \\ e^{-at} & -e^{-3at} \end{pmatrix},$$

and the homogeneous solution can be rewritten as

$$x_h(t) = \Phi(t)c.$$

Note that $\Phi(t) \to 0_2$ as $t \to +\infty$. By Corollary 3.3, we know that the general solution to the full problem is the sum of the homogeneous and particular solutions:

$$x(t) = \Phi(t)c + x_p(t).$$

If the initial condition is $x(0) = x_0$, then the solution to the IVP is

$$x(t) = \Phi(t)\Phi(0)^{-1}\left[x_0 - x_p(0)\right] + x_p(t).$$

Since $\Phi(t) \to 0_2$ as $t \to +\infty$, it is the case that for large time,

$$x(t) \sim x_p(t).$$

Consequently, the concentration in each tank for large times will be described by the particular solution and will not depend on the initial concentration in each tank.

Now consider the particular solution. We will follow the one-tank case study of section 2.6.1 and assume that the incoming concentration is sinusoidal with mean c_0:

$$c(t) = c_0(1 - \cos(\omega t)).$$

Since the forcing vector is then

$$\begin{pmatrix} aV_0 c(t) \\ 0 \end{pmatrix} = \begin{pmatrix} aV_0 c_0 \\ 0 \end{pmatrix} - \cos(\omega t)\begin{pmatrix} aV_0 c_0 \\ 0 \end{pmatrix},$$

we know that the particular solution will have the form

$$x_p(t) = a_0 + \cos(\omega t)a_1 + \sin(\omega t)a_2.$$

Plugging this guess into the system (3.7.1) and equating the constant vector term gives the system

$$Aa_0 + aV_0 c_0 \begin{pmatrix} 1 \\ 0 \end{pmatrix} = 0 \quad \rightsquigarrow \quad a_0 = \frac{1}{3}V_0 c_0 \begin{pmatrix} 2 \\ 1 \end{pmatrix},$$

and equating the cosine and sine terms yields the pair of linear systems

$$Aa_1 - \omega a_2 = \begin{pmatrix} aV_0 c_0 \\ 0 \end{pmatrix}, \quad \omega a_1 + Aa_2 = \begin{pmatrix} 0 \\ 0 \end{pmatrix}.$$

3.7. Case studies

In block matrix form, this pair of linear systems can be rewritten as the system

$$\begin{pmatrix} A & -\omega I_2 \\ \omega I_2 & A \end{pmatrix} \begin{pmatrix} a_1 \\ a_2 \end{pmatrix} = \begin{pmatrix} aV_0c_0 \\ 0 \\ 0 \\ 0 \end{pmatrix}.$$

We solve the algebraic system using SAGE. On setting

$$\Delta := (9a^2 + \omega^2)(a^2 + \omega^2) > 0,$$

which is the determinant of the coefficient matrix, the (undetermined) vectors are

$$a_1 = -\frac{a^2 V_0 c_0}{\Delta} \begin{pmatrix} 6a^2 + 2\omega^2 \\ 3a^2 - \omega^2 \end{pmatrix}, \quad a_2 = -\frac{aV_0 c_0 \omega}{\Delta} \begin{pmatrix} 5a^2 + \omega^2 \\ 4a^2 \end{pmatrix}.$$

On collecting terms, we see that the particular solution is

$$x_p(t) = \frac{1}{3} V_0 c_0 \begin{pmatrix} 2 \\ 1 \end{pmatrix} - \frac{a^2 V_0 c_0}{\Delta} \begin{pmatrix} 6a^2 + 2\omega^2 \\ 3a^2 - \omega^2 \end{pmatrix} \cos(\omega t)$$
$$- \frac{aV_0 c_0 \omega}{\Delta} \begin{pmatrix} 5a^2 + \omega^2 \\ 4a^2 \end{pmatrix} \sin(\omega t).$$

We now determine the concentration in each tank for $t \gg 0$. The concentration in each tank is

$$c_1(t) = \frac{x_1(t)}{V_0}, \quad c_2(t) = \frac{x_2(t)}{V_0} \quad \rightsquigarrow \quad c(t) = \frac{x(t)}{V_0}.$$

Using the above analysis for the solution, the concentration in each tank for large time is approximately

$$c(t) \sim \frac{x_p(t)}{V_0}.$$

In other words, using the individual components for the particular solution, we have

$$c_1(t) \sim c_0 \left\{ \frac{2}{3} - \frac{a}{\Delta} \left[a(6a^2 + 2\omega^2) \cos(\omega t) + \omega(5a^2 + \omega^2) \sin(\omega t) \right] \right\},$$
$$c_2(t) \sim c_0 \left\{ \frac{1}{3} - \frac{a}{\Delta} \left[a(3a^2 - \omega^2) \cos(\omega t) + 4\omega a^2 \sin(\omega t) \right] \right\}.$$

In order to better interpret these expressions, we will use the identity

$$e_1 \cos(\omega t) + e_2 \sin(\omega t) = \sqrt{e_1^2 + e_2^2} \cos(\omega t - \phi), \quad \tan \phi = \frac{e_2}{e_1}.$$

After some simplification, this identity yields

$$\frac{aV_0 c_0}{\Delta} \left[a(6a^2 + 2\omega^2) \cos(\omega t) + \omega(5a^2 + \omega^2) \sin(\omega t) \right]$$
$$= V_0 c_0 A_1^*(\omega) \cos(\omega t - \phi_1^*(\omega)),$$
$$\frac{aV_0 c_0}{\Delta} \left[a(3a^2 - \omega^2) \cos(\omega t) + 4\omega a^2 \sin(\omega t) \right]$$
$$= V_0 c_0 A_2^*(\omega) \cos(\omega t - \phi_2^*(\omega)),$$

with the amplitudes being

$$A_1^*(\omega) := \frac{a\sqrt{a^2(6a^2+2\omega^2)^2 + \omega^2(5a^2+\omega^2)^2}}{\Delta},$$

$$A_2^*(\omega) := \frac{a\sqrt{a^2(3a^2-\omega^2)^2 + 16a^4\omega^2}}{\Delta}$$

and the phases satisfying

$$\tan\phi_1^*(\omega) := \frac{\omega(5a^2+\omega^2)}{a(6a^2+2\omega^2)}, \quad \tan\phi_2^*(\omega) := \frac{4\omega a}{3a^2-\omega^2}.$$

It can be checked that

$$A_1^*(0) = \frac{2}{3}, \quad A_2^*(0) = \frac{1}{3},$$

and moreover, for fixed a, each amplitude function is a strictly decreasing function of the frequency ω. In conclusion, after a long time, the concentration in each tank is

$$c_1(t) \sim c_0\left[\frac{2}{3} - A_1^*\cos(\omega t - \phi_1^*)\right], \quad c_2(t) \sim c_0\left[\frac{1}{3} - A_2^*\cos(\omega t - \phi_2^*)\right].$$

The mean concentration in tank 1 is $2c_0/3$, and the mean concentration in tank 2 is $c_0/3$. Just as in the one-tank example of section 2.6.1, there is a variation about the mean; moreover, the variation decreases as the frequency increases. A sample plot of the amplitudes and phases is given in Figure 3.12 for the case that $a = 0.45$. Note that in tank 1, there will more of a variation about the mean concentration in that tank but that in tank 2, the larger phase value implies that there will be more lag between the variation in the incoming solute and that in the tank.

A different perspective of the amplitudes and phases is given in Figure 3.13 when $a = 0.45$. In the left panel, we compare the amplitude $A_1^*(\omega)$ with the amplitude ratio $A_2^*(\omega)/A_1^*(\omega)$, and in the right panel, we compare the phase $\phi_1^*(\omega)$ with the phase difference $\phi_2^*(\omega) - \phi_1^*(\omega)$. The curves in each panel practically coincide for $\omega \geq 4$; moreover, they are quite similar to those given for the one-tank problem (illustrated in Figure 2.6). We can conclude that the amplitudes and phases vary primarily as a function of the amplitude and phase variation of the incoming concentration from the left (at least in the case of two coupled tanks). The interested reader will explore in Group Project 3.2 if this pattern continues when a third tank is added to the system.

Finally, a solution curve for the concentration in each tank for $t \gg 0$ is given in Figure 3.14 when $a = 0.45, \omega = 3.0$, and $c_0 = 1.0$. Here we clearly see that the concentration in each tank varies about its mean and that the amount of variation is markedly different for each tank. Further note the phase lag: the maximum concentration in tank 2 occurs at a later time than that in tank 1, which in turn occurs at a later time than that for the incoming solute. For large frequency, the lag in tank 1 is approximately $T/4$, whereas the lag for tank 2 is approximately $T/2$. Here T refers to the period of oscillation, $T = 2\pi/\omega$.

3.7. Case studies 187

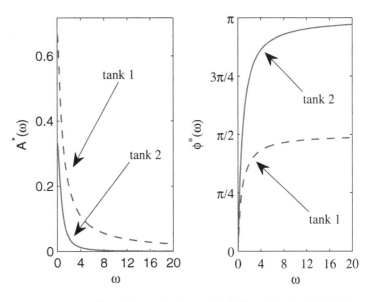

Figure 3.12. *A plot of the amplitudes $A_j^*(\omega)$ (left panel) and $\phi_j^*(\omega)$ (right panel) for the variation about the mean concentration in each tank when $a = 0.45$. The curves for tank 1 are given by a dashed curve and those for tank 2 by a solid curve. Note that the variation about the mean in the first tank—that closest to the incoming solute—is always greater than that in the second tank. However, there is more of a phase lag for the concentration in the tank that is farthest from the incoming solute.*

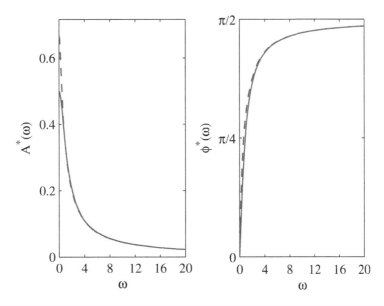

Figure 3.13. *A plot of the amplitudes $A_j^*(\omega)$ (left panel) and $\phi_j^*(\omega)$ (right panel) for the variation about the mean concentration in each tank when $a = 0.45$. In the left panel, the curve for $A_1^*(\omega)$ is given by a dashed curve, and the curve for the ratio $A_2^*(\omega)/A_1^*(\omega)$ is given by a solid curve. Note that the two curves almost coincide. In the right panel, the curve for $\phi_1^*(\omega)$ is given by a dashed curve, and the curve for the phase difference $\phi_2^*(\omega) - \phi_1^*(\omega)$ is given by a solid curve. Again note that the two curves almost coincide.*

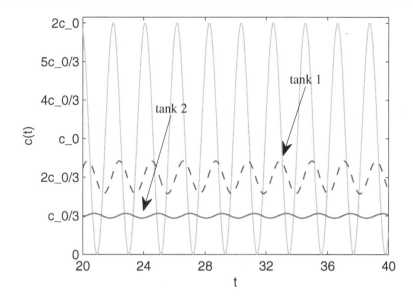

Figure 3.14. *A plot of the solution to* (3.7.1) *for large t when* $a = 0.45, \omega = 3.0,$ *and* $c_0 = 1.0$. *The incoming concentration is given by the solid curve, the concentration in tank 1 is given by the dashed curve, and the concentration in tank 2 is denoted by the solid curve. Note that the incoming concentration varies about its mean* c_0, *the concentration in tank 1 varies about its mean* $2c_0/3$, *and the concentration in tank 2 varies about its mean* $c_0/3$.

3.7.2 ▪ Lead in the human body

Lead poisoning is a medical condition that is caused by increased levels of the heavy metal lead in the body. Lead interferes with a variety of body processes and is toxic to many organs and tissues, including the heart, bones, intestines, kidneys, and reproductive and nervous systems. It interferes with the development of the nervous system; therefore, it is particularly toxic to children, as it causes potentially permanent learning and behavior disorders. Routes of exposure to lead include contaminated air, water, soil, food, and consumer products. One of the largest threats to children is lead paint that exists in many homes, especially older ones. No safe threshold for lead exposure has been discovered. As far as is known, any amount of ingested lead whatsoever is harmful.

Lead from the environment may enter the body through inhalation or ingestion. In either case, it ends up in the bloodstream, and from there it moves into the tissues and bones. The behavior is captured by a three-compartment model (see Figure 3.15). Note that this model has the same qualitative features as a three-tank mixing problem; indeed, mixing problems are a particular example of compartment models. The variables are defined as follows:

(a) x_1: the amount of lead in the blood,

(b) x_2: the amount of lead in tissue,

(c) x_3: the amount of lead in the bones, and

(d) $I_L(t)$: the rate of ingestion of lead into the blood.

3.7. Case studies

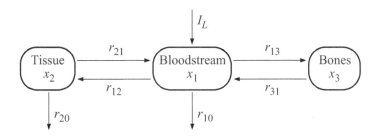

Figure 3.15. *A graphic of the compartment model for the ingestion of lead into the body. The lead enters through the bloodstream at rate I_L. The lead then enters into tissue and bone via transfer from the blood. Lead is filtered out of the bloodstream from the kidneys and leaves tissue loss (e.g., hair loss). Lead is filtered out from the bone only after first being carried away by the blood.*

The unit of measurement for time is days. Lead enters the bloodstream at rate I_L and is transferred from compartment j to compartment k at rate r_{jk}. Lead leaves the body through urination (bloodstream to kidneys) or tissue loss (e.g., hair loss).

The lead ingestion compartment model yields a linear system,

$$x_1' = -(r_{10} + r_{12} + r_{13})x_1 + r_{21}x_2 + r_{31}x_3 + I_L(t),$$
$$x_2' = r_{12}x_1 - r_{20}x_2,$$
$$x_3' = r_{13}x_1 - r_{31}x_3.$$

The implicit assumption underlying the linearity of the model is that the rate constants are independent of the amount of lead in the blood, tissue, and/or bone. The initial amount of lead in the blood is $x_1(0)$, in the tissue is $x_2(0)$, and in the bone is $x_3(0)$. In Borelli and Coleman [8, Chapter 10.6] the rate constants are given, and the equations become

$$x_1' = -\frac{13}{360}x_1 + \frac{272}{21,875}x_2 + \frac{7}{200,000}x_3 + I_L(t),$$
$$x_2' = \frac{1}{90}x_1 - \frac{1}{35}x_2, \quad (3.7.2)$$
$$x_3' = \frac{7}{1800}x_1 - \frac{7}{200,000}x_3.$$

The goal is to determine how the amount of lead in each component of the body changes as a function of time. In particular, we wish to understand how the lead concentrates in the body of someone who is initially lead-free.

Written as a first-order system, the ODEs become

$$x' = Ax + \begin{pmatrix} I_L(t) \\ 0 \\ 0 \end{pmatrix}, \quad A = \begin{pmatrix} -13/360 & 272/21,875 & 7/200,000 \\ 1/90 & -1/35 & 0 \\ 7/1800 & 0 & -7/200,000 \end{pmatrix}.$$

In order to solve the ODE, first consider the homogeneous system. Using SAGE, the eigenvalues and associated eigenvectors of A are

$$\lambda_1 \sim -4.47 \times 10^{-2},\ v_1 \sim \begin{pmatrix} 1.00 \\ -0.69 \\ -0.09 \end{pmatrix};\quad \lambda_2 \sim -2.00 \times 10^{-2},\ v_2 \sim \begin{pmatrix} 1.00 \\ 1.30 \\ -0.19 \end{pmatrix};$$

and

$$\lambda_3 \sim -3.06 \times 10^{-5},\ v_3 \sim \begin{pmatrix} 1.00 \\ 0.39 \\ 892.56 \end{pmatrix}.$$

Since all of the eigenvalues are negative, on arguing as in the previous example of the two-tank mixing problem, it will be the case that the homogeneous solution satisfies $x_h(t) \to 0$ as $t \to +\infty$. However, the smallness of the eigenvalues in absolute value implies in this case that "large" means times of $\mathcal{O}(10^6)$ days, which is $\mathcal{O}(10^3)$ years! Thus, while it is mathematically the case that for large times the amount of lead in each part of the body will not depend on the initial amount of lead in the body, from a practical perspective, this is simply not true.

Now consider the particular solution under the assumption of a constant ingestion rate of $I_L(t) = 49.3$. The particular solution will be of the form

$$x_p(t) = a_0 \quad \rightsquigarrow \quad A a_0 + \begin{pmatrix} 49.3 \\ 0 \\ 0 \end{pmatrix} = 0,$$

so that on solving using SAGE,

$$x_p(t) = \begin{pmatrix} 1{,}848{,}750/1027 \\ 2{,}156{,}875/3081 \\ 616{,}250{,}000/3081 \end{pmatrix} \sim \begin{pmatrix} 1800.14 \\ 700.06 \\ 200{,}016.23 \end{pmatrix}.$$

From zero initial data and after approximately 1 year, the amount of lead in the blood and tissue, respectively, is given by

$$x_1(t) \sim 1{,}577.78,\quad x_2(t) \sim 613.00.$$

These values are relatively close to the equilibrium values. On the other hand, even after 81 years, $x_3(t) \sim 119{,}086.27$: the amount of lead in the blood at this time is roughly only 60% of the equilibrium value. In other words, the amount of lead in the bones continually increases until death and never reaches an equilibrium value.

Now let us briefly consider the problem of what happens to the amount of lead in the body after no more lead is ingested. In particular, suppose that

$$I_L(t) = \begin{cases} 49.3, & 0 \le t < 365, \\ 0, & 365 \le t, \end{cases}$$

and further suppose that at time $t = 0$, the body contains no lead. In other words, we are supposing that the body is continually exposed to lead for 1 year and is afterward placed in a lead-free environment. The solution is plotted in Figure 3.16. Here we see that after 1 year of living in a lead-free environment, there will no longer be any significant amounts of lead in the blood or tissue. On the other

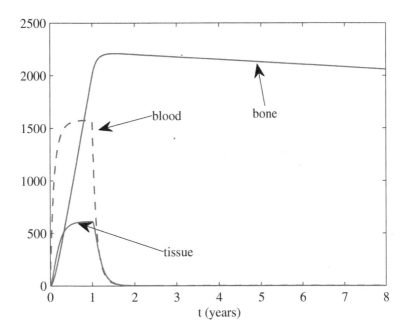

Figure 3.16. *A plot of the solution to the system (3.7.2) with the initial condition $x_1(0) = x_2(0) = x_3(0) = 0$. The ingestion rate is given by $I_L = 49.3$ for $0 \le t < 365$ and $I_L = 0$ for $t \ge 365$. The amount of lead in the bone is given by the solid curve, the amount of lead in the blood is given by the dashed curve, and the amount of lead in the tissue is given by the solid curve. Note that the lead in the blood and tissue clears out after a year or so of living in a lead-free environment but that there is still a significant amount of lead in the bone.*

hand, there will still be a significant amount of lead in the bone. Indeed, while it is not shown in this figure, there will be 910.3 micrograms of lead in the bone after 80 years of living in a lead-free environment. In conclusion, exposure to lead for a relatively short time means that (according to this mathematical model) you will carry that lead with you for the rest of your life.

Why does it take so long for the lead to clear out of the bones? The disparity in the size of the eigenvalues means that for large time, the solution will approximately be

$$x(t) \sim c_3 e^{\lambda_3 t} v_3, \quad t \gg 0.$$

In this case, since

$$t > 580 \quad \leadsto \quad e^{\lambda_1 t}, e^{\lambda_2 t} < 10^{-5},$$

large time is on the order of about 2 years. Since an eigenvector is unique only up to scalar multiplication, we can rescale it so that each entry corresponds to a percentage (this was done in section 1.13.2 in the voter registration study, section 1.13.3 in the discrete SIR study, and section 1.13.4 in the study of the Northern spotted owl). Rescaling this eigenvector so that the sum of the components is one gives the eigenvector:

$$v_3 \sim \begin{pmatrix} 0.0011 \\ 0.0004 \\ 0.9984 \end{pmatrix}.$$

Thus, of the remaining lead in the body, approximately 0.11% is in the blood, 0.04% is in the tissue, and 99.84% is in the bones. As a final remark, the size of the eigenvalue λ_3 means that it will take at least $t = \mathcal{O}(10^6)$ days (which is $\mathcal{O}(10^3)$ years!) before this final term becomes negligible. Once the lead has been ingested, it will never really leave your body.

3.7.3 ▪ Linear SIR model for zoonotic microorganisms

Zoonosis is the process where a disease is transmitted from animals to humans (or vice versa). In particular, regular exposure to a zoonotic pathogen such as *Campylobacter jejuni*—a species of bacteria found in animal feces—can occur through both food and environmental routes, but person-to-person transmission is considered rare. Ingestion of *C. jejuni* is a form of food poisoning, and symptoms (which may last up to 1 week or more) include abdominal pain, diarrhea, and fever. Rural workers in regular contact with livestock have a greater and more frequent exposure to zoonotic pathogens and consequently develop a greater immunity than the bulk of the population. On the other hand, those who live in an urban environment may develop an immunity after illness but then lose it afterward because of infrequent exposure. Any mathematical model for this disease must take into account the differential immunity status between various subgroups within a population.

McBride and French [29] proposed an *age-dependent* SIR (*Susceptible-Infected-Recovered*) model for the transmission of zoonotic pathogens. The three distinct subgroups are

(a) susceptible (S): those who are able to get a disease but have not yet been infected;

(b) infected (I): those who are currently fighting the disease; and

(c) recovered (R): those who have had the disease or are immune to the disease.

In formulating the model, it is assumed that the number of people in each group does not depend on location; hence, there are no spatial effects. The model is given by

$$\begin{aligned} S' &= -cK_1 S + \alpha I + (\alpha + \delta)R + b(a), \\ I' &= cK_1 K_2 S - (\alpha + \gamma)I, \\ R' &= cK_1(1 - K_2)S + \gamma I - (\alpha + \delta)R. \end{aligned} \tag{3.7.3}$$

Here $' = d/da$, where a represents the age (in years). The relevant initial condition is that when $a = 0$ (time of birth),

$$(S, I, R) = (N_0, 0, 0),$$

where N_0 is the total population at the time of birth. In other words, all people are born susceptible. Moreover, it is assumed that

(a) person-to-person transmission can be ignored, and

(b) individuals may be become infected and ill and then recover to become immune or on exposure may pass directly into the immune class.

3.7. Case studies

The linearity of the model arises because person-to-person transmission is ignored. Set the total population to be $N = S + I + R$. Since

$$N' = b(a) \quad \leadsto \quad N = \int b(a)\,da, \qquad (3.7.4)$$

(see Exercise 3.7.3) the function $b(a)$ describes the rate at which the total population changes as a function of age. As for the other parameters,

(a) c is the specific rate of contact with the pathogen,

(b) K_1 is the probability of infection given contact,

(c) K_2 is the probability of illness given infection,

(d) γ is the inverse of the shedding period,

(e) α is the specific death rate in the population, and

(f) δ is the specific immunity loss rate.

Setting

$$x = \begin{pmatrix} S \\ I \\ R \end{pmatrix}, \quad A = \begin{pmatrix} -cK_1 & \alpha & \alpha+\delta \\ cK_1K_2 & -\alpha-\gamma & 0 \\ cK_1(1-K_2) & \gamma & -\alpha-\delta \end{pmatrix},$$

the system (3.7.3) with the appropriate initial condition becomes the IVP:

$$x' = Ax + \begin{pmatrix} b(a) \\ 0 \\ 0 \end{pmatrix}, \quad x(0) = \begin{pmatrix} N_0 \\ 0 \\ 0 \end{pmatrix}. \qquad (3.7.5)$$

We solve this IVP under the assumption that $b(a) \equiv 0$, which by (3.7.4) means that the total population is constant as it ages.

We consider what happens for two distinct population groups: an urban group and a rural group. The goal is to understand the SIR distribution for an "old" population in each group, where old is appropriately defined. Following [29], for both groups, we set

$$\alpha = \frac{1}{80}, \quad \gamma = 12.$$

For the urban population, we have

$$c = 1, \, K_1 = \frac{1}{2}, \, K_2 = \frac{1}{2}, \, \delta = \frac{9}{10},$$

and for the rural population, we set

$$c = 10, \, K_1 = \frac{1}{10}, \, K_2 = \frac{1}{2}, \, \delta = \frac{1}{20}.$$

The parameters reflect the fact that

(a) the rate of contact with the pathogen is much higher for the rural group than for the urban group,

(b) the rural group is more resistant to infection, and

(c) the rural group is much more likely to remain immune than is the urban group.

First, consider the urban population, where on substitution of the appropriate parameter values, the matrix is

$$A = \begin{pmatrix} -1/2 & 1/80 & 73/80 \\ 1/4 & -961/80 & 0 \\ 1/4 & 12 & -73/80 \end{pmatrix}.$$

Using SAGE, the eigenvalues and associated eigenvectors for A are

$$\lambda_1 \sim -11.99, \; v_1 \sim \begin{pmatrix} 1.00 \\ 11.75 \\ -12.75 \end{pmatrix}; \quad \lambda_2 \sim -1.43, \; v_2 \sim \begin{pmatrix} 1.00 \\ 0.02 \\ -1.02 \end{pmatrix};$$

and

$$\lambda_3 = 0, \; v_3 = \begin{pmatrix} 1 \\ 20/961 \\ 38,420/70,153 \end{pmatrix}.$$

The general solution is

$$x(a) = c_1 e^{\lambda_1 a} v_1 + c_2 e^{\lambda_2 a} v_2 + c_3 v_3.$$

Since $N(a) = N_0$ for all $a \geq 0$ and since $\lambda_1, \lambda_2 < 0$ implies that

$$\lim_{a \to +\infty} x(a) = c_3 v_3,$$

the sum of the entries of the vector $c_3 v_3$ must equal N_0. This yields

$$c_3 v_3 = \frac{N_0}{110,033} \begin{pmatrix} 70,153 \\ 1460 \\ 38,420 \end{pmatrix} \sim N_0 \begin{pmatrix} 0.638 \\ 0.013 \\ 0.349 \end{pmatrix}.$$

In other words, for an old urban population, 63.8% of the people will be susceptible, 1.3% will be infected, and 34.9% will be recovered or immune. Moreover, since

$$a > 7 \quad \leadsto \quad e^{\lambda_1 a}, e^{\lambda_2 a} < 10^{-4},$$

we know that in this context, "old" means at least 7 years.

Next consider the rural population, where on substitution of the appropriate parameter values,

$$A = \begin{pmatrix} -1 & 1/80 & 1/16 \\ 1/2 & -961/80 & 0 \\ 1/2 & 12 & -1/16 \end{pmatrix}.$$

Using SAGE, the eigenvalues and associated eigenvectors for A are

$$\lambda_1 \sim -12.01, \; v_1 \sim \begin{pmatrix} 1.00 \\ 218.95 \\ -219.95 \end{pmatrix}; \quad \lambda_2 \sim -1.06, \; v_2 \sim \begin{pmatrix} 1.00 \\ 0.05 \\ -1.05 \end{pmatrix};$$

and
$$\lambda_3 = 0, \quad v_3 = \begin{pmatrix} 1 \\ 40/961 \\ 15,368/961 \end{pmatrix}.$$

The general solution is
$$x(a) = c_1 e^{\lambda_1 a} v_1 + c_2 e^{\lambda_2 a} v_2 + c_3 v_3.$$

Since $N(a) = N_0$ for all $a \geq 0$ and since $\lambda_1, \lambda_2 < 0$ implies that
$$\lim_{a \to +\infty} x(a) = c_3 v_3,$$

the sum of the entries of the vector $c_3 v_3$ must equal N_0. This yields
$$c_3 v_3 = \frac{N_0}{16,369} \begin{pmatrix} 961 \\ 40 \\ 15,368 \end{pmatrix} \sim N_0 \begin{pmatrix} 0.059 \\ 0.002 \\ 0.939 \end{pmatrix}.$$

In other words, for an old rural population, 5.9% of the people will be susceptible, 0.2% will be infected, and 93.9% will be recovered or immune. Moreover, since
$$a > 9 \quad \leadsto \quad e^{\lambda_1 a}, e^{\lambda_2 a} < 10^{-4},$$

we know that in this context, "old" means at least 9 years.

In conclusion, we see that the model says that there are important differences between the two populations regarding the manner in which they respond to the disease. After the age of 9, there is a much higher percentage of the urban population (63.8%) that is susceptible to the disease than there is of the rural population (5.9%). Furthermore, the vast majority of the rural population is recovered from or immune to the disease (93.9%), whereas only approximately 1/3 (34.9%) of the urban population enjoys a similar property.

--- **Exercises** ---

Exercise 3.7.1. *Consider the system of two interconnected tanks given below.*

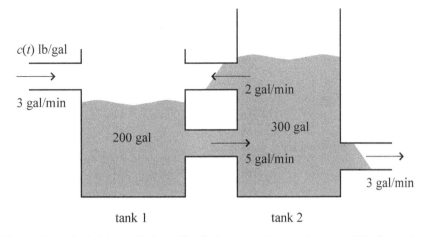

The tank on the left initially has 5 lb of salt present in its solution, while the tank on the right has 12 lb in its solution.

(a) *Set up the initial value problem whose solution will determine the amount of salt in each tank.*

(b) *After a very long time, does the amount of salt in each tank substantively depend on the initial amount of salt in each tank?*

(c) *Suppose that $c(t) = 10$. Find the limiting concentration in each tank.*

Exercise 3.7.2. *Consider the system of two interconnected tanks given below:*

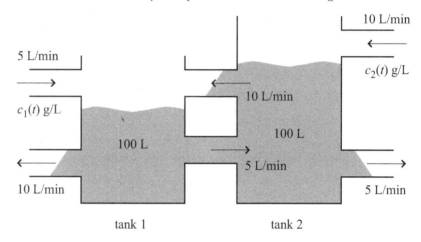

The tank on the left initially has 40 g of salt present in its solution, while the tank on the right has 50 g in its solution.

(a) *Set up the initial value problem whose solution will determine the amount of salt in each tank.*

(b) *After a very long time, does the amount of salt in each tank substantively depend on the initial amount of salt in each tank?*

(c) *Suppose that*
$$c_1(t) = 15, \quad c_2(t) = 30.$$
Find the limiting concentration in each tank.

Exercise 3.7.3. *Consider the SIR model given in (3.7.3). If $N = S + I + R$, show that*
$$N' = b(a).$$

Exercise 3.7.4. *Consider the SIR model of given in (3.7.3) with $b(a) \equiv 0$. The primary source for Campylobacteriosis is animals that are farmed for meat. For this bacterial infection, the parameters are*
$$\alpha = \frac{1}{80}, \ \gamma = 26, \ c = 2, \ K_1 = \frac{1}{10}, \ K_2 = \frac{2}{10}, \ \delta = \frac{35}{100}.$$

(a) *For an old population, what percentage of the original population is susceptible? Infected? Recovered or immune?*

(b) *Using the error criteria of 10^{-4}, determine what is "old" in (a).*

Exercise 3.7.5. *Consider the SIR model of given in (3.7.3) with $b(a) \equiv 0$. The primary symptom associated with Cryptosporidiosis is a self-contained diarrhea. For this bacterial infection, the parameters are*

$$\alpha = \frac{1}{80}, \quad \gamma = 26, \quad c = 1, \quad K_1 = \frac{2}{10}, \quad K_2 = \frac{2}{10}, \quad \delta = \frac{25}{1000}.$$

(a) *For an old population, what percentage of the original population is susceptible? Infected? Recovered or immune?*

(b) *Using the error criteria of 10^{-4}, determine what is "old" in (a).*

Group projects

3.1. Consider the following variation on the two-tank system considered in section 3.7.1:

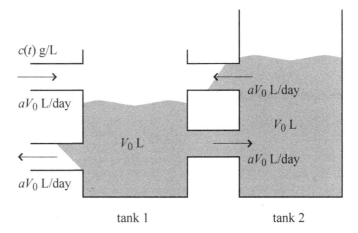

(a) Let $x_{1,0}$ and $x_{2,0}$ represent the amount of salt in tank 1 and tank 2, respectively, when $t = 0$. Set up the initial value problem whose solution will determine the amount of salt in each tank.

(b) After a very long time, does the amount of salt in each tank substantively depend on the initial amount of salt in each tank?

(c) Suppose that
$$c(t) = c_0(1 - \cos(\omega t)), \quad \omega > 0.$$
Find the limiting concentration in each tank, i.e., the approximate concentration in each tank for $t \gg 0$.

(d) For $t \gg 0$, what is the mean concentration in each tank?

(e) In each tank, there is a variation about the mean of amplitude $c_0 A_j^*(\omega)$ for $j = 1, 2$. Produce a plot of each $A_j^*(\omega)$ on the same graph. In which tank is there more variation about the mean? Does the answer depend on the frequency?

(f) Plot $A_1^*(\omega)$ and $A_2^*(\omega)/A_1^*(\omega)$ on the same graph. What can you conclude?

(g) In each tank, the asymptotic concentration has a phase shift $\phi_j^*(\omega)$ on the same graph. Produce a plot of each phase shift. Is there more of a phase shift in one tank than the other? Does the answer depend on the frequency?

(h) Plot $\phi_1^*(\omega)$ and $\phi_2^*(\omega) - \phi_1^*(\omega)$ on the same graph. What can you conclude?

(i) Discuss the similarities and differences between results for this problem and those for the two-tank problem given in section 3.7.1.

3.2. Consider the variant of Figure 3.11 in which there is an additional tank between the two pictured. We have three identical tanks with volume V_0, each of which is coupled to its nearest neighbor only. The flow rate between each tank is aV_0 L/day, so the total volume in each tank is fixed. For the tank on the left, there is entering a briny solution with concentration $c(t)$ g/L. For the tank on the right, there is entering freshwater. Going from left to right, let x_j denote the amount of salt in tank j. The governing equation for the system is

$$x_1' = -2ax_1 + ax_2 + aV_0 c(t),$$
$$x_2' = ax_1 - 2ax_2 + ax_3,$$
$$x_3' = ax_2 - 2ax_3.$$

(a) Write the system in matrix/vector form:

$$x' = Ax + f(t).$$

Clearly identify the matrix A.

(b) After a very long time, does the amount of salt in each tank substantively depend on the initial amount of salt in each tank?

(c) Suppose that $c(t) = c_0(1 - \cos(\omega t))$. Show that the concentration in each tank for $t \gg 0$ has the form

$$c_j(t) \sim c_0\left(\alpha_j - A_j^* \cos(\omega t - \phi_j^*)\right), \quad j = 1, 2, 3.$$

Write expressions for the means, α_j; amplitudes, $A_j^*(\omega)$; and phases, $\phi_j^*(\omega)$.

(d) What is the mean concentration in each tank? What can you conclude?

(e) Plot $A_1^*(\omega)$, $A_2^*(\omega)/A_1^*(\omega)$, and $A_3^*(\omega)/A_2^*(\omega)$ on the same graph. What can you conclude?

(f) Plot $\phi_1^*(\omega)$, $\phi_2^*(\omega) - \phi_1^*(\omega)$, and $\phi_3^*(\omega) - \phi_2^*(\omega)$ on the same graph. What can you conclude?

(g) Based on your answer in (d) and the result for two tanks in section 3.7.1, if there were a system of $N \geq 4$ identical tanks, provide a conjecture as to what would be the average concentration in tank j for $j = 1, 2, \ldots, N$.

3.3. In section 3.7.2, we considered the manner in which lead propagates throughout the body under the assumption of a constant rate of ingestion. Now let us

consider what happens if the ingestion rate decays. In particular, now suppose that the ingestion rate is
$$I_L(t) = 49.3e^{-at}.$$

The constant $a > 0$ reflects how quickly you stop taking the lead. In particular, at time $t_{1/2} = \ln(2)/a$, the rate of ingestion is half what it was at the beginning. We will assume that $a \geq 0.05$, so $t_{1/2} < 14$ (2 weeks). We further suppose there is initially no lead in the body.

(a) For a given value of a, how much lead is ingested in 1 year?

(b) Suppose that 2 or more years have passed. Does the value of a have any influence on the conclusion that 99.84% of the lead in the body is trapped in the bones?

(c) Suppose that $a = 0.1$, which means that you essentially stop ingesting lead after approximately 20 weeks. How much lead was ingested in 20 weeks? How much lead is in the bones after 5 years? After 10 years? After 30 years?

(d) Suppose that $a = 1$, which means that you essentially stop ingesting lead after approximately 2 weeks. How much lead was ingested in 2 weeks? How much lead is in the bones after 5 years? After 10 years? After 30 years?

(e) Suppose that $a = 4$, which means that you essentially stop ingesting lead after approximately 4 days. How much lead was ingested in 4 days? How much lead is in the bones after 5 years? After 10 years? After 30 years?

3.4. The equations for a particular armature-controlled DC motor are given by

$$x_1' = -2x_1 - 0.5x_2 + v(t),$$
$$x_2' = 100x_1 - 1.5x_2,$$

where x_1 represents the motor's current, x_2 represents the motor's rotational velocity, and $v(t)$ is the applied voltage.

(a) After a very long time, does the current substantively depend on the initial current in the motor?

(b) Suppose that
$$v(t) = V_0(1 - \cos(\omega t));$$

i.e., the applied current varies periodically in time. Find the approximate current and rotational velocity for $t \gg 0$.

(c) For $t \gg 0$, what is the mean current and mean rotational velocity?

(d) There is a variation about the mean of the current given by $V_0 A_1^*(\omega)$ and a variation about the mean of the rotational velocity given by $V_0 A_2^*(\omega)$. Produce a plot of $A_1^*(\omega)$ and $A_2^*(\omega)$ on the same graph.

(e) Is there more variation about the mean of the current or of the rotational velocity? Does the answer depend on the frequency?

(f) The asymptotic current and rotational velocity have associated with them a phase shift $\phi_j^*(\omega)$ for $j = 1, 2$. Produce a plot of each phase shift on the same graph. Is there more of a phase shift for one quantity than the other? Does the answer depend on the frequency?

3.5. Consider the SIR model of (3.7.3). Suppose that when $a = 0$, the population is N_0. Further suppose that the size of the population slowly decays as it ages and that this decay is modeled by

$$N(a) = e^{-\mu a} N_0, \quad \mu = \frac{\ln(10/9)}{60} \sim 1.756 \times 10^{-3}.$$

Suppose that for a particular bacterial infection, the parameters are

$$\alpha = \frac{1}{80}, \ \gamma = 26, \ c = 1, \ K_1 = \frac{3}{200}, \ K_2 = \frac{2}{10}, \ \delta = \frac{35}{100}.$$

(a) Show that the group of 60-year-old people is 90% of the original population.

(b) Find the function $b(a)$ to be used in the model. (*Hint*: Consider Exercise 3.7.3.)

(c) For a group of 5-year-old children, what percentage of the population is susceptible? Infected? Recovered or immune?

(d) For a group of 20-year-old adults, what percentage of the population is susceptible? Infected? Recovered or immune?

(e) For a group of 60-year-old people, what percentage of the population is susceptible? Infected? Recovered or immune?

(f) Is there an age for which the percentages in each group are essentially fixed thereafter (use an error bound of 10^{-3})? If so, what is it? Is it reasonable to expect that this age can be reached?

Chapter 4

Scalar higher-order linear differential equations

> *Truth is ever to be found in the simplicity, and not in the multiplicity and confusion of things.*
>
> —Isaac Newton

In the previous chapter, we learned how to find the general solution to systems of ODEs of the form $x' = Ax + f(t)$. We used the eigenvalues and eigenvectors of A in order to construct the homogeneous solution, and then we constructed the particular solution using either the method of undetermined coefficients or variation of parameters.

In this chapter, we concern ourselves with the study of nth-order scalar ODEs:

$$y^{(n)} + a_{n-1}y^{(n-1)} + \cdots + a_1 y' + a_0 y = f(t). \tag{4.0.1}$$

As we saw in the last chapter, such a problem readily arises in many applications. For example, if $n = 2$, it is a model used in both circuit theory and the damped-forced mass spring; if $n = 3$, it is a model used in the theory of water waves (see Drazin and Johnson [14, Chapter 1]); and if $n = 4$, it is a model used to study the deflection of a beam from its equilibrium state (see Meirovitch [30, Chapter 10.2]).

In the last chapter, we showed that we could convert (4.0.1) into a first-order linear system. On setting

$$x_1 = y, \ x_2 = y', \ x_3 = y'', \ldots, x_n = y^{(n-1)}, \tag{4.0.2}$$

we get

$$x' = \underbrace{\begin{pmatrix} 0 & 1 & 0 & 0 & 0 & \cdots & 0 \\ 0 & 0 & 1 & 0 & 0 & \cdots & 0 \\ 0 & 0 & 0 & 1 & 0 & \cdots & 0 \\ \vdots & \vdots & \vdots & \vdots & \vdots & \vdots & \vdots \\ 0 & 0 & 0 & 0 & 0 & \cdots & 1 \\ -a_0 & -a_1 & -a_2 & -a_3 & -a_4 & \cdots & -a_{n-1} \end{pmatrix}}_{A} x + \underbrace{\begin{pmatrix} 0 \\ 0 \\ 0 \\ \vdots \\ 0 \\ f(t) \end{pmatrix}}_{f(t)}. \tag{4.0.3}$$

The matrix $A \in \mathcal{M}_n(\mathbb{R})$ is sometimes known as the *companion matrix* for the nth-order ODE (4.0.1). Thus, the problem (4.0.1) is amenable to the ideas and techniques presented in the last chapter.

We wish to solve the problem (4.0.1) without first converting it into the system (4.0.3). Because of the equivalence between the two problems, this will be possible. In particular, the first component of the solution to the vectorized problem (4.0.3) is a solution to the original problem (4.0.1). Since the general solution to (4.0.3) is the sum of the homogeneous and particular solutions, the same can then be said for the original problem (4.0.1).

4.1 ▪ The homogeneous solution

In order to solve the original problem (4.0.1), we first proceed to solve the system (4.0.3). First, consider the vectorized homogeneous problem

$$x' = Ax,$$

which is equivalent to the scalar problem,

$$y^{(n)} + a_{n-1} y^{(n-1)} + \cdots + a_1 y' + a_0 y = 0. \tag{4.1.1}$$

As already noted, the first component of the solution to the vectorized problem is the solution to the scalar problem. While we will not prove it here (see Meyer [31, p. 648] for the proof), the characteristic polynomial for the companion matrix is

$$p_A(\lambda) = \lambda^n + a_{n-1} \lambda^{n-1} + \cdots + a_1 \lambda + a_0; \tag{4.1.2}$$

in other words, it is simply the ODE (4.0.1) with the derivative of kth-order being replaced by the monomial λ^k. The interested student should verify this statement for the case of $n = 2, 3$ (see Exercise 4.1.2). It further turns out to be the case that an eigenvector associated with an eigenvalue $\lambda = \lambda_0$ is given by

$$\lambda = \lambda_0, \quad v = \begin{pmatrix} 1 \\ \lambda_0 \\ \lambda_0^2 \\ \vdots \\ \lambda_0^{n-2} \\ \lambda_0^{n-1} \end{pmatrix}. \tag{4.1.3}$$

The jth entry of the eigenvector is the $(j-1)$st power of the eigenvalue.

With this form of the eigenvector, we can easily write down the solutions to the scalar problem (4.1.1). If λ_0 is real valued, then the associated solution to the vector homogeneous system (4.1.1) is

$$x(t) = c e^{\lambda_0 t} \begin{pmatrix} 1 \\ \lambda_0 \\ \lambda_0^2 \\ \vdots \\ \lambda_0^{n-2} \\ \lambda_0^{n-1} \end{pmatrix}.$$

4.1. The homogeneous solution

Since a solution to the scalar homogeneous system (4.1.1) is given by the first component of the vector-valued solution, we have the solution

$$y(t) = ce^{\lambda_0 t}.$$

Now suppose that the eigenvalue is complex valued, i.e., $\lambda_0 = a + ib$. The associated eigenvector will then be of the form

$$v = \begin{pmatrix} 1 \\ \lambda_0 \\ \lambda_0^2 \\ \vdots \\ \lambda_0^{n-2} \\ \lambda_0^{n-1} \end{pmatrix} = \begin{pmatrix} 1 \\ p_1 \\ p_2 \\ \vdots \\ p_{n-2} \\ p_{n-1} \end{pmatrix} + i \begin{pmatrix} 0 \\ q_1 \\ q_2 \\ \vdots \\ q_{n-2} \\ q_{n-1} \end{pmatrix},$$

where we are using the notation

$$\lambda_0^k = p_k + iq_k, \quad k = 1, \ldots, n-1.$$

The two solutions associated with this eigenvalue are

$$x_1(t) = c_1 e^{at} \left[\cos(bt) \begin{pmatrix} 1 \\ p_1 \\ p_2 \\ \vdots \\ p_{n-2} \\ p_{n-1} \end{pmatrix} - \sin(bt) \begin{pmatrix} 0 \\ q_1 \\ q_2 \\ \vdots \\ q_{n-2} \\ q_{n-1} \end{pmatrix} \right],$$

$$x_2(t) = c_2 e^{at} \left[\sin(bt) \begin{pmatrix} 1 \\ p_1 \\ p_2 \\ \vdots \\ p_{n-2} \\ p_{n-1} \end{pmatrix} + \cos(bt) \begin{pmatrix} 0 \\ q_1 \\ q_2 \\ \vdots \\ q_{n-2} \\ q_{n-1} \end{pmatrix} \right].$$

Two solutions to the scalar homogeneous system (4.1.1) are given by the first component of each vector-valued solution:

$$y_1(t) = c_1 e^{at} \cos(bt), \quad y_2(t) = c_2 e^{at} \sin(bt).$$

Example 4.1. Consider the second-order ODE

$$y'' - 5y' + 6y = 0.$$

The characteristic polynomial is

$$p(\lambda) = \lambda^2 - 5\lambda + 6 = (\lambda - 2)(\lambda - 3),$$

and the roots are $\lambda = 2, 3$. The general solution is

$$y(t) = c_1 e^{2t} + c_2 e^{3t}. \quad \blacksquare$$

Example 4.2. Consider the second-order ODE

$$y'' + 4y' + 29y = 0.$$

The characteristic polynomial is

$$p(\lambda) = \lambda^2 + 4\lambda + 29 = (\lambda + 2)^2 + 25,$$

and the roots are $\lambda = -2 \pm i5$. The general solution is

$$y(t) = c_1 e^{-2t} \cos(5t) + c_2 e^{-2t} \sin(5t).$$

Now suppose that there is the initial condition $y(0) = 2$, $y'(0) = -5$. This leads to a linear system of equations:

$$2 = y(0) = c_1, \quad -5 = y'(0) = -2c_1 + 5c_2 \quad \rightsquigarrow \quad c_1 = 2, \; c_2 = -\frac{1}{5}.$$

The solution to the IVP is

$$y(t) = 2e^{-2t} \cos(5t) - \frac{1}{5} e^{-2t} \sin(5t). \quad \blacksquare$$

Example 4.3. Consider the fourth-order ODE

$$y^{(4)} - 3y'' + 6y' + 9y = 0.$$

The associated characteristic polynomial is

$$p(\lambda) = \lambda^4 - 3\lambda^2 + 6\lambda + 9.$$

The roots cannot easily be found analytically. However, they can be found numerically using SAGE, and we find the four roots to be

$$\lambda_1 \sim -1.92, \quad \lambda_2 \sim -1.13, \quad \lambda_3 \sim 1.53 - i1.35, \quad \lambda_4 \sim 1.53 + i1.35.$$

As expected, the roots with nonzero imaginary part come in complex-conjugate pairs. The general solution is

$$y(t) = c_1 e^{-1.92t} + c_2 e^{-1.13t} + c_3 e^{1.53t} \cos(1.35t) + c_4 e^{1.53t} \sin(1.35t). \quad \blacksquare$$

As we can see from the last example, the most difficult part in solving the homogeneous problem is factoring the characteristic polynomial. We now introduce some notation that will help us write an nth-order ODE in "factored" form. Set

$$D := \frac{d}{dt} \quad \rightsquigarrow \quad D^k = \frac{d^k}{dt^k}, \quad k = 1, 2 \ldots.$$

The homogeneous problem

$$y^{(n)} + a_{n-1} y^{(n-1)} + \cdots + a_1 y' + a_0 y = 0$$

4.1. The homogeneous solution

can then be rewritten as

$$(D^n + a_{n-1}D^{n-1} + \cdots + a_1 D + a_0)y = 0.$$

The characteristic polynomial is found simply via the substitution $D \mapsto \lambda$. Suppose that the characteristic polynomial has been factored, i.e.,

$$p(\lambda) = \lambda^n + a_{n-1}\lambda^{n-1} + \cdots + a_1\lambda + a_0 = (\lambda - \lambda_1)(\lambda - \lambda_2)\cdots(\lambda - \lambda_n),$$

where $\lambda_1, \lambda_2, \ldots, \lambda_n$ are the roots. On replacing λ with D in the factored form of the characteristic polynomial, it is not difficult to check that

$$(D^n + a_{n-1}D^{n-1} + \cdots + a_1 D + a_0)y = (D - \lambda_1)(D - \lambda_2)\cdots(D - \lambda_n)y.$$

(Here we implicitly use the fact that the coefficients are constants.) This factored form of the ODE allows us to easily read off the roots of the characteristic polynomial and thus solve ODEs of high order without doing the difficult work of factoring a polynomial of large order.

Example 4.4. Consider the ODE

$$(D+2)(D-4)(D^2+2D+17)(D^2-6D+13)y = 0.$$

This is a sixth-order ODE. The characteristic polynomial is

$$p(\lambda) = (\lambda + 2)(\lambda - 4)(\lambda^2 + 2\lambda + 17)(\lambda^2 - 6\lambda + 13).$$

Using

$$\lambda^2 + 2\lambda + 17 = 0 \quad \leadsto \quad \lambda = -1 \pm i4$$

and

$$\lambda^2 - 6\lambda + 13 = 0 \quad \leadsto \quad \lambda = 3 \pm i2,$$

the general solution is

$$y(t) = c_1 e^{-2t} + c_2 e^{4t} + c_3 e^{-t}\cos(4t) + c_4 e^{-t}\sin(4t) + c_5 e^{3t}\cos(2t) + c_6 e^{3t}\sin(2t). \blacksquare$$

We finally must consider the problem where the zero of the characteristic polynomial is not simple. For an illuminating sequence of examples, start with

$$Dy = 0 \quad \leadsto \quad p(\lambda) = \lambda.$$

Clearly, $\lambda = 0$ is a simple root of the characteristic polynomial. The solution can be found simply by integrating once:

$$y(t) = c_1.$$

Next consider the ODE

$$D^2 y = 0 \quad \leadsto \quad p(\lambda) = \lambda^2.$$

Here $\lambda = 0$ is a root of the characteristic polynomial of multiplicity two. This ODE is solved on integrating twice:

$$y(t) = c_1 + c_2 t.$$

Similarly, for the ODE,
$$D^3 y = 0 \quad \leadsto \quad p(\lambda) = \lambda^3,$$
$\lambda = 0$ is a root of the characteristic polynomial of multiplicity three, and integrating three times yields the solution:
$$y(t) = c_1 + c_2 t + c_3 t^2.$$
In general, we have
$$D^k y = 0 \quad \leadsto \quad y = c_1 + c_2 t + c_3 t^2 + \cdots + c_k t^{k-1}.$$

Now we show that these examples act as a template for solving the general problem,
$$(D - \lambda_0)^k y = 0 \quad \leadsto \quad p(\lambda) = (\lambda - \lambda_0)^k,$$
where now $\lambda = \lambda_0$ is a root of multiplicity k. The key is to rewrite the ODE to make it look like the previous example. Using the product rule, we first note that
$$\begin{aligned} D\left(e^{-\lambda_0 t} y\right) &= -\lambda_0 e^{-\lambda_0 t} y + e^{-\lambda_0 t} y' \\ &= e^{-\lambda_0 t}\left(-\lambda_0 y + y'\right) \\ &= e^{-\lambda_0 t}(D - \lambda_0) y. \end{aligned}$$

Again, using the product rule and the above calculation yields for the second derivative
$$\begin{aligned} D^2\left(e^{-\lambda_0 t} y\right) &= D\left[D\left(e^{-\lambda_0 t} y\right)\right] \\ &= D\left[e^{-\lambda_0 t}(D - \lambda_0) y\right] \\ &= e^{-\lambda_0 t}(D - \lambda_0)(D - \lambda_0) y \\ &= e^{-\lambda_0 t}(D - \lambda_0)^2 y. \end{aligned}$$

Continuing with this argument leads to the general rule:
$$e^{-\lambda_0 t}(D - \lambda_0)^k y = D^k\left(e^{-\lambda_0 t} y\right) \quad \leadsto \quad (D - \lambda_0)^k y = e^{\lambda_0 t} D^k\left(e^{-\lambda_0 t} y\right).$$

In conclusion, since the exponential is never zero,
$$(D - \lambda_0)^k y = 0 \quad \leadsto \quad D^k\left(e^{-\lambda_0 t} y\right) = 0.$$
We have already solved the latter ODE, so
$$e^{-\lambda_0 t} y = c_1 + c_2 t + c_3 t^2 + \cdots + c_k t^{k-1}.$$
Multiplication by the exponential leads to the solution for the original ODE:
$$y(t) = \left(c_1 + c_2 t + c_3 t^2 + \cdots + c_k t^{k-1}\right) e^{\lambda_0 t}.$$
If λ_0 is real, we are finished. If $\lambda_0 = a + ib$ for some nonzero b, then on taking the real and imaginary parts, we arrive at two linearly independent sets of solutions:
$$\left(c_1 + c_2 t + c_3 t^2 + \cdots + c_k t^{k-1}\right) e^{at} \cos(bt)$$

4.1. The homogeneous solution

and
$$\left(c_{k+1}+c_{k+2}t+c_{k+3}t^2+\cdots+c_{2k}t^{k-1}\right)e^{at}\sin(bt).$$

A real zero of order k yields k linearly independent solutions to the ODE, while a complex-valued zero of order k yields $2k$ linearly independent solutions.

The case of complex-valued roots with higher multiplicity, i.e., $k \geq 2$, can happen only if $n \geq 4$.

Theorem 4.5. *Consider the scalar nth-order homogeneous ODE*
$$y^{(n)} + a_{n-1}y^{(n-1)} + \cdots + a_1 y' + a_0 y = 0.$$

The characteristic polynomial is given by
$$p(\lambda) = \lambda^n + a_{n-1}\lambda^{n-1} + \cdots + a_1 \lambda + a_0.$$

If λ_0 is a real root of order k of the characteristic polynomial, then the associated solution to the ODE is given by
$$y(t) = (c_1 + c_2 t + c_3 t^2 + \cdots + c_k t^{k-1})e^{\lambda_0 t},$$

where the constants $c_1, c_2, \ldots, c_k$ are arbitrary. The general solution associated with the complex-valued roots $\lambda_0 = a \pm ib$ of order k is given by
$$y(t) = (c_1 + c_2 t + \cdots + c_k t^k)e^{at}\cos(bt) + (c_{k+1} + c_{k+2}t + \cdots + c_{2k}t^k)e^{at}\sin(bt),$$

where the constants $c_1, c_2, \ldots, c_{2k}$ are arbitrary. The homogeneous solution is a linear combination of these n linearly independent solutions.

Example 4.6. Suppose that the ODE is
$$(D-4)(D-5)(D+3)^2 y = 0.$$
This is a fourth-order ODE, and the associated characteristic polynomial is
$$p(\lambda) = (\lambda - 4)(\lambda - 5)(\lambda + 3)^2.$$
On using Theorem 4.5, the general solution is
$$y(t) = c_1 e^{4t} + c_2 e^{5t} + (c_3 + c_4 t)e^{-3t}. \qquad \blacksquare$$

Example 4.7. Suppose that the ODE is
$$(D+1)(D-6)^3(D^2+4D+13)y = 0.$$
This is a sixth-order ODE, and the associated characteristic polynomial is
$$p(\lambda) = (\lambda + 1)(\lambda - 6)^3(\lambda^2 + 4\lambda + 13).$$
Since
$$\lambda^2 + 4\lambda + 13 = 0 \quad \rightsquigarrow \quad \lambda = -2 \pm i3,$$
on using Theorem 4.5, the general solution is
$$y(t) = c_1 e^{-t} + (c_2 + c_3 t + c_4 t^2)e^{6t} + c_5 e^{-2t}\cos(3t) + c_6 e^{-2t}\sin(3t). \qquad \blacksquare$$

Example 4.8. Suppose that the ODE is
$$(D+7)(D^2+8D+25)^2 y = 0.$$
This is a fifth-order ODE, and the associated characteristic polynomial is
$$p(\lambda) = (\lambda+7)(\lambda^2+8\lambda+25).$$
Since
$$\lambda^2 + 8\lambda + 25 = 0 \rightsquigarrow \lambda = -4 \pm i3,$$
on using Theorem 4.5, the general solution is
$$y(t) = c_1 e^{-7t} + (c_2 + c_3 t)e^{-4t}\cos(3t) + (c_4 + c_5 t)e^{-4t}\sin(3t). \blacksquare$$

Exercises

Exercise 4.1.1. *Consider the set of functions*
$$\{e^{at}, te^{at}, t^2 e^{at}, \ldots, t^{k-1} e^{at}\}, \quad a \in \mathbb{R}.$$

(a) *If $k = 2$, show that the set is linearly independent.*

(b) *If $k = 3$, show that the set is linearly independent.*

(c) *Show that the set is linearly independent for any positive integer k.*

(Hint: Recall the discussion following Corollary 1.75 and the accompanying Exercise 1.6.8 regarding the linear dependence and independence of a set of functions.)

Exercise 4.1.2. *Consider the companion matrix A given in (4.0.3).*

(a) *If $n = 2$, verify that $p_A(\lambda) = \lambda^2 + a_1 \lambda + a_0$.*

(b) *If $n = 3$, verify that $p_A(\lambda) = \lambda^3 + a_2 \lambda^2 + a_1 \lambda + a_0$.*

Exercise 4.1.3. *Find the general solution for each of the following second-order ODEs.*

(a) $y'' - 2y' + 5y = 0$

(b) $y'' - 3y' - 28y = 0$

(c) $y'' + 7y' + 10y = 0$

(d) $y'' - 9y' + 18y = 0$

(e) $y'' + 64y = 0$

(f) $y'' - 10y' + 25y = 0$

(g) $y'' + 6y' + 13y = 0$

Exercise 4.1.4. *Solve the following IVPs.*

(a) $y'' - 2y' + 5y = 0; \quad y(0) = -2,\ y'(0) = -4$

(b) $y'' - 3y' - 28y = 0; \quad y(0) = 4,\ y'(0) = 1$

(c) $y'' + 7y' + 10y = 0; \quad y(0) = -5, \; y'(0) = 4$

(d) $y'' - 9y' + 18y = 0; \quad y(0) = -1, \; y'(0) = 8$

(e) $y'' + 64y = 0; \quad y(0) = 3, \; y'(0) = -1$

(f) $y'' - 10y' + 25y = 0; \quad y(0) = 7, \; y'(0) = 2$

(g) $y'' + 6y' + 13y = 0; \quad y(0) = 9, \; y'(0) = -2$

Exercise 4.1.5. *For each of the following problems, state the order of the ODE and then find the general solution.*

(a) $(D-1)(D+3)^2(D-5)y = 0$

(b) $(D^2 + 8D + 41)(D-2)^3(D+4)^2 y = 0$

(c) $(D-7)(D^2-9)^2(D^2+9)^3 y = 0$

(d) $(D^2 + 10D + 34)^4 (D+5)y = 0$

4.2 • The particular solution

The general solution is the sum of the homogeneous and particular solutions. There will again be two techniques for finding a particular solution: variation of parameters and undetermined coefficients. Regarding the method of variation of parameters, we will assume that $n = 2$. The case of $n \geq 3$ is considered in the discussion leading to Corollary 5.24. We will make no restriction on the size of n when discussing the method of undetermined coefficients.

4.2.1 • Variation of parameters

Let us first consider the method of variation of parameters for

$$y'' + a_1 y' + a_0 y = f(t). \tag{4.2.1}$$

In order to derive a particular solution using this method, it is best if we first consider the vectorized form of the above (4.2.1):

$$x' = \begin{pmatrix} 0 & 1 \\ -a_0 & -a_1 \end{pmatrix} x + \begin{pmatrix} 0 \\ f(t) \end{pmatrix}. \tag{4.2.2}$$

In this manner, we can use the result of Theorem 3.25, which states that the particular solution is given by

$$x_p(t) = \Phi(t) \int \Phi(t)^{-1} \begin{pmatrix} 0 \\ f(t) \end{pmatrix} dt, \tag{4.2.3}$$

where $\Phi(t)$ is the matrix-valued solution to the homogeneous problem. Now, by using Theorem 4.5, we can find two solutions, say, $y_1(t)$ and $y_2(t)$, to the homogeneous problem associated with (4.2.1). Since the vectorized form (4.2.2) is found via $x_1 = y$, $x_2 = y'$, this means that a matrix-valued solution to the homogeneous problem is given by

$$\Phi(t) = \begin{pmatrix} y_1(t) & y_2(t) \\ y_1'(t) & y_2'(t) \end{pmatrix}.$$

Since

$$\Phi(t)^{-1} = \frac{1}{\det \Phi(t)} \begin{pmatrix} y_2'(t) & -y_2(t) \\ -y_1'(t) & y_1(t) \end{pmatrix},$$

The matrix $\Phi(t)$ is the Wronskian of the two solutions (see section 1.6.3).

we can rewrite (4.2.3) in terms of the solutions y_1 and y_2 as

$$x_p(t) = \Phi(t) \int \frac{1}{\det \Phi(t)} \begin{pmatrix} y_2'(t) & -y_2(t) \\ -y_1'(t) & y_1(t) \end{pmatrix} \begin{pmatrix} 0 \\ f(t) \end{pmatrix} dt$$

$$= \begin{pmatrix} y_1(t) & y_2(t) \\ y_1'(t) & y_2'(t) \end{pmatrix} \int \frac{1}{\det \Phi(t)} \begin{pmatrix} -f(t)y_2(t) \\ f(t)y_1(t) \end{pmatrix} dt.$$

After performing one more matrix/vector multiplication, we see that

$$x_p(t) = -\left(\int \frac{f(t)y_2(t)}{\det \Phi(t)} dt \right) \begin{pmatrix} y_1(t) \\ y_1'(t) \end{pmatrix} + \left(\int \frac{f(t)y_1(t)}{\det \Phi(t)} dt \right) \begin{pmatrix} y_2(t) \\ y_2'(t) \end{pmatrix}.$$

Taking the first component of this vector-valued solution then yields the following result:

Variation of parameters

Theorem 4.9. *A particular solution for the second-order ODE*

$$y'' + a_1 y' + a_0 y = f(t)$$

is given by

$$y_p(t) = -\left(\int \frac{f(t)y_2(t)}{\det \Phi(t)} dt \right) y_1(t) + \left(\int \frac{f(t)y_1(t)}{\det \Phi(t)} dt \right) y_2(t).$$

Here $y_1(t)$ and $y_2(t)$ are homogeneous solutions found via Theorem 4.5, and $\Phi(t)$ is the matrix

$$\Phi(t) = \begin{pmatrix} y_1(t) & y_2(t) \\ y_1'(t) & y_2'(t) \end{pmatrix}.$$

Example 4.10. Let us find the general solution to the second-order ODE

$$y'' + y = \sec(t).$$

Since the characteristic polynomial is $p(\lambda) = \lambda^2 + 1$, by using Theorem 4.5, two solutions to the homogeneous problem are given by

$$y_1(t) = \cos(t), \quad y_2(t) = \sin(t).$$

Using Theorem 4.9, the matrix $\Phi(t)$ is given by

$$\Phi(t) = \begin{pmatrix} \cos(t) & \sin(t) \\ -\sin(t) & \cos(t) \end{pmatrix} \rightsquigarrow \det \Phi(t) \equiv 1;$$

4.2. The particular solution

hence, the particular solution is

$$y_p(t) = -\cos(t)\int \sec(t)\sin(t)\,dt + \sin(t)\int \sec(t)\cos(t)\,dt$$
$$= \cos(t)\ln(\cos(t)) + t\sin(t).$$

The general solution is the sum of the homogeneous and particular solutions:

$$y(t) = c_1 \cos(t) + c_2 \sin(t) + \cos(t)\ln(\cos(t)) + t\sin(t). \blacksquare$$

Example 4.11. Consider

$$y'' + 4y' + 3y = 2\sin(2t).$$

Since the characteristic polynomial is

$$p(\lambda) = \lambda^2 + 4\lambda + 3 = (\lambda+1)(\lambda+3),$$

the two homogeneous solutions are

$$y_1(t) = e^{-t}, \quad y_2(t) = e^{-3t}.$$

Using Theorem 4.9, the matrix-valued solution is

$$\Phi(t) = \begin{pmatrix} e^{-t} & e^{-3t} \\ -e^{-t} & -3e^{-3t} \end{pmatrix} \rightsquigarrow \det \Phi(t) = -2e^{-4t}.$$

The particular solution has the form

$$y_p(t) = -e^{-t}\int \frac{2\sin(2t)e^{-3t}}{-2e^{-4t}}\,dt + e^{-3t}\int \frac{2\sin(2t)e^{-t}}{-2e^{-4t}}\,dt$$
$$= e^{-t}\int e^{t}\sin(2t)\,dt - e^{-3t}\int e^{3t}\sin(2t)\,dt$$
$$= -\frac{16}{65}\cos(2t) - \frac{2}{65}\sin(2t).$$

(The evaluations were done using SAGE.) The general solution is the sum of homogeneous and particular solutions:

$$y(t) = c_1 e^{-t} + c_2 e^{-3t} - \frac{16}{65}\cos(2t) - \frac{2}{65}\sin(2t). \blacksquare$$

The argument leading to Theorem 4.9 does not require that the solutions to the homogeneous problem be explicitly given as in Theorem 4.5. In other words, it can also be used to solve the problem

$$y'' + a_1(t)y' + a_0(t)y = f(t).$$

The problem, of course, is that in general, we do not know how to explicitly solve the homogeneous problem

$$y'' + a_1(t)y' + a_0(t)y = 0.$$

However, if the homogeneous solutions are available, we can proceed (e.g., see Exercise 4.2.6).

Example 4.12. Consider the second-order ODE

$$t^2 y'' - 6y = t^5 \ln(t).$$

It can be checked that two solutions to the homogeneous problem are

$$y_1(t) = \frac{1}{t^2}, \quad y_2(t) = t^3.$$

The matrix-valued solution is then

$$\Phi(t) = \begin{pmatrix} 1/t^2 & t^3 \\ -2/t^3 & 3t^2 \end{pmatrix} \rightsquigarrow \det \Phi(t) \equiv 5.$$

In order to use the variation-of-parameters formula, we must first write the ODE in standard form:

$$y'' - \frac{6}{t^2} y = t^3 \ln(t).$$

The particular solution has the form

$$\begin{aligned} y_p(t) &= -\frac{1}{t^2} \int \frac{t^6 \ln(t)}{5} \, dt + t^3 \int \frac{t \ln(t)}{5} \, dt \\ &= -\frac{1}{5t^2} \int t^6 \ln(t) \, dt + \frac{1}{5} t^3 \int t \ln(t) \, dt \\ &= \frac{1}{14} t^5 \ln(t) - \frac{9}{196} t^5. \end{aligned}$$

(The evaluations were done using SAGE.) The general solution is the sum of the homogeneous and particular solutions:

$$y(t) = \frac{c_1}{t^2} + c_2 t^3 + \frac{1}{14} t^5 \ln(t) - \frac{9}{196} t^5.$$

Finally, suppose that there is the initial condition $y(1) = -3$, $y'(1) = 7$. This leads to a linear system of equations:

$$-3 = y(1) = c_1 + c_2 - \frac{9}{196},$$

$$7 = y'(1) = -2c_1 + 3c_2 + \frac{1}{14} - \frac{45}{196} \rightsquigarrow c_1 = -\frac{157}{49}, \ c_2 = \frac{1}{4}.$$

The solution to the IVP is

$$y(t) = -\frac{157/49}{t^2} + \frac{1}{4} t^3 + \frac{1}{14} t^5 \ln(t) - \frac{9}{196} t^5. \quad \blacksquare$$

4.2.2 ▪ Undetermined coefficients

Let us now find the particular solution using the method of undetermined coefficients. The idea is exactly the same as in section 2.5.3 (first-order scalar ODEs) and section 3.6.2 (first-order systems of ODEs), namely, guess the solution form based on the functional form of the forcing function. As in section 2.5.3, in order for the

4.2. The particular solution

method to work, the forcing function must be a linear combination of functions of the form $p(t)e^{at}\cos(bt)$ and $p(t)e^{at}\sin(bt)$, where $p(t)$ is a polynomial. We will again illustrate the method via a couple of examples. In all of the examples, the homogeneous problem will be

$$y'' + 4y' + 3y = 0 \quad \leadsto \quad y_h(t) = c_1 e^{-t} + c_2 e^{-3t}.$$

Example 4.13. First consider

$$y'' + 4y' + 3y = 5e^t.$$

The guess for the particular solution is

$$y_p(t) = a_0 e^t,$$

which, after plugging into the ODE and simplifying, yields the algebraic equation

$$8a_0 = 5 \quad \leadsto \quad a_0 = \frac{5}{8}.$$

The particular solution is

$$y_p(t) = \frac{5}{8} e^t,$$

and the general solution is the sum of the homogeneous and particular solutions:

$$y(t) = c_1 e^{-t} + c_2 e^{-3t} + \frac{5}{8} e^t. \quad \blacksquare$$

Example 4.14. For our second example, consider

$$y'' + 4y' + 3y = 7e^{-t}.$$

We would like to guess $y_p(t) = a_0 e^{-t}$. However, this guess is also a homogeneous solution, so we know that it will not work. Using the idea first presented in section 2.5.3, we modify the initial guess by multiplying it by t to get

$$y_p(t) = a_0 t e^{-t}.$$

Plugging this guess into the ODE and simplifying yields the algebraic equation

$$2a_0 = 7 \quad \leadsto \quad a_0 = \frac{7}{2}.$$

The particular solution is

$$y_p(t) = \frac{7}{2} t e^{-t},$$

and the general solution is the sum of the homogeneous and particular solutions:

$$y(t) = c_1 e^{-t} + c_2 e^{-3t} + \frac{7}{2} t e^{-t}.$$

Now suppose that there is the initial condition $y(0) = -4$, $y'(0) = 5$. This leads to the linear system of equations:

$$-4 = y(0) = c_1 + c_2, \quad 5 = y'(0) = -c_1 - 3c_2 + \frac{7}{2} \quad \rightsquigarrow \quad c_1 = -\frac{21}{4}, \quad c_2 = \frac{5}{4}.$$

The solution to the IVP is

$$y(t) = -\frac{21}{4} e^{-t} + \frac{5}{4} e^{-3t} + \frac{7}{2} t e^{-t}. \quad \blacksquare$$

Example 4.15. For our last example in which we will actually find the solution, consider

$$y'' + 4y' + 3y = 4\cos(2t).$$

This problem is closely related to that presented in Example 4.11. Our guess for the particular solution is

$$y_p(t) = a_0 \cos(2t) + a_1 \sin(2t).$$

Plugging this guess into the ODE, equating the coefficients associated with $\sin(2t)$ and $\cos(2t)$, and simplifying yields the algebraic system

$$-a + 8b = 4, \quad -8a - b = 0 \quad \rightsquigarrow \quad \begin{pmatrix} -1 & 8 \\ -8 & -1 \end{pmatrix} a = \begin{pmatrix} 4 \\ 0 \end{pmatrix}.$$

The solution to the linear system is

$$a = \begin{pmatrix} -1 & 8 \\ -8 & -1 \end{pmatrix}^{-1} \begin{pmatrix} 4 \\ 0 \end{pmatrix} = \frac{1}{65} \begin{pmatrix} -4 \\ 32 \end{pmatrix},$$

so that the particular solution is

$$y_p(t) = -\frac{4}{65} \cos(2t) + \frac{32}{65} \sin(2t).$$

The general solution is the sum of the homogeneous and particular solutions:

$$y(t) = c_1 e^{-t} + c_2 e^{-3t} - \frac{4}{65} \cos(2t) + \frac{32}{65} \sin(2t). \quad \blacksquare$$

Let us finally consider the problem of simply determining the form of the particular solution for a more complicated homogeneous problem. For all of the given forcing functions, it will be assumed that the homogeneous problem is the ninth-order ODE

$$(D^2 + 4)(D - 1)^3 (D^2 + 6D + 25)^2 y = 0 \quad \rightsquigarrow$$
$$y_h(t) = c_1 \cos(2t) + c_2 \sin(2t) + (c_3 + c_4 t + c_5 t^2) e^t$$
$$+ (c_6 + c_7 t) e^{-3t} \cos(4t) + (c_8 + c_9 t) e^{-3t} \sin(4t).$$

For each forcing function $f(t)$, we will simply write the form of the corresponding particular solution $y_p(t)$. The key in the end is that the guess must have nothing in common with the homogeneous solution:

(a) $f(t) = 7e^{-3t} + t^2 \rightsquigarrow y_p(t) = a_0 e^{-3t} + a_1 + a_2 t + a_3 t^2,$

(b) $f(t) = 5t\sin(4t) \rightsquigarrow y_p(t) = (a_0 + a_1 t)\cos(4t) + (a_2 + a_3 t)\sin(4t),$

(c) $f(t) = 17t^2 e^t \rightsquigarrow y_p(t) = t^3(a_0 + a_1 t + a_2 t^2)e^t,$ and

(d) $f(t) = 22te^{-3t}\sin(4t) \rightsquigarrow y_p(t) = t^2(a_0 + a_1 t)e^{-3t}\cos(4t) + t^2(a_2 + a_3 t)e^{-3t}\sin(4t).$

Regarding examples (c) and (d), the power of t that is seen as a multiplicative prefactor was chosen to be the minimal power so that no part of the guess corresponded to any part of a homogeneous solution.

Exercises

Exercise 4.2.1. *Find the general solution for each of the following second-order ODEs.*

(a) $y'' - 2y' + 5y = 8\cos(3t)$

(b) $y'' - 3y' - 28y = 3t + e^t$

(c) $y'' + 7y' + 10y = 6e^{-2t}$

(d) $y'' - 9y' + 18y = -7e^{-t}\cos(2t)$

(e) $y'' + 64y = -4\sin(8t)$

(f) $y'' - 10y' + 25y = 17te^{5t}$

(g) $y'' + 6y' + 13y = 5e^{-3t}$

Exercise 4.2.2. *Solve the following IVPs.*

(a) $y'' - 2y' + 5y = 8\cos(3t); \quad y(0) = -2, \; y'(0) = -4$

(b) $y'' - 3y' - 28y = 3t + e^t; \quad y(0) = 3, \; y'(0) = 2$

(c) $y'' + 7y' + 10y = 6e^{-2t}; \quad y(0) = 5, \; y'(0) = -1$

(d) $y'' - 9y' + 18y = -7e^{-t}\cos(2t); \quad y(0) = -1, \; y'(0) = 5$

(e) $y'' + 64y = -4\sin(8t); \quad y(0) = 6, \; y'(0) = 9$

(f) $y'' - 10y' + 25y = 17te^{5t}; \quad y(0) = 2, \; y'(0) = 3$

(g) $y'' + 6y' + 13y = 5e^{-3t}; \quad y(0) = -5, \; y'(0) = 7$

Exercise 4.2.3. *For each of the following problems, state the general form of the particular solution that you would guess using the method of undetermined coefficients (do not actually find the particular solution).*

(a) $y'' + 2y' + y = 5te^{-t} + 6\sin(2t)$

(b) $y'' - 3y' - 4y = 3t\cos(4t) + 4e^{2t}\sin(3t)$

(c) $y'' + 2y' + 2y = 7e^{2t} + 5e^{-t}\cos(t)$

Exercise 4.2.4. *For each of the following problems, find the particular solution using SAGE.*

(a) $y'' + 2y' + y = 5te^{-t} + 6\sin(2t)$

(b) $y'' - 3y' - 4y = 3t\cos(4t) + 4e^{2t}\sin(3t)$

(c) $y'' + 2y' + 2y = 7e^{2t} + 5e^{-t}\cos(t)$

Exercise 4.2.5. *For each of the following problems, state the general form of the particular solution that you would guess using the method of undetermined coefficients (do not actually find the particular solution).*

(a) $(D-1)^2(D+3)(D+5)(D^2+2D+5)y = 5t^2 e^t$

(b) $(D+1)^3(D+4)(D^2+25)(D^2+4D+13)^4 y = 3t\cos(3t) - 9e^{-2t}\sin(3t)$

(c) $(D-5)^2(D-3)(D+5)^4(D^2-8D+41)y = -4t^2 + 3e^{5t}$

Exercise 4.2.6. *Find the general solution for each of the following second-order ODEs.*

(a) $y'' - y = 6/(1+e^t)$

(b) $t^2 y'' - ty' + y = 4t \ln(t)$ *(two solutions to the homogeneous problem are given by $y_1(t) = t$ and $y_2(t) = t\ln(t)$)*

(c) $t^2 y'' - 2ty' + 2y = t^3/(1+t^2)$ *(two solutions to the homogeneous problem are given by $y_1(t) = t$ and $y_2(t) = t^2$)*

Exercise 4.2.7. *The second-order Cauchy–Euler equation is given by*

$$t^2 y'' + a_1 t y' + a_0 y = f(t).$$

On setting

$$s = \ln(t) \quad \Leftrightarrow \quad t = e^s,$$

the ODE becomes the second-order constant coefficient problem

$$\frac{d^2 y}{ds^2} + (a_1 - 1)\frac{dy}{ds} + a_0 y = f(e^s).$$

(a) *Verify the above equivalence between the two equations.*

(b) *If the initial values for the Cauchy–Euler equation are*

$$y(1) = y_0, \quad y'(1) = y_1,$$

verify that the corresponding initial values for the transformed problem are

$$y(0) = y_0, \quad \frac{dy}{ds}(0) = y_1.$$

4.2. The particular solution

(c) If the homogeneous solution to the transformed problem is
$$y_h(s) = c_1 e^{\lambda_1 s} + c_2 e^{\lambda_2 s},$$
what is the homogeneous solution, $y_h(t)$, to the Cauchy–Euler equation?

(d) If the homogeneous solution to the transformed problem is
$$y_h(s) = c_1 e^{as} \cos(bs) + c_2 e^{as} \sin(bs),$$
what is the homogeneous solution, $y_h(t)$, to the Cauchy–Euler equation?

Exercise 4.2.8. *Find the general solution for each of the following second-order ODEs. (Hint: Use the result of Exercise 4.2.7.)*

(a) $t^2 y'' - t y' + y = 0$

(b) $t^2 y'' - 2t y' + 2y = 0$

(c) $t^2 y'' + 4t y' + 2y = 0$

(d) $t^2 y'' - 7t y' + 16y = 0$

(e) $t^2 y'' + t y' + 9y = 0$

(f) $t^2 y'' - 3t y' + 29y = 0$

Exercise 4.2.9. *Find the general solution for each of the following second-order ODEs. (Hint: Use the result of Exercise 4.2.7.)*

(a) $t^2 y'' - t y' + y = 5t^2$

(b) $t^2 y'' - 2t y' + 2y = -8/t^3$

(c) $t^2 y'' + 4t y' + 2y = 3\ln(t)$

(d) $t^2 y'' - 7t y' + 16y = 3 - 2t$

(e) $t^2 y'' + t y' + 9y = 4\cos(2\ln(t))$

(f) $t^2 y'' - 3t y' + 29y = -8\sin(\ln(t))$

Exercise 4.2.10. *Solve the following IVPs. (Hint: Use the result of Exercise 4.2.7.)*

(a) $t^2 y'' - t y' + y = 0; \quad y(1) = 3,\ y'(1) = -2$

(b) $t^2 y'' - 2t y' + 2y = 0; \quad y(1) = 5,\ y'(1) = 4$

(c) $t^2 y'' + 4t y' + 2y = 0; \quad y(1) = 8,\ y'(1) = -1$

(d) $t^2 y'' - 7t y' + 16y = 3 - 2t; \quad y(1) = 1,\ y'(1) = 6$

(e) $t^2 y'' + t y' + 9y = 4\cos(2\ln(t)); \quad y(1) = -3,\ y'(1) = -5$

(f) $t^2 y'' - 3t y' + 29y = -8\sin(\ln(t)); \quad y(1) = -7,\ y'(1) = -9$

4.3 • Case studies

4.3.1 • Classification result, revisited

The physical system to be studied in the first two case studies is the damped mass-spring system (see section 3.1.3). The equation of motion is

$$y'' + 2by' + \omega_0^2 y = f(t), \qquad (4.3.1)$$

where $b > 0$, and the coefficients and functions in the ODE are related to the physical terms via

$$2b = \frac{c}{m}, \quad \omega_0^2 = \frac{k}{m}, \quad f(t) = \frac{1}{m} F(t).$$

This equation also appears when studying RLC circuits, glucose and diabetes, and the simple pendulum, among others. Our goal here is to classify the dynamics for the homogeneous (unforced) problem as a function of the parameters.

The homogeneous problem is

$$y'' + 2by' + \omega_0^2 y = 0,$$

which on setting $x_1 = y$, $x_2 = y'$ is equivalent to the system

$$x' = \begin{pmatrix} 0 & 1 \\ -\omega_0^2 & -2b \end{pmatrix}.$$

The characteristic polynomial is

$$p(\lambda) = \lambda^2 + 2b\lambda + \omega_0^2,$$

which has the zeros

$$\lambda = \lambda_\pm = -b \pm \sqrt{b^2 - \omega_0^2}.$$

The solution behavior depends on the relationship between b and ω_0. A summary cartoon is provided in Figure 4.1.

First suppose that $b > \omega_0$, so the zeros are real. The homogeneous solution is

$$y_h(t) = c_1 e^{\lambda_- t} + c_2 e^{\lambda_+ t}.$$

The zeros satisfy

$$\lambda_- < b < \lambda_+ < 0,$$

so $y_h(t) \to 0$ as $t \to +\infty$. Since the zeros are real and negative, in the phase plane the origin is a stable node. Generally, we expect solutions to decay at the slower rate, $y_h(t) \sim c_2 e^{\lambda_+ t}$ for $t \gg 0$. We say that the system is *overdamped*.

Now suppose that $0 \leq b < \omega_0$ so that the zeros have nonzero imaginary part. Setting $\omega_1^2 = \omega_0^2 - b^2$, we have

$$\lambda_\pm = -b \pm i\omega_1.$$

The homogeneous solution is

$$y_h(t) = c_1 e^{-bt} \cos(\omega_1 t) + c_2 e^{-bt} \sin(\omega_1 t).$$

If $b > 0$, the solution decays to zero as $t \to +\infty$ at exponential rate $|y_h(t)| = \mathcal{O}(e^{-bt})$. In the phase plane, the origin is a stable spiral. We say that the system

4.3. Case studies

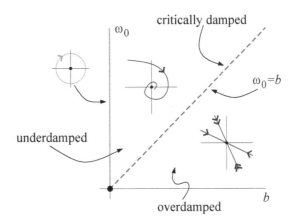

Figure 4.1. *A cartoon of the dynamics in the $x_1 x_2$-plane for the homogeneous problem associated with (4.3.1). In the context of the mass-spring problem, $x_1 = y$ is the position of the mass, and $x_2 = y'$ is the velocity. When there is no damping ($b = 0$), the origin is a linear center, and the solutions are periodic with period $2\pi/\omega_0$. When there is damping ($b > 0$), the origin is stable: underdamping with oscillatory decay for $0 < b < \omega_0$ and overdamping with no oscillations for $b > \omega_0$. Solutions generically decay most quickly when the damping is critical, $b = \omega_0$.*

is *underdamped*. On the other hand, if $b = 0$ (no damping), the origin is a linear center. In this case, $\omega_1 = \omega_0$, so the period of the solution is $2\pi/\omega_0$. We say that ω_0 is the *natural frequency* of the undamped system.

Finally, suppose that $b = \omega_0$, so $\lambda_- = \lambda_+ = -b$. This is the case of *critical damping*. The homogeneous solution is now

$$y_h(t) = (c_1 + c_2 t) e^{-bt}.$$

Since $b > 0$, the solution still decays to zero as $t \to +\infty$.

In conclusion, if $b > 0$, the homogeneous solution decays to zero as $t \to +\infty$; i.e., the homogeneous solution is *transient*. This means that if we are interested in the dynamics for large t, then the initial condition is unimportant. Moreover, a comparison of the decay rates in each case shows that the solution will generically decay most quickly in the case of critical damping, $b = \omega_0$ (see Figure 4.2 for a plot of the decay rates as a function of b for fixed ω). In practice, the decay rate will be optimized as long as the damping constant, $b = c/(2m)$, is sufficiently close to the natural frequency, $\omega_0 = \sqrt{k/m}$. On the other hand, if $b = 0$, the homogeneous solution is not transient, and the initial condition will play a role in the long time dynamics.

4.3.2 ▪ Undamped mass-spring system

We first consider the undamped problem, $b = 0$, so the equation of motion is this system is equivalent to

$$y'' + \omega_0^2 y = f(t).$$

The constant ω_0 is the natural frequency for the undamped system. We will solve this system under the assumption that the forcing is sinusoidal,

$$f(t) = f_0 \cos(\omega t),$$

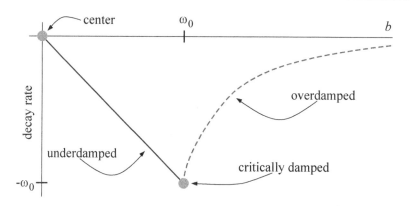

Figure 4.2. *A cartoon of the plot of the exponential decay rate (if $y(t) = e^{at}$ for $a < 0$, the decay rate is a) of the homogeneous solution as a function of the damping constant b for fixed natural frequency ω_0. Note that the rate is maximized for the case of critical damping.*

where f_0 is the amplitude of the forcing and ω is the frequency. In conclusion, we will study the solution behavior to the ODE

$$y'' + \omega_0^2 y = f_0 \cos(\omega t). \tag{4.3.2}$$

As we saw in the previous subsection, the homogeneous solution is

$$y_h(t) = c_1 \cos(\omega_0 t) + c_2 \sin(\omega_0 t).$$

The homogeneous solution is periodic with frequency, ω_0, hence the moniker "natural frequency." Since the solution is not transient, the initial condition will be important in determining solution behavior. Now consider the particular solution. Using the method of undetermined coefficients—and assuming that $\omega \neq \omega_0$!—we know that the particular solution is of the form

$$y_p(t) = a_0 \cos(\omega t) + a_1 \sin(\omega t).$$

We will later worry about the case that the forcing frequency is the same as the natural frequency, i.e., $\omega = \omega_0$. Plugging this guess into (4.3.2) and collecting terms yields

$$\left(-\omega^2 a_0 + \omega_0^2 a_0 - f_0\right)\cos(\omega t) + \left(-\omega^2 a_1 + \omega_0^2 a_1\right)\sin(\omega t) = 0,$$

from which we get the linear system

$$(\omega_0^2 - \omega^2) a_0 = f_0, \quad (\omega_0^2 - \omega^2) a_1 = 0.$$

Thus,

$$a_0 = \frac{f_0}{\omega_0^2 - \omega^2}, \quad a_1 = 0,$$

so the particular solution is

$$y_p(t) = \frac{f_0}{\omega_0^2 - \omega^2} \cos(\omega t).$$

4.3. Case studies

We clearly see here that the assumption $\omega \neq \omega_0$ is necessary for our solution guess. The general solution is the sum of the homogeneous and particular solutions:

$$y(t) = c_1 \cos(\omega_0 t) + c_2 \sin(\omega_0 t) + \frac{f_0}{\omega_0^2 - \omega^2} \cos(\omega t).$$

We will now focus solely on the effect of forcing on the solution behavior, which means that we will consider the IVP with initial condition $y(0) = y'(0) = 0$. Physically, the mass is initially at its equilibrium position and is not moving when the forcing is applied. We use this initial condition to find the constants c_1 and c_2:

$$0 = y(0) = c_1 + \frac{f_0}{\omega_0^2 - \omega^2}, \quad 0 = y'(0) = \omega_0 c_2.$$

We find

$$c_1 = -\frac{f_0}{\omega_0^2 - \omega^2}, \quad c_2 = 0,$$

so on collecting terms, the solution to the IVP is

$$y(t) = \frac{f_0}{\omega_0^2 - \omega^2} [\cos(\omega t) - \cos(\omega_0 t)]. \tag{4.3.3}$$

While we have a formula for the solution, we do not yet really understand the solution behavior. Let us first consider the solution behavior for $\omega \gg \omega_0$. Rewrite the solution as

$$y(t) = \underbrace{\frac{f_0}{\omega^2 - \omega_0^2} \cos(\omega_0 t)}_{y_{\text{slow}}(t)} - \underbrace{\frac{f_0}{\omega^2 - \omega_0^2} \cos(\omega t)}_{y_{\text{rapid}}(t)}.$$

The solution, which overall is small with maximum bounded above to leading order by $2f_0/\omega^2 \ll 1$, is the sum of two period functions: the first, $y_{\text{slow}}(t)$, is a slowly varying cosine with period $2\pi/\omega_0$, and the second, $y_{\text{rapid}}(t)$, is a rapidly varying cosine wave with period $2\pi/\omega \ll 2\pi/\omega_0$. The sum of these two functions is essentially the rapidly oscillating function $y_{\text{rapid}}(t)$ with a slowly varying mean $y_{\text{slow}}(t)$ (see Figure 4.3). Since the forcing is $f_0 \cos(\omega t)$ and the rapidly varying part of the solution is a positive multiple of $-\cos(\omega t)$, the solution is out of phase with the forcing.

Now let us consider the solution behavior when the forcing frequency is close to the natural frequency. Unfortunately, in its current form, the solution formula (4.3.3) does not give a clear explanation as to the motion of the mass. We now find another equivalent version of the solution formula (4.3.3). Using the trigonometric identity

$$\cos(\theta_1) - \cos(\theta_2) = 2 \sin\left(\frac{\theta_2 - \theta_1}{2}\right) \sin\left(\frac{\theta_1 + \theta_2}{2}\right),$$

we can rewrite the solution as

$$y(t) = \frac{2f_0}{\omega_0^2 - \omega^2} \sin\left(\frac{\omega_0 - \omega}{2} t\right) \sin\left(\frac{\omega_0 + \omega}{2} t\right). \tag{4.3.4}$$

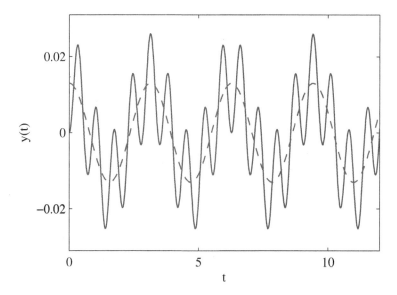

Figure 4.3. *A plot of the solution to the sinusoidally forced mass-spring for the initial condition $y(0) = y'(0) = 0$ with $f_0 = 1$, $\omega_0 = 2.0$, and $\omega = 9.0$. The dashed curve is the slow modulation of the mean of the solution, and the solid curve is the actual solution.*

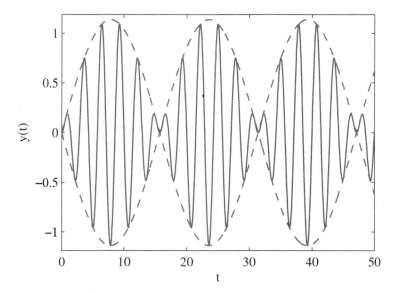

Figure 4.4. *A plot of the solution to the sinusoidally forced mass spring for the initial condition $y(0) = y'(0) = 0$ with $f_0 = 1$, $\omega_0 = 2.0$, and $\omega = 2.4$. The dashed curve is the amplitude of the solution (see Figure 4.5), and the solid curve is the actual solution. The beating phenomena is pronounced here.*

The solution is a modulated sinusoid (see Figure 4.4). The amplitude is given by the sinusoidal function

$$A(t) = \frac{2f_0}{\omega_0^2 - \omega^2} \sin\left(\frac{\omega_0 - \omega}{2} t\right).$$

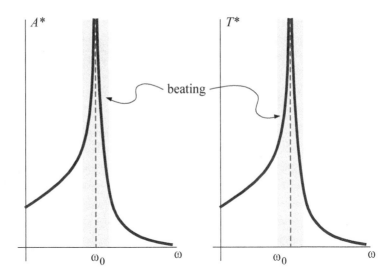

Figure 4.5. *A cartoon of the plot of the amplitude $A^*(\omega)$ is given by the thick curve in the left figure, and a cartoon of the plot of the period of the amplitude function $T^*(\omega)$ is given by the thick curve in the right figure. In both figures, the domain where the beating phenomena is pronounced is shaded.*

The amplitude function has amplitude

$$A^*(\omega) := \frac{2f_0}{|\omega_0^2 - \omega^2|}$$

and period

$$T^*(\omega) := \frac{4\pi}{|\omega_0 - \omega|}.$$

The plots of these two functions are given in Figure 4.5. Note that both the amplitude and the period become arbitrarily

(a) large as the forcing frequency approaches the natural frequency, and

(b) small as the forcing frequency becomes large.

Moreover, when the forcing frequency is close to the natural frequency, the solution operates on two time scales:

(a) the fast time scale $|\omega_0 + \omega|t/2 \sim \omega_0 t$, which is seen in the relatively rapid internal oscillations of the solution; and

(b) the slow time scale $|\omega_0 - \omega|t/2$, which is seen in the slow variation of the amplitude function $A(t)$.

If $|\omega - \omega_0|$ is small, the solution will exhibit a *beating phenomena*. The mass will oscillate with a frequency close to the natural frequency, but the amplitude of the oscillation will change periodically and slowly over a much longer time scale.

What is the solution when $\omega = \omega_0$, i.e., when there is *resonant forcing*? We find it via a limiting process. Setting

$$h = \frac{\omega_0 - \omega}{2} t,$$

we can rewrite the amplitude function as

$$A(t) = \frac{f_0}{\omega_0 + \omega} t \frac{\sin(h)}{h}.$$

Since $\omega \to \omega_0$ implies that $h \to 0$, we have

$$\lim_{\omega \to \omega_0} A(t) = \frac{f_0}{2\omega_0} t \lim_{h \to 0} \frac{\sin(h)}{h} = \frac{f_0}{2\omega_0} t.$$

Thus, for the solution as written in (4.3.4), we have

$$\lim_{\omega \to \omega_0} y(t) = \lim_{\omega \to \omega_0} A(t) \lim_{\omega \to \omega_0} \sin\left(\frac{\omega_0 + \omega}{2} t\right) = \frac{f_0}{2\omega_0} t \sin(\omega_0 t).$$

The amplitude of the sinusoidal wave, $f_0 t / 2\omega_0$, becomes large as t increases, which leads to an eventual failure of the spring (unless there are nonlinear effects that make the model invalid for large oscillations).

Of course, in an actual system the forcing need not be exactly at resonance in order for the spring to fail. If ω is chosen so that $A^*(\omega)$ is larger than the spring can stretch and still recover, then the spring will fail. Referring again to Figure 4.5, we then see that for a given spring, there is a number δ such that if forcing frequency satisfies $|\omega - \omega_0| < \delta$, then the spring will fail.

Remark 4.16. *The Tacoma Narrows Bridge collapse in 1940 is a famous example of resonant forcing leading to system failure (see the YouTube video[9]). On the day that the bridge collapsed, there was a relatively steady and strong wind transverse to the length of the bridge, so the structure was seemingly NOT subjected to a periodic resonant forcing. However, when the wind blew over the bridge, it produced vortices through a process called vortex shedding. The shedding of these vortices produced an alternating and periodic vertical forcing that was both strong and in resonance with the normal modes of the bridge. The interested reader should consult Billah and Scanlan [4] and the online PDF[10] by J. Koughan for more details.*

4.3.3 ▪ Damped mass-spring system

We have considered the dynamics of the mass-spring system under the assumption that there is no damping. We now consider the problem in which damping is present. Since the homogeneous solution is *transient*, i.e., it decays to zero as $t \to +\infty$, we will not worry about the effect of the initial condition on the solution. We will, as in the previous section 4.3.2, assume that the forcing is sinusoidal. In conclusion, we will focus solely on finding and analyzing the particular solution to

$$y'' + 2by' + \omega_0^2 y = f_0 \cos(\omega t), \tag{4.3.5}$$

as this solution will give us the behavior of the full solution for $t \gg 0$.

Using the method of undetermined coefficients, we know that the particular solution is of the form

$$y_p(t) = a_0 \cos(\omega t) + a_1 \sin(\omega t).$$

---

[9]http://www.youtube.com/watch?v=3mclp9QmCGs.
[10]http://www.aapt.org/Store/upload/tacoma_narrows2.pdf.

Plugging this guess into (4.3.5) and collecting the cosine and sine terms yields

$$[(\omega_0^2-\omega^2)a_0+2b\omega a_1]\cos(\omega t)+[-2b\omega a_0+(\omega_0^2-\omega^2)a_1]\sin(\omega t)=f_0\cos(\omega t).$$

Equating the coefficients yields the linear system for the coefficients,

$$(\omega_0^2-\omega^2)a_0+2b\omega a_1=f_0, \quad -2b\omega a_0+(\omega_0^2-\omega^2)a_1=0,$$

which in matrix form is

$$\begin{pmatrix} \omega_0^2-\omega^2 & 2b\omega \\ -2b\omega & \omega_0^2-\omega^2 \end{pmatrix} a = \begin{pmatrix} f_0 \\ 0 \end{pmatrix} \rightsquigarrow a = \frac{f_0}{\Delta}\begin{pmatrix} \omega_0^2-\omega^2 \\ 2b\omega \end{pmatrix}.$$

Here the positive constant Δ is defined by

$$\Delta=(\omega_0^2-\omega^2)^2+4b^2\omega^2.$$

The particular solution is then

$$y_p(t) = \frac{(\omega_0^2-\omega^2)f_0}{\Delta}\cos(\omega t) + \frac{2b\omega f_0}{\Delta}\sin(\omega t)$$

$$= \frac{f_0}{\sqrt{\Delta}}\cos(\omega t - \phi), \quad \tan\phi = \frac{2b\omega}{\omega_0^2-\omega^2}.$$

We know that for $t \gg 0$, the solution satisfies $y(t) \sim y_p(t)$. The amplitude of oscillation for the asymptotic solution is

$$A^*(\omega) = \frac{f_0}{\sqrt{\Delta}},$$

and although the frequency of oscillation is exactly that of the forcing frequency, there is the phase shift $\phi^*(\omega)$, which satisfies

$$\phi^*(\omega) = \tan^{-1}\left(\frac{2b\omega}{\omega_0^2-\omega^2}\right).$$

The amplitude is maximized when

$$\frac{dA^*}{d\omega}(\omega) = 0 \rightsquigarrow \omega = \omega^* := \sqrt{\omega_0^2 - 2b^2} < \omega_0.$$

If $b \geq \omega_0/\sqrt{2}$, then the amplitude is maximized at $\omega = 0$. Note that with respect to the homogeneous problem, at the transition point $\omega_0 = \sqrt{2}b$, the spring is underdamped (recall Figure 4.1). Since

$$A^*(\omega^*) = \frac{f_0}{2b\sqrt{\omega_0^2-b^2}}, \quad \omega_0^2 > 2b^2,$$

the maximal amplitude becomes arbitrarily large as $b \to 0^+$. It is also true that $\omega^* \to \omega_0$ as $b \to 0^+$; hence, as $b \to 0^+$, the plot of $A^*(\omega)$ closely approximates that given for the undamped problem in Figure 4.5, and the maximal amplitude occurs roughly for the case of resonant forcing. For $\omega > \omega^*$, the amplitude decreases

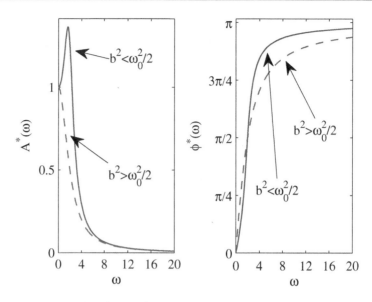

Figure 4.6. *Plots of the amplitude $A^*(\omega)$ (left figure) and the phase $\phi^*(\omega)$ (right figure) for the damped and forced harmonic oscillator. For these plots, we set $f_0 = 4.0$ and $\omega_0 = 2.0$. The solid curve corresponds to weak damping, $b^2 < \omega_0^2/2$ (we set $b = 0.8$ to generate this specific curve), and the dashed curve corresponds to relatively strong damping, $b^2 > \omega_0^2/2$ (we set $b = 2.0$ to generate this curve). Note that the amplitude decreases as the forcing frequency increases. Further note that when ω is small, the phase is roughly zero, so that the mass roughly oscillates in-phase with the forcing. On the other hand, if ω is large, then the phase is roughly π, so that the mass is oscillating out-of-phase with the forcing.*

monotonically as the frequency increases and satisfies $A^*(\omega) \to 0$ as $\omega \to +\infty$. As for the phase shift, it is a monotone increasing function that satisfies $\phi^*(0) = 0$, $\phi^*(\omega_0) = \pi/2$, and $\phi^*(\omega) \to \pi$ as $\omega \to +\infty$. The graphs of each of these functions are given in Figure 4.6.

We conclude by noting here that the addition of damping suppresses two effects that are present for the undamped problem. No matter the value of the forcing frequency relative to the natural frequency,

(a) the solution will not exhibit beating behavior; in other words, damping kills beating; and

(b) the amplitude of the solution is bounded; in particular, damping kills the notion that resonant forcing leads to solutions growing without bound.

Remark 4.17. *In the episode* **Shattering Subwoofer**[11] *(August 16, 2006) of the TV show* Mythbusters, *there is a discussion regarding how fast you should drive over a bumpy road in order to have the smoothest ride. The analysis given above provides a mathematical answer to the question. Suppose that we regard the suspension of the car as a (stiff) spring and the car as the mass at the end of the spring. The forcing corresponds to driving on the bumpy road, and the frequency of the forcing is determined by the speed at which the car is traveling. If the car is going slow, then the frequency is small, while if the car is moving quickly, then the frequency is large. We have seen in our*

---

[11]http://www.discovery.com/tv-shows/mythbusters/2006-episodes/.

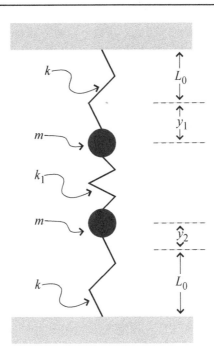

Figure 4.7. *A cartoon that depicts the coupled mass-spring problem. Two identical undamped mass-spring systems—whose motion was studied in section 4.3.2—are connected by a spring.*

analysis that as the frequency (speed) increases, the amplitude response of the mass to the forcing decreases; indeed, the amplitude response goes to zero as the frequency gets very large. Since amplitude response corresponds to how much the car is bouncing, we can conclude that for a smoother ride, it is best if the car travels as fast as possible. This is precisely what was concluded in that TV episode.

4.3.4 ▪ Coupled and undamped mass-spring system

Consider the coupled mass-spring system depicted in Figure 4.7. Here we are taking two identical mass-spring systems (spring constant k and mass m) and coupling them via a spring with a different spring constant (spring constant k_1). If the systems are uncoupled, i.e., $k_1 = 0$, then we saw in section 4.3.2 that the motion of the mass at the end of each spring is sinusoidal with frequency $\omega_0 = \sqrt{k/m}$. One can then intuit that with $k_1 > 0$, each mass will act as a resonant periodic forcing on the other mass, which by the analysis presented in section 4.3.2 means that the oscillation of each mass becomes unbounded. However, as we see in numerical simulations, this is not the case. Indeed, it appears to be the case that for small k_1, there is near-resonant forcing in that each mass exhibits the beating phenomena associated with near-resonant periodic forcing. However, the beating behavior oscillates between the two masses (see Figure 4.9). For larger values of the coupling constant k_1, the simulations suggest that the beating phenomena vanishes. Thus, our intuition is not correct, and we must analyze the equations of motion in order to understand the solution behavior.

While we will not go through the derivation of the equations of motion here, it turns out that the governing ODE is given by

$$my_1'' = -ky_1 + k_1(y_2 - y_1), \quad my_2'' = -ky_2 - k_1(y_2 - y_1). \qquad (4.3.6)$$

See Fay and Graham [17] for a nice discussion of the derivation. In matrix/vector form, we can write (4.3.6) in the form

$$y'' + Ky = 0, \quad K = \frac{1}{m}\begin{pmatrix} k+k_1 & -k_1 \\ -k_1 & k+k_1 \end{pmatrix}. \qquad (4.3.7)$$

To solve this system, we will write down a Fourier expansion for the solution $y(t)$ using the eigenvectors of the matrix K (see section 1.12).

The eigenvalues and associated eigenvectors of K are

$$\lambda_1 = \omega_n^2,\ u_1 = \begin{pmatrix} 1 \\ 1 \end{pmatrix}; \quad \lambda_2 = \omega_c^2,\ u_2 = \begin{pmatrix} 1 \\ -1 \end{pmatrix},$$

where

$$\omega_n^2 = \frac{k}{m}, \quad \omega_c^2 = \frac{k + 2k_1}{m}.$$

The first eigenvalue is the square of the natural frequency, ω_n, for the uncoupled mass-spring oscillator. The square root of the second (larger) eigenvalue, ω_c, will be denoted the frequency induced by the coupling.

We next write the vector y in terms of the eigenvectors through the Fourier expansion:

$$y(t) = z_1(t)u_1 + z_2(t)u_2.$$

By the linearity of differentiation, we have

$$y'' = z_1'' u_1 + z_2'' u_2,$$

and by the expansion result of Lemma 1.99, we have

$$Ky = \omega_n^2 z_1 u_1 + \omega_c^2 z_2 u_2.$$

Substituting these expansions into (4.3.7) and collecting terms yields

$$\left(z_1'' + \omega_n^2 z_1\right) u_1 + \left(z_2'' + \omega_c^2 z_2\right) u_2 = 0. \qquad (4.3.8)$$

We also arrive at (4.3.9) by setting $z_1 = y_1 + y_2$ and $z_2 = y_1 - y_2$ in (4.3.6) and then deriving the appropriate ODE for each new variable.

The original ODE (4.3.7) has now been written as a Fourier expansion (4.3.8). Each Fourier coefficient is a second-order scalar ODE. Since the eigenvectors are linearly independent, each time-dependent coefficient must be identically zero. We then arrive at two *uncoupled* second-order ODEs:

$$z_1'' + \omega_n^2 z_1 = 0,$$
$$z_2'' + \omega_c^2 z_2 = 0. \qquad (4.3.9)$$

These two problems were solved in the previous example, and the general solution is given by

$$z_1(t) = c_1 \cos(\omega_n t) + c_2 \sin(\omega_n t), \quad z_2(t) = c_3 \cos(\omega_c t) + c_4 \sin(\omega_c t).$$

4.3. Case studies

It is not difficult to check that the solution to a generic IVP is

$$z_1(t) = z_1(0)\cos(\omega_n t) + \frac{z_1'(0)}{\omega_n}\sin(\omega_n t), \quad z_2(t) = z_2(0)\cos(\omega_c t) + \frac{z_2'(0)}{\omega_c}\sin(\omega_c t). \tag{4.3.10}$$

Suppose we choose the initial conditions

$$y_1(0) = y_1'(0) = 0; \quad y_2(0) = F, \; y_2'(0) = 0.$$

In other words, suppose that the top mass is started at its equilibrium position and that the bottom mass is moved F units but is given no initial velocity. We first need to translate these initial conditions into initial conditions for z_1 and z_2. The Fourier expansion gives

$$y = z_1 u_1 + z_2 u_2 \quad \leadsto \quad y = Uz, \; U = (u_1 \; u_2).$$

Since

$$U^{-1} = \frac{1}{2}\begin{pmatrix} 1 & 1 \\ 1 & -1 \end{pmatrix} = \frac{1}{2}U,$$

we have

$$z = \frac{1}{2}Uy.$$

In particular,

$$y(0) = \begin{pmatrix} 0 \\ F \end{pmatrix}, \quad y'(0) = \begin{pmatrix} 0 \\ 0 \end{pmatrix}$$

means that

$$z(0) = \frac{1}{2}Uy(0) = \frac{1}{2}F\begin{pmatrix} 1 \\ -1 \end{pmatrix}$$

and

$$z'(0) = \frac{1}{2}Uy'(0) = \begin{pmatrix} 0 \\ 0 \end{pmatrix}.$$

For the given initial condition, the solution (4.3.10) becomes

$$z_1(t) = \frac{1}{2}F\cos(\omega_n t), \quad z_2(t) = -\frac{1}{2}F\cos(\omega_c t). \tag{4.3.11}$$

We are now (finally!) ready to find the solution to the problem in the original variables. Since $y = Uz$ means

$$y_1 = z_1 + z_2, \quad y_2 = z_1 - z_2,$$

we can use (4.3.11) to write

$$y_1(t) = \frac{1}{2}F\left[\cos(\omega_n t) - \cos(\omega_c t)\right]$$
$$= -F\sin\left(\frac{\omega_c - \omega_n}{2}t\right)\sin\left(\frac{\omega_c + \omega_n}{2}t\right)$$

and

$$y_2(t) = \frac{1}{2}F\left[\cos(\omega_n t) + \cos(\omega_c t)\right]$$
$$= F\cos\left(\frac{\omega_c - \omega_n}{2}t\right)\cos\left(\frac{\omega_c + \omega_n}{2}t\right).$$

We saw the equality for $y_1(t)$ in the previous example, and the second equality follows from the identity

$$\cos(\theta_1) + \cos(\theta_2) = 2\cos\left(\frac{\theta_2 - \theta_1}{2}\right)\cos\left(\frac{\theta_1 + \theta_2}{2}\right).$$

The solution for $y_1(t)$ and $y_2(t)$ are now in the form described by the forced mass-spring problem, and the plot for each one will be similar to that described in Figure 4.4. The time-dependent amplitude for each mass is given by

$$A_1(t) = F\sin\left(\frac{\omega_c - \omega_n}{2}t\right), \quad A_2(t) = F\cos\left(\frac{\omega_c - \omega_n}{2}t\right).$$

Since for any Ω

$$\cos(\Omega t) = \sin(\Omega t + \pi/2) = \sin\left(\Omega\left(t + \frac{\pi}{2\Omega}\right)\right),$$

we have that

$$A_1\left(t + \frac{\pi}{\omega_c - \omega_n}\right) = A_2(t);$$

in other words, the amplitude for $y_1(t)$ is $\pi/(\omega_c - \omega_n)$ out of phase with that for $y_2(t)$. Now the period of each amplitude function is given by

$$T^* = \frac{4\pi}{\omega_c - \omega_n}.$$

Since the phase difference between the two amplitude functions is precisely one-fourth the period, we see that when one amplitude is at a maximum or minimum, the other must be at a zero.

There are two distinguished limits for the strength of the coupling spring. First, suppose that k_1 is large. This implies that the frequency induced by the coupling is much larger than the natural frequency, $\omega_c \gg \omega_n$. The situation is similar to what was seen for the periodically forced mass-spring system in which the forcing frequency was large. Going to our solution formula for $y_1(t)$ and $y_2(t)$, we see that each is the sum of a rapidly oscillating function (frequency ω_c) with relatively slowly varying mean (frequency ω_n). The mean for each function is identical; however, the rapidly varying portion of each is out of phase with the other (see Figure 4.8).

Now suppose that k_1 is small. The frequency induced by the coupling will now be approximately the same as the natural frequency. The situation is now similar as to what was the seen for the periodically forced mass-spring system in which there was almost-resonant forcing. Going to our solution formula for $y_1(t)$ and $y_2(t)$, we see that each is the product of a slowly varying amplitude function and a relatively rapidly varying trigonometric function with frequency of almost the natural frequency. Consequently, each function exhibits a beating phenomena. As we have already discussed, the phase difference for the amplitudes is one-fourth the period so that when one amplitude is maximized in absolute value, the other is at a zero. Thus, the beating alternates between the two masses. This is precisely what is seen in the numerical simulations (see Figure 4.9).

4.3. Case studies

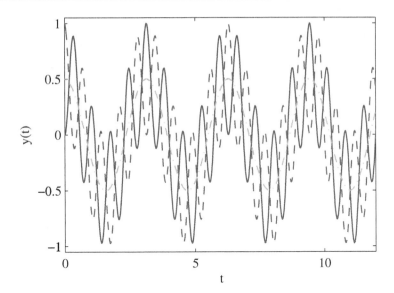

Figure 4.8. *A plot of the solution to the coupled mass-springs for the initial condition $y_1(0) = y_1'(0) = 0$ and $y_2(0) = 1$, $y_2'(0) = 0$. Here $\omega_n = 2.0$ and $\omega_c = 9.0$. The dashed-dotted curve is the slowly varying mean, the solid curve is the solution $y_1(t)$, and the dashed curve is the solution $y_2(t)$.*

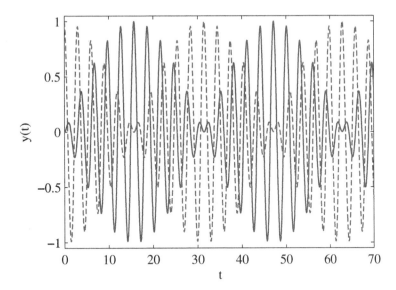

Figure 4.9. *A plot of the solution to the coupled mass-springs for the initial condition $y_1(0) = y_1'(0) = 0$ and $y_2(0) = 1$, $y_2'(0) = 0$. Here $\omega_n = 2.0$ and $\omega_c = 2.2$. The solid curve is the solution $y_1(t)$, and the dashed curve is the solution $y_2(t)$. Note that each solution exhibits the beating phenomena, and that moreover the beating alternates between the two masses.*

4.3.5 • Pendulum driven by constant torque

Recall the derivation of the equation of motion for the nonlinear pendulum as given in section 3.1.4. There we assumed that the suspension point was being moved horizontally. If we assume that the suspension point is fixed but instead is driven by a constant torque, then the rescaled equation of motion is

$$\theta'' + 2b\theta' + \omega_0^2 \sin\theta = \Gamma.$$

See Strogatz [37, Chapter 4.4]. If $\Gamma > 0$, then the applied torque drives the pendulum counterclockwise, while $\Gamma < 0$ means that the pendulum is being driven clockwise. If, as in section 3.1.4, we assume that the oscillations about the straight-down equilibrium are small, then we can write $\sin\theta \sim \theta$ to get the second-order linear ODE

$$\theta'' + 2b\theta' + \omega_0^2 \theta = \Gamma.$$

We now solve this ODE. First suppose that $b > 0$; i.e., the system experiences damping. The homogeneous problem is exactly the same as that for the damped mass-spring system. The homogeneous solution is transient, and the behavior for large t is determined by the particular solution. Using the method of undetermined coefficients, we know the particular solution is of the form

$$\theta_p(t) = a_0.$$

Plugging this solution into the ODE yields

$$\omega_0^2 a_0 = \Gamma \quad \leadsto \quad a_0 = \frac{\Gamma}{\omega_0^2}.$$

Thus, for $t > 0$ sufficiently large, we have

$$\theta(t) \sim \theta_p(t) = \frac{\Gamma}{\omega_0^2}.$$

For this solution to be physically valid, our small angle approximation requires that

$$\left|\frac{\Gamma}{\omega_0^2}\right| \ll 1;$$

i.e., the applied torque is small relative to the natural frequency of the undamped system. Under this assumption, we conclude that the pendulum is essentially motionless and tilts off to the side at an angle that is proportional to the applied torque.

Now suppose that the system is undamped, i.e., $b = 0$. We consider the solution under the assumption that the unforced system is at rest:

$$\theta(0) = 0, \quad \theta'(0) = 0.$$

The homogeneous solution is

$$\theta_h(t) = a_0 \cos(\omega_0 t) + a_1 \sin(\omega_0 t),$$

and the particular solution is that as given above. The general solution is the sum of these two solutions:

$$\theta(t) = a_0 \cos(\omega_0 t) + a_1 \sin(\omega_0 t) + \frac{\Gamma}{\omega_0^2}.$$

The initial conditions yield the linear system

$$0 = \theta(0) = a_0 + \frac{\Gamma}{\omega_0^2}, \quad 0 = \theta'(0) = \omega_0 a_1.$$

On solving the first equation for a_0 and using $a_1 = 0$, we can conclude that the solution to the IVP is

$$\theta(t) = \frac{\Gamma}{\omega_0^2}(1 - \cos(\omega_0 t)).$$

The mean of this solution is Γ/ω_0^2, which is the large time solution when damping is present. However, the absence of damping means that there are now oscillations about the mean. Again assuming that the applied torque is small relative to the natural frequency, we conclude that

(a) the pendulum oscillates about its mean with the natural frequency of the system, and

(b) the amplitude of the oscillation is precisely the mean value.

Exercises

Exercise 4.3.1. *Give an explanation as to why the intuition that coupling two identical mass-spring systems should lead to something akin to periodic resonant forcing is incorrect (see the first paragraph of section 4.3.4).*

Exercise 4.3.2. *Consider the coupled mass-spring system with the initial condition*

$$y_1(0) = F, \quad y_1'(0) = y_2(0) = y_2'(0) = 0.$$

How does this solution compare to that given in section 4.3.4?

Exercise 4.3.3. *Suppose that the initial condition for the coupled oscillator problem is*

$$y_1'(0) = F, \quad y_1(0) = y_2(0) = y_2'(0) = 0.$$

(a) *Solve the IVP.*

(b) *Suppose that the coupling spring is weak, $k_1 \ll 1$. Is there still an alternating of the beating phenomena? Explain.*

(c) *Suppose that the coupling spring is strong, $k_1 \gg 1$. Describe the solution behavior.*

Group projects

4.1. In section 3.1.4, we showed that the equation of motion for the horizontally forced pendulum can be written in the form

$$\theta'' + 2b\theta' + \omega_0^2 \sin\theta = f(t)\cos\theta,$$

where θ represents the angle from the equilibrium position in which the mass is hanging straight down ($\theta = 0$). We wish to consider the dynamics near the (unstable) straight-up equilibrium position ($\theta = \pi$).

(a) Setting $\phi = \theta - \pi$, show that the equation of motion becomes
$$\phi'' + 2b\phi' - \omega_0^2 \sin\phi = -f(t)\cos\phi.$$
(See Exercise 3.1.4.)

(b) Show that if ϕ is small, then the approximating second-order ODE is
$$\phi'' + 2b\phi' - \omega_0^2 \phi = -f(t).$$

(c) Henceforth consider only the linear approximating problem. Show that for the homogeneous problem, $\phi = 0$ is an unstable saddle point for any $b \geq 0$.

(d) Henceforth assume that there is no damping, i.e., $b = 0$. Further suppose that
$$f(t) = f_0 \cos(\omega t).$$
If the initial conditions are
$$\phi(0) = \phi_0, \quad \phi'(0) = \phi_1,$$
then find a function $f_0 = f_0^s(\phi_1, \phi_2, \omega_0, \omega)$ such that if $f_0 = f_0^s$, then the solution $\phi(t)$ is uniformly bounded.

(e) Consider the uniformly bounded solution found in (d). Is it approximately periodic for $t \gg 0$? If so, what is its frequency? Is there a phase difference between the solution and the forcing? If so, how is this phase difference a function of the forcing amplitude and forcing frequency?

(f) Now suppose that there is damping, i.e., $b > 0$. Find a function $f_0 = f_0^s(\phi_1, \phi_2, \omega_0, b, \omega)$ such that if $f_0 = f_0^s$, then the solution $\phi(t)$ is uniformly bounded.

(g) Consider the uniformly bounded solution found in (f). Is it approximately periodic for $t \gg 0$? If so, what is its frequency? Is there a phase difference between the solution and the forcing? If so, how is this phase difference a function of the forcing amplitude, forcing frequency, damping, and initial data?

4.2. Recall that in section 3.1.2, we provided the linear Ackerman model for the concentrations of glucose and insulin in the blood:
$$G' = aG + bI + J(t), \quad G(0) = G_0,$$
$$I' = cG + dI, \quad I(0) = I_0.$$
We will suppose that there is a $2\pi/\omega$-periodic ingestion of insulin:
$$J(t) = J_0(1 - \cos(\omega t)).$$
One can think of this model as representing having a large meal every $2\pi/\omega$ time units with snacking between the large meals.

(a) What is the average amount of insulin being consumed over one period?

(b) Show that the concentration of insulin satisfies the second-order linear ODE
$$I'' - (a+d)I' + (ad-bc)I = cJ(t); \quad I(0) = I_0, \; I'(0) = cG_0 + dI_0.$$

(c) If $I(t)$ solves the ODE in (b), how do we find $G(t)$ without solving another ODE?

(d) Under the model assumptions,
$$a, b, c < 0, \quad d > 0,$$
show that the concentration of insulin in the blood for $t \gg 0$ does not depend on the initial amount of insulin and glucose in the body.

(e) Show that for $t \gg 0$, the concentration of insulin in the body is modeled by an ω-periodic function.

(f) Now use the experimentally determined coefficient values:
$$a = -2.92, \quad b = -4.34, \quad c = 0.21, \quad d = -0.78.$$
Assume that $t \gg 0$. What is the average concentration of insulin that is present in the body over the period $2\pi/\omega$?

(g) Assume that $t \gg 0$. Carefully describe the solution behavior in the two distinguished limits, $0 < \omega \ll 1$ and $\omega \gg 1$. Use amplitude and phase diagrams (if appropriate).

4.3. Consider three identical masses with mass m, each of which is coupled to its nearest neighbor by a spring with spring constant k. The outer two masses are connected to a fixed wall with a spring with spring constant k. In other words, imagine the scenario depicted in Figure 4.7, where $k_1 = k$, and there is an additional mass between the two pictured. Letting y_j denote the distance from equilibrium for mass j, the governing equation for the system is
$$my_1'' = -ky_1 + k(y_2 - y_1),$$
$$my_2'' = -k(y_2 - y_1) + k(y_3 - y_2),$$
$$my_3'' = -k(y_3 - y_2) - ky_3.$$

(a) Write the system in matrix/vector form:
$$y'' + Ky = 0.$$
Clearly identify the matrix K.

(b) Find the eigenvalues and associated eigenvectors for the matrix K.

(c) Find vectors u_1, u_2, u_3 such that if we write
$$y(t) = z_1(t)u_1 + z_2(t)u_2 + z_3(t)u_3,$$
then the equation of motion of each $z_j(t)$ is
$$z_j'' + \omega_j^2 z_j = 0.$$
What are the natural frequencies ω_j, and how do they relate to $\omega_0 = \sqrt{k/m}$?

(d) Describe the motion of each mass if $z_1(t) = z_3(t) \equiv 0$. For each mass, plot the curve that describes its motion.

(e) Describe the motion of each mass if $z_2(t) = z_3(t) \equiv 0$. For each mass, plot the curve that describes its motion.

(f) Solve the IVP with initial data

$$y_2(0) = F, \quad y_1(0) = y_3(0) = y_1'(0) = y_2'(0) = y_3'(0) = 0.$$

Plot the curve describing the motion of each mass.

Chapter 5

Discontinuous forcing and the Laplace transform

> *If people do not believe that mathematics is simple, it is only because they do not realize how complicated life is.*
>
> — John von Neumann

The goal of this chapter is to understand how to find the general solution for (primarily) second-order ODEs of the form

$$y'' + a_1 y' + a_0 y = f(t), \tag{5.0.1}$$

where the function $f(t)$ is *piecewise continuous*. While we focus on the second-order problem, the ideas and techniques can be easily generalized to higher-order problems. We already know how to find the homogeneous solution, so we will focus on finding the particular solution. We will not talk about general forcing functions; instead, we will focus our discussion on two functions that often appear in applications: the *Heaviside (unit) step function* and the *(Dirac) delta function*. The first function models the application of an "on-off" switch, whereas the second is used to model an instantaneous application of force (such as hitting a nail with a hammer).

5.1 • The Heaviside (unit) step function and the (Dirac) delta function

The Heaviside (unit) step function is defined as

$$H(t) = \begin{cases} 0, & -\infty < t < 0, \\ 1, & 0 \le t. \end{cases} \tag{5.1.1}$$

Using the property that the graph of the function $f(t-a)$ for $a > 0$ is simply the graph of $f(t)$ shifted a units to the right, we have

$$H(t-a) = \begin{cases} 0, & -\infty \le t < a, \\ 1, & a \le t. \end{cases}$$

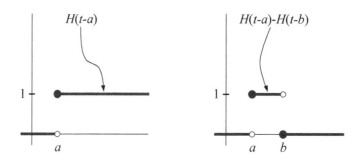

Figure 5.1. *The plot of the step function $H(t-a)$ is given in the left panel, and the plot of the linear combination $H(t-a)-H(t-b)$ is given in the right panel.*

The graph of the shifted step function is given in the left panel of Figure 5.1. The step function $H(t-a)$ is a mathematical model for an instantaneous "on-off" switch: the switch is turned off until time $t = a$, and afterward it is turned on. More generally, we have for $0 < a < b$

$$H(t-a)-H(t-b) = \begin{cases} 0, & -\infty \leq t < a, \\ 1, & a \leq t < b, \\ 0, & b \leq t. \end{cases} \qquad (5.1.2)$$

The graph of this linear combination of step functions is given in the right panel of Figure 5.1. The step function is useful because it allows us to write a piecewise-defined function as an appropriately weighted linear combination of shifted step functions.

Example 5.1. For our first example, suppose that

$$f(t) = \begin{cases} 2, & 0 \leq t < 3, \\ -5, & 3 \leq t < 6, \\ 3, & 6 \leq t. \end{cases}$$

Using the formulation of (5.1.2), we can rewrite the function as the sum

$$f(t) = 2[H(t-0)-H(t-3)] - 5[H(t-3)-H(t-6)] + 3H(t-6)$$
$$= 2H(t-0) - 7H(t-3) + 8H(t-6). \quad \blacksquare$$

Example 5.2. For our second example, suppose that

$$f(t) = \begin{cases} 6, & 0 \leq t < 3, \\ t^2 + 2, & 3 \leq t < 5, \\ 27 - t, & 5 < t. \end{cases}$$

As in the previous example, we can rewrite the function as the sum

$$f(t) = 6[H(t-0)-H(t-3)] + (t^2+2)[H(t-3)-H(t-5)] + (27-t)H(t-5)$$
$$= 6H(t-0) + (t^2-4)H(t-3) - (t^2+t-25)H(t-5).$$

5.1. The Heaviside (unit) step function and the (Dirac) delta function

It will later be useful to rewrite the function in the following manner. Using the algebraic identities,

$$t^2 - 4 = [(t-3)+3]^2 - 4 = [(t-3)^2 + 6(t-3) + 9] - 4 = (t-3)^2 + 6(t-3) + 5$$

and, similarly,

$$t^2 + t - 25 = (t-5)^2 + 11(t-5) + 5,$$

we can rewrite the function as

$$f(t) = 6H(t-0) + [(t-3)^2 + 6(t-3) + 5]H(t-3) - [(t-5)^2 + 11(t-5) + 5]H(t-5).$$

As we will see, it is this form of the function that turns out to be more useful when finding its Laplace transform. ∎

We define the (Dirac) delta function, $\delta(t-a)$ for $a \geq 0$ in the following manner:

Definition 5.3. *Suppose that $f(t)$ is continuous at the point $t = a$, where $a > 0$. The delta function, $\delta(t-a)$, satisfies the integral property*

$$\int_0^{+\infty} \delta(t-a)f(t)\,dt = f(a).$$

A physical interpretation of the delta function is provided in section 5.4.1. Therein it is shown that for mass-spring systems, the forcing function $v_0 \delta(t-a)$ models an instantaneous change of velocity by magnitude v_0 at time $t = a$. This can be accomplished, e.g., by striking the oscillating mass with a hammer with some force (which depends on the physical parameters of the system) at the appropriate time.

Example 5.4. We have

$$\int_0^{+\infty} \delta(t-1)\sin(5t)\,dt = \sin(5)$$

and

$$\int_0^{+\infty} \delta(t-9)e^{-3t}\,dt = e^{-27}. \quad \blacksquare$$

Exercises

Exercise 5.1.1. *Write the following piecewise continuous functions as a linear combination of step functions.*

(a) $f(t) = \begin{cases} 3, & 0 \leq t < 5, \\ 6\sin(4t), & 5 \leq t, \end{cases}$

(b) $f(t) = \begin{cases} t^2, & 0 \leq t < 2, \\ 4e^t, & 2 \leq t < 5, \\ 6, & 5 \leq t, \end{cases}$

(c) $f(t) = \begin{cases} 5\cos(2t), & 0 \leq t < \pi, \\ t-5, & \pi \leq t < 7, \\ 2t^2, & 7 \leq t. \end{cases}$

5.2 • The Laplace transform

We now provide a new solution technique to find the particular solution. The general idea behind this technique is similar to what we have seen before: transform the ODE into an algebraic problem, solve the algebraic problem, and then invert the algebraic solutions to ODE solutions. For one example, for nth-order ODEs, the homogeneous problem was converted to an nth-order characteristic polynomial, the zeros of the polynomial were found, and then the zeros were used to determine the general solution. For a second example, when finding the particular solution using the method of undetermined coefficients, the solution is found by guessing appropriately and then solving a linear system of equations where the variables are the coefficients associated with the guess. For the technique using the *Laplace transform*, we will see that the solution to the ODE will be realized as a rational function and that the solution will be found by first expanding the rational function using partial fractions and then using the table (5.2.1) to determine the function associated with each term in the partial fraction expansion.

We often write $\mathscr{L}[f](s) = F(s)$.

Laplace transform

Definition 5.5. *Let $f(t)$ be piecewise continuous and satisfy $|f(t)| \leq Ce^{bt}$ for some $M, b > 0$. The Laplace transform of $f(t)$, denoted $\mathscr{L}[f](s)$, is given by*

$$\mathscr{L}[f](s) := \int_0^{+\infty} e^{-st} f(t)\, dt.$$

The Laplace transform takes a function, $f(t)$, which is a function of t, and creates a new function, $F(s)$, which is a function of a new variable s. The exponential growth bound on $|f(t)|$ implies that the $F(s)$ is well defined for $s > b$ (see Exercise 5.3.1). Throughout the rest of this chapter, we implicitly use the fact that s can be complex valued and that the transform is well defined for $\mathrm{Re}(s) > b$.

5.2.1 • Computing the Laplace transform

From this point forward, it will be assumed that any functions for which the Laplace transform is being computed satisfies the exponential growth bound. As a consequence of the fact that the Laplace transform is defined via an improper integral, it is a linear function.

Proposition 5.6. *The Laplace transform is linear,*

$$\mathscr{L}[c_1 f + c_2 g](s) = c_1 \mathscr{L}[f](s) + c_2 \mathscr{L}[g](s).$$

5.2. The Laplace transform

Example 5.7. Let us compute the Laplace transform of the exponential function. We have

$$\mathscr{L}[e^{at}](s) = \int_0^{+\infty} e^{-st} e^{at}\, dt = \int_0^{+\infty} e^{-(s-a)t}\, dt = \frac{1}{s-a}, \quad s > a.$$

The inequality $s > a$ follows from an evaluation of the improper integral and

$$\lim_{t \to +\infty} e^{-(s-a)t} = 0 \quad \Leftrightarrow \quad s > a. \quad \blacksquare$$

Example 5.8. Let us compute the Laplace transform of the step function. We have

$$\mathscr{L}[H(t-a)](s) = \int_0^{+\infty} e^{-st} H(t-a)\, dt = \int_a^{+\infty} e^{-st}\, dt = \frac{e^{-as}}{s}, \quad s > 0.$$

The second equality follows from the definition of the step function in (5.1.1). $\blacksquare$

Example 5.9. Let us compute the Laplace transform of the delta function. We have

$$\mathscr{L}[\delta(t-a)](s) = \int_0^{+\infty} e^{-st} \delta(t-a)\, dt = e^{-st}\big|_{t=a} = e^{-as}, \quad s > 0.$$

The second equality follows from Definition 5.3. $\blacksquare$

Since the Laplace transform is defined via an integration against an exponential function, as in Example 5.7, it is not terribly difficult to compute when $f(t)$ is some linear combination of polynomials, exponential functions, and sines and cosines. We present the results of these calculations for the functions that appear most commonly in applications in the table of (5.2.1). All but the last entry in the table follow from the definition of the Laplace transform and direct integration. The last entry in the table of (5.2.1) requires just a bit more work, which we do in Proposition 5.10.

> **Proposition 5.10.** $\mathscr{L}[f(t-a)H(t-a)](s) = e^{-as} \mathscr{L}[f](s).$

Proof. By definition,

$$\mathscr{L}[f(t-a)H(t-a)](s) = \int_0^{+\infty} e^{-st} f(t-a) H(t-a)\, dt$$

$$= \int_a^{+\infty} e^{-st} f(t-a)\, dt = e^{-as} \int_0^{+\infty} e^{-st} f(t)\, dt.$$

The result follows from the fact that the last integral on the right is precisely $\mathscr{L}[f](s)$. $\square$

$f(t)$	$\mathscr{L}[f](s)$	Domain
1	$\dfrac{1}{s}$	$s > 0$
t^n	$\dfrac{n!}{s^{n+1}}$	$s > 0$
e^{at}	$\dfrac{1}{s-a}$	$s > a$
$t^n e^{at}$	$\dfrac{n!}{(s-a)^{n+1}}$	$s > a$
$\cos(bt)$	$\dfrac{s}{s^2+b^2}$	$s > 0$
$\sin(bt)$	$\dfrac{b}{s^2+b^2}$	$s > 0$
$e^{at}\cos(bt)$	$\dfrac{s-a}{(s-a)^2+b^2}$	$s > a$
$e^{at}\sin(bt)$	$\dfrac{b}{(s-a)^2+b^2}$	$s > a$
$H(t-a)$	$\dfrac{e^{-as}}{s}$	$s > 0$
$\delta(t-a)$	e^{-as}	$s > 0$
$f(t-a)H(t-a)$	$e^{-as}\mathscr{L}[f](s)$	same as for $\mathscr{L}[f](s)$

(5.2.1)

Example 5.11. Recall that we showed in Example 5.1 the function $f(t)$ given by

$$f(t) = \begin{cases} 2, & 0 \le t < 3, \\ -5, & 3 \le t < 6, \\ 3, & 6 \le t \end{cases}$$

can be rewritten as

$$f(t) = 2H(t-0) - 7H(t-3) + 8H(t-6).$$

As a consequence of linearity of the Laplace transform (Proposition 5.6), we have

$$\mathscr{L}[f](s) = 2\mathscr{L}[H(t-0)](s) - 7\mathscr{L}[H(t-3)](s) + 8\mathscr{L}[H(t-6)](s).$$

We already computed for any $a > 0$,

$$\mathscr{L}[H(t-a)](s) = \frac{e^{-as}}{s}, \quad s > 0.$$

Substitution gives

$$\mathscr{L}[f](s) = 2\frac{1}{s} - 7e^{-3s}\frac{1}{s} + 8e^{-6s}\frac{1}{s}. \quad \blacksquare$$

5.2. The Laplace transform

Example 5.12. Recall that we showed in Example 5.2 the function $f(t)$ given by

$$f(t) = \begin{cases} 6, & 0 \leq t < 3, \\ t^2 + 2, & 3 \leq 5, \\ 27 - t, & 5 < t, \end{cases}$$

can be rewritten as

$$f(t) = 6H(t-0) + [(t-3)^2 + 6(t-3) + 5]H(t-3) - [(t-5)^2 + 11(t-5) + 5]H(t-5).$$

As a consequence of linearity of the Laplace transform (Proposition 5.6), we can consider each term separately. For the sake of clarity, let us focus on the middle term:

$$h(t) = [(t-3)^2 + 6(t-3) + 5]H(t-3).$$

Using linearity gives

$$\mathscr{L}[h](s) = \mathscr{L}[(t-3)^2 H(t-3)](s) + 6\mathscr{L}[(t-3)H(t-3)](s) + 5\mathscr{L}[H(t-3)](s).$$

From the table in (5.2.1), we have

$$\mathscr{L}[t](s) = \frac{1}{s^2}, \quad \mathscr{L}[t^2](s) = \frac{2}{s^3}.$$

Using Proposition 5.10, we have

$$\mathscr{L}[(t-3)H(t-3)](s) = e^{-3s}\frac{1}{s^2}, \quad \mathscr{L}[(t-3)^2 H(t-3)](s) = e^{-3s}\frac{2}{s^3}.$$

Thus, we can conclude that

$$\mathscr{L}[h](s) = e^{-3s}\frac{2}{s^3} + 6e^{-3s}\frac{1}{s^2} + 5e^{-3s}$$
$$= e^{-3s}\left[\frac{2}{s^3} + 6\frac{1}{s^2} + 5\frac{1}{s}\right].$$

Applying a similar argument to the other terms gives the Laplace transform for the full function to be

$$\mathscr{L}[f](s) = 6\frac{1}{s} + e^{-3s}\left[\frac{2}{s^3} + 6\frac{1}{s^2} + 5\frac{1}{s}\right] - e^{-5s}\left[\frac{2}{s^3} + 11\frac{1}{s^2} + 5\frac{1}{s}\right]. \blacksquare$$

Example 5.13. Suppose that

$$f(t) = 3 - 6t + 5\delta(t-2) + 7e^{2t}\cos(t-\pi)H(t-\pi).$$

As a consequence of linearity of the Laplace transform (Proposition 5.6), we have

$$\mathscr{L}[f](s) = 3\mathscr{L}[1](s) - 6\mathscr{L}[t](s) + 5\mathscr{L}[\delta(t-2)](s) + 7\mathscr{L}[e^{2t}\cos(t-\pi)H(t-\pi)](s).$$

In order to use Proposition 5.10, we must first rewrite the last term,

$$e^{2t}\cos(t-\pi)H(t-\pi) = e^{2\pi}e^{2(t-\pi)}\cos(t-\pi)H(t-\pi),$$

so

$$\mathscr{L}[f](s) = 3\mathscr{L}[1](s) - 6\mathscr{L}[t](s) + 5\mathscr{L}[\delta(t-2)](s)$$
$$+ 7e^{2\pi}\mathscr{L}[e^{2(t-\pi)}\cos(t-\pi)H(t-\pi)](s).$$

Using the fact that from the table in (5.2.1)

$$\mathscr{L}[e^{2t}\cos(t)](s) = \frac{s-2}{(s-2)^2+1},$$

we have from Proposition 5.10

$$\mathscr{L}[e^{2(t-\pi)}\cos(t-\pi)H(t-\pi)](s) = e^{-\pi s}\frac{s-2}{(s-2)^2+1}.$$

For the other terms, we use the table in (5.2.1) and get

$$\mathscr{L}[f](s) = \frac{3}{s} - \frac{6}{s^2} + 5e^{-2s} + 7e^{2\pi}e^{-\pi s}\frac{s-2}{(s-2)^2+1}. \quad\blacksquare$$

Example 5.14. Now let us compute the Laplace transform directly without using the table. Suppose that

$$f(t) = \begin{cases} t, & 0 \leq t < 2 \\ t^2 - 2, & 2 \leq t. \end{cases}$$

Using the definition of the Laplace transform,

$$\mathscr{L}[f](s) = \int_0^{+\infty} e^{-st} f(t)\,dt$$
$$= \int_0^2 e^{-st} \cdot t\,dt + \int_2^{+\infty} e^{-st} \cdot (t^2-2)\,dt$$
$$= \frac{1}{s^2} + e^{-2s}\left(\frac{3}{s^2} + \frac{2}{s^3}\right).$$

The integrals were evaluated using SAGE. $\blacksquare$

5.2.2 • Inverting the Laplace transform

When using the Laplace transform to solve ODEs, we will need to invert the Laplace transform in order to find the original function. This leads to the following question: given a transform $\mathscr{L}[f](s)$, what is the generating function $f(t)$? For the problems of interest, the table, as well as the result of Proposition 5.10, will be sufficient in order to answer this question. What we will need to do is algebraically massage the original $\mathscr{L}[f](s)$ so that we see expressions in the second column of the table: once we see these expressions, the generating function can be found simply by appealing to the corresponding function $f(t)$ in the first column. The algebraic massaging takes place through the method of partial fractions. We will not review that material here; instead, when necessary, we will use SAGE in order to properly decompose the original transform.

5.2. The Laplace transform

Example 5.15. For our first example, suppose that

$$\mathcal{L}[f](s) = \frac{2s-5}{(s^2+4)(s+2)}.$$

Since

$$\frac{2s-5}{(s^2+4)(s+2)} = \frac{1}{8}\frac{9s-2}{s^2+4} - \frac{9}{8}\frac{1}{s+2},$$

after some algebraic manipulation, we can rewrite the Laplace transform as

$$\mathcal{L}[f](s) = \frac{9}{8}\frac{s}{s^2+4} - \frac{1}{8}\frac{2}{s^2+4} - \frac{9}{8}\frac{1}{s+2}.$$

Comparing each term in the sum with the table in (5.2.1) and using linearity gives

$$f(t) = \frac{9}{8}\cos(2t) - \frac{1}{8}\sin(2t) - \frac{9}{8}e^{-2t}. \quad \blacksquare$$

Example 5.16. For our next example, suppose that

$$\mathcal{L}[f](s) = e^{-5s}\frac{2s-5}{(s^2+4)(s+2)}.$$

Recall the result of Example 5.15:

$$\mathcal{L}[g](s) = \frac{2s-5}{(s^2+4)(s+2)} \quad \rightsquigarrow \quad g(t) = \frac{9}{8}\cos(2t) - \frac{1}{8}\sin(2t) - \frac{9}{8}e^{-2t}.$$

By Proposition 5.10 (also see the last row in the table of (5.2.1)), the inverse Laplace transform is a shift of the above followed by a multiplication of the appropriate step function:

$$f(t) = \left[\frac{9}{8}\cos(2(t-5)) - \frac{1}{8}\sin(2(t-5)) - \frac{9}{8}e^{-2(t-5)}\right]H(t-5). \quad \blacksquare$$

Example 5.17. For our last example, suppose that

$$\mathcal{L}[f](s) = \frac{s^4+6s^3-15s^2-10s+50}{s^3(s^2-2s+10)}.$$

Using SAGE, we have the partial fraction expansion

$$\mathcal{L}[f](s) = \frac{5}{s^3} - \frac{2}{s} + \frac{3s+2}{s^2-2s+10}.$$

Since

$$s^2-2s+10 = (s-1)^2+9,$$

in order to use the table, we must replace s with $s-1$ in the last term. Doing so and simplifying yields

$$\mathcal{L}[f](s) = \frac{5}{s^3} - \frac{2}{s} + \frac{3(s-1)+5}{(s-1)^2+9} = \frac{5}{2}\frac{2}{s^3} - 2\frac{1}{s} + 3\frac{s-1}{(s-1)^2+9} + \frac{5}{3}\frac{3}{(s-1)^2+9}.$$

Comparing with the table in (5.2.1) and using linearity, we see that
$$f(t) = \frac{5}{2}t^2 - 2 + 3e^t \cos(3t) + \frac{5}{3}e^t \sin(3t). \quad \blacksquare$$

Exercises

Exercise 5.2.1. *If $f(t)$ is piecewise continuous and satisfies the estimate $|f(t)| \leq Ce^{bt}$ for some $C, b > 0$, show that*
$$|\mathcal{L}[f](s)| \leq \frac{C}{s-b}, \quad s > b.$$

Exercise 5.2.2. *Find the Laplace transform of the following functions using the table in (5.2.1).*

(a) $f(t) = 3e^{2t} + 9\cos(4t)$

(b) $f(t) = 7t^2 e^{5t} - 12H(t-4)$

(c) $f(t) = -4e^{-5t} \sin(3t) + 15\delta(t-6)$

(d) $f(t) = -4te^{-3t} + 6H(t-1) + 4 - 8t$

(e) $f(t) = -5\delta(t-2) + e^{3t} H(t-2)$

(f) $f(t) = (t^2 - 3)H(t-1) + 3t^3$

(g) $f(t) = \begin{cases} 4\sin(t), & 0 \leq t < 2\pi, \\ 3(t-2\pi), & 2\pi \leq t < 2\pi + 4, \\ 12, & 2\pi + 4 \leq t. \end{cases}$

Exercise 5.2.3. *Find the inverse Laplace transform of the following functions using the table in (5.2.1).*

(a) $\mathcal{L}[f](s) = \dfrac{3s+7}{s^2+s-2}$

(b) $\mathcal{L}[f](s) = \dfrac{6s-5}{s^2+6s+13}$

(c) $\mathcal{L}[f](s) = \dfrac{4s^3 - 6s + 1}{(s+3)(s-2)(s^2+25)}$

(d) $\mathcal{L}[f](s) = e^{-3s} \dfrac{s+6}{s^2+6s+8}$

(e) $\mathcal{L}[f](s) = e^{-5s} \dfrac{2s-3}{s^2+4s+13}$

(f) $\mathcal{L}[f](s) = e^{-4s} \dfrac{5s^2 - 7}{(s+4)(s-1)^2}$

(g) $\mathcal{L}[f](s) = e^{-2s} \dfrac{3s^2 + 6s - 8}{(s+2)(s-5)(s^2-6s+13)}$

5.3 • Application: Solving second-order scalar ODEs

We now use the Laplace transform to solve the second-order ODE

$$y'' + a_1 y' + a_0 y = f(t).$$

On taking the Laplace transform of both sides, we can use linearity to write

$$\mathscr{L}[y''](s) + a_1 \mathscr{L}[y'](s) + a_0 \mathscr{L}[y](s) = \mathscr{L}[f](s).$$

At the moment, it appears that we have one equation in three variables, which are the Laplace transform of the solution, its derivative, and its second derivative. We need to make this into one equation with one variable—the Laplace transform of the solution. We have the following important result, which tells us the manner in which the Laplace transform of a derivative of a function is related to the Laplace transform of a function.

Using induction, a formula for $\mathscr{L}[y^{(n)}](s)$, $n \geq 3$, in terms of $\mathscr{L}[y](s)$ can be derived (see Exercise 5.3.1).

> **Theorem 5.18.** *Assume that $|y(t)|, |y'(t)|, |y''(t)| \leq Ce^{bt}$ for some $b > 0$. The Laplace transform then satisfies for $s > b$*
>
> $$\mathscr{L}[y'](s) = s\mathscr{L}[y](s) - y(0), \quad \mathscr{L}[y''](s) = s^2 \mathscr{L}[y](s) - sy(0) - y'(0).$$

Proof. The result follows from an integration by parts. By definition, we have

$$\mathscr{L}[y'](s) = \int_0^{+\infty} e^{-st} y'(t) \, dt = \lim_{A \to +\infty} \left(e^{-st} y(t) \big|_{t=0}^{t=A} \right) + s \int_0^{+\infty} e^{-st} y(t) \, dt.$$

The second term is $s\mathscr{L}[y](s)$, and assuming that $s > b$, we have

$$\lim_{A \to +\infty} \left(e^{-st} y(t) \big|_{t=0}^{t=A} \right) = -y(0).$$

The first result now follows. As for the second result, we start with an identity:

$$\mathscr{L}[y''](s) = \mathscr{L}[(y')'](s).$$

We can then use the expression for the Laplace transform of a derivative to get

$$\mathscr{L}[y''](s) = s\mathscr{L}[y'](s) - y'(0) = s\left[s\mathscr{L}[y](s) - y(0)\right] - y'(0),$$

which simplifies to the desired result. □

We are now ready to use the Laplace transform to solve second-order IVPs:

$$y'' + a_1 y' + a_0 y = f(t); \quad y(0) = y_0, \ y'(0) = y_1. \tag{5.3.1}$$

There are two pieces to the general solution: the homogeneous solution and the particular solution. The homogeneous solution can be found using the techniques from the previous Chapter 4 (see Theorem 4.5), so there is little reason to introduce a new technique to find it. On the other hand, while the particular solution can always be found using variation of parameters, it is cumbersome to use it when dealing with forcing functions that are piecewise continuous. The Laplace transform will be ideal for these cases; indeed, it will be the preferred solution technique when the forcing function can be written as some linear combination of step functions and delta functions.

If we are to use the Laplace transform to find the particular solution, then we see in Theorem 5.18 that we must specify the initial data. It will be easiest if we assume zero initial data,

$$y_p(0) = y_p'(0) = 0,$$

for in this case

$$\mathscr{L}[y_p'](s) = s\mathscr{L}[y_p](s), \quad \mathscr{L}[y_p''](s) = s^2\mathscr{L}[y_p](s).$$

As we will see in the examples, this assumption also makes it possible to deal with the initial data only by using the homogeneous solution. Under this assumption of zero initial data, the Laplace transform of the left-hand side of the ODE (5.3.1) satisfies

$$\mathscr{L}[y_p'' + a_1 y_p' + a_0 y_p](s) = \mathscr{L}[y_p''](s) + a_1 \mathscr{L}[y_p'](s) + a_0 \mathscr{L}[y_p](s)$$
$$= s^2 \mathscr{L}[y_p](s) + a_1 s \mathscr{L}[y_p](s) + a_0 \mathscr{L}[y_p](s)$$
$$= (s^2 + a_1 s + a_0) \mathscr{L}[y_p](s).$$

The Laplace transform of the left-hand side of the ODE becomes the product of the characteristic polynomial with the Laplace transform of the particular solution. This will always be the case no matter the order of the ODE.

In conclusion, on taking the Laplace transform of both sides of (5.3.1) and under the assumption of zero initial data, the Laplace transform of the particular solution is

$$(s^2 + a_1 s + a_0)\mathscr{L}[y_p](s) = \mathscr{L}[f](s) \quad \rightsquigarrow \quad \mathscr{L}[y_p](s) = \frac{\mathscr{L}[f](s)}{s^2 + a_1 s + a_0}. \quad (5.3.2)$$

The particular solution is found on inverting the right-hand side using the table in (5.2.1). The general solution, as always, is the sum of the homogeneous solution and the particular solution.

> **Lemma 5.19.** *The Laplace transform of the particular solution to the second-order IVP,*
>
> $$y'' + a_1 y' + a_0 y = f(t); \quad y(0) = y'(0) = 0,$$
>
> *is given by*
>
> $$(s^2 + a_1 s + a_0)\mathscr{L}[y_p](s) = \mathscr{L}[f](s).$$

Example 5.20. Let us find the solution to the IVP

$$y'' + 4y' + 3y = 5H(t-2); \quad y(0) = 2, \ y'(0) = 2.$$

The characteristic polynomial is

$$p(\lambda) = \lambda^2 + 4\lambda + 3 = (\lambda + 1)(\lambda + 3),$$

5.3. Application: Solving second-order scalar ODEs

so the homogeneous solution is

$$y_h(t) = c_1 e^{-t} + c_2 e^{-3t}.$$

In order to find the particular solution corresponding to zero initial data, we apply the Laplace transform to both sides as in Lemma 5.19 and use the table in (5.2.1) to get

$$(s^2 + 4s + 3)\mathscr{L}[y_p](s) = 5e^{-2s}\frac{1}{s}.$$

The Laplace transform of the particular solution is then

$$\mathscr{L}[y_p](s) = e^{-2s}\frac{5}{s(s^2+4s+3)} = e^{-2s}\left(\frac{5}{3}\frac{1}{s} - \frac{5}{2}\frac{1}{s+1} + \frac{5}{6}\frac{1}{s+3}\right).$$

Using the table in (5.2.1), we invert to get

$$y_p(t) = \left[\frac{5}{3} - \frac{5}{2}e^{-(t-2)} + \frac{5}{6}e^{-3(t-2)}\right]H(t-2).$$

The general solution is the sum of the homogeneous and particular solutions:

$$y(t) = c_1 e^{-t} + c_2 e^{-3t} + \left[\frac{5}{3} - \frac{5}{2}e^{-(t-2)} + \frac{5}{6}e^{-3(t-2)}\right]H(t-2).$$

As for the initial condition, on using the fact that $y_p(0) = y_p'(0) = 0$, we see that

$$2 = y(0) = c_1 + c_2, \quad 2 = y'(0) = -c_1 - 3c_2.$$

This linear system can be written in matrix/vector form as

$$\begin{pmatrix} 1 & 1 \\ -1 & -3 \end{pmatrix}\begin{pmatrix} c_1 \\ c_2 \end{pmatrix} = \begin{pmatrix} 2 \\ 2 \end{pmatrix} \rightsquigarrow \begin{pmatrix} c_1 \\ c_2 \end{pmatrix} = \begin{pmatrix} 4 \\ -2 \end{pmatrix}.$$

The solution to the IVP is

$$y(t) = 4e^{-t} - 2e^{-3t} + \left[\frac{5}{3} - \frac{5}{2}e^{-(t-2)} + \frac{5}{6}e^{-3(t-2)}\right]H(t-2). \blacksquare$$

Example 5.21. Let us find the solution to the IVP

$$y'' + 4y' + 3y = 5\delta(t-2); \quad y(0) = 5, \; y'(0) = -2.$$

From the previous example, we know the homogeneous solution is

$$y_h(t) = c_1 e^{-t} + c_2 e^{-3t}.$$

In order to find the particular solution corresponding to zero initial data, we apply the Laplace transform to both sides as in Lemma 5.19 and use the table in (5.2.1) to get

$$(s^2 + 4s + 3)\mathscr{L}[y_p](s) = 5e^{-2s}.$$

Thus,

$$\mathscr{L}[y_p](s) = e^{-2s}\frac{5}{s^2+4s+3} = e^{-2s}\left(\frac{5}{2}\frac{1}{s+1} - \frac{5}{2}\frac{1}{s+3}\right),$$

and on using the table in (5.2.1), we invert to get

$$y_p(t) = \left[\frac{5}{2}e^{-(t-2)} - \frac{5}{2}e^{-3(t-2)}\right]H(t-2).$$

The general solution is the sum of the homogeneous and particular solutions:

$$y(t) = c_1 e^{-t} + c_2 e^{-3t} + \left[\frac{5}{2}e^{-(t-2)} - \frac{5}{2}e^{-3(t-2)}\right]H(t-2).$$

Since $y_p(0) = y_p'(0) = 0$,

$$5 = y(0) = c_1 + c_2, \quad -2 = y'(0) = -c_1 - 3c_2 \quad \leadsto \quad c_1 = \frac{13}{2}, \; c_2 = -\frac{3}{2}.$$

The solution to the IVP is

$$y(t) = \frac{13}{2}e^{-t} - \frac{3}{2}e^{-3t} + \left[\frac{5}{2}e^{-(t-2)} - \frac{5}{2}e^{-3(t-2)}\right]H(t-2). \quad \blacksquare$$

Exercises

Exercise 5.3.1. *For the following ODEs, find the particular solution that satisfies* $y_p(0) = y_p'(0) = 0$.

(a) $y'' + 7y' + 12y = 3e^{-2t}$

(b) $y'' + 8y' + 20y = 4\delta(t-3) - 12\delta(t-5)$

(c) $y'' + 25y = 2\delta(t-1) + 5H(t-3)$

(d) $y'' + 8y' + 15y = -4\delta(t-7)$

(e) $y'' - 4y' + 29y = 3H(t-2)$

Exercise 5.3.2. *Solve the following IVPs.*

(a) $y'' + 7y' + 12y = 3e^{-2t}; \quad y(0) = 1, y'(0) = 3$

(b) $y'' + 8y' + 20y = 4\delta(t-3) - 12\delta(t-5); \quad y(0) = -2, y'(0) = 6$

(c) $y'' + 25y = 2\delta(t-1) + 5H(t-3); \quad y(0) = 4, y'(0) = -1$

(d) $y'' + 8y' + 15y = -4\delta(t-7); \quad y(0) = 2, y'(0) = 3$

(e) $y'' - 4y' + 29y = 3H(t-2); \quad y(0) = -1, y'(0) = -6$

Exercise 5.3.3. *Show that*

(a) $\mathscr{L}[y'''](s) = s^3 \mathscr{L}[y](s) - s^2 y(0) - s y'(0) - y''(0)$, *and*

(b) $\mathscr{L}[y^{(n)}](s) = s^n \mathscr{L}[y](s) - s^{n-1} y(0) - s^{n-2} y'(0) - s^{n-3} y''(0) - \cdots - y^{(n-1)}(0)$
for $n = 4, 5, \ldots$.

(Hint: Generalize the result of Theorem 5.18.)

5.4 ▪ Case studies

We are now ready to consider physical problems in which the forcing is not continuous. We start by comparing nonhomogeneous systems with delta function forcing with homogeneous systems with initial velocity. We then look at two mass-spring problems—one with no damping and one with strong damping—and a one-tank mixing problem. As part of our analysis, we show that the strongly damped mass spring is formally equivalent to a one-tank mixing problem.

5.4.1 ▪ Physical interpretation of delta function forcing

Using the delta function as an example of a discontinuous forcing function is mathematically convenient. We now wish to provide a physical interpretation of this forcing function for second-order equations. An interpretation for higher-order equations is presented following Corollary 5.24.

We start with the second-order nonhomogeneous IVP

$$y'' + a_1 y' + a_0 y = f_0 \delta(t); \quad y(0) = y'(0) = 0.$$

The system is initially at rest but is subject to an instantaneous force. The solution is referred to as the *impulse response*. On applying the Laplace transform to both sides and using Theorem 5.18, we see that the transform of the solution to this IVP is

$$(s^2 + a_1 s + a_0)\mathscr{L}[y](s) = f_0 \quad \rightsquigarrow \quad \mathscr{L}[y](s) = \frac{f_0}{s^2 + a_1 s + a_0}.$$

Alternatively, consider the second-order homogeneous IVP

$$y'' + a_1 y' + a_0 y = 0; \quad y(0) = y_0, \ y'(0) = y_1.$$

In the mass-spring context, we are modeling an unforced system where the mass is not necessarily at the equilibrium position and is also initially moving with velocity y_1. On applying the Laplace transform to both sides and using Theorem 5.18, the transform of the solution to the IVP is

$$(s^2 + a_1 s + a_0)\mathscr{L}[y](s) - (s + a_1)y_0 - y_1 = 0 \quad \rightsquigarrow \quad \mathscr{L}[y](s) = \frac{(s + a_1)y_0 + y_1}{s^2 + a_1 s + a_0}.$$

On comparing the Laplace transforms of the solutions to the two problems, we see that they coincide if and only if $y_0 = 0$ and $f_0 = y_1$. In other words, the solution to the two problems is the same if and only if $y_0 = 0$ and $f_0 = y_1$. Physical interpretations of this mathematical result are as follows. In the mass-spring context, applying an external instantaneous force of magnitude $f_0 > 0$ is equivalent to instantaneously increasing the velocity of the mass, which is at the equilibrium position by amount f_0. In the electrical circuit context, applying an external instantaneous voltage of magnitude $V_0 > 0$ is equivalent to instantaneously increasing the current by amount V_0.

5.4.2 ▪ Undamped mass-spring system

Here we will revisit the problem discussed in section 4.3.2: the forced but undamped mass-spring system, which is modeled by

$$y'' + \omega_0^2 y = f(t).$$

We will assume that the mass is initially at equilibrium:
$$y(0) = y'(0) = 0.$$
We will suppose that the forcing function is given by
$$f(t) = f_0 \delta(t-1) + f_1 \delta(t-a), \quad a > 1. \tag{5.4.1}$$
This forcing function is meant to model the striking of the system with, say, a hammer once at $t=1$ with force f_0 and yet again at $t = a > 1$ with force f_1. Given an f_0, our goal is to determine f_1 and a such that there is no motion for $t \geq a$. The mass starts oscillating for $t \geq 1$, and the amplitude of oscillation is determined by f_0. Our goal is to find value(s) of f_1 and a such that the mass is again at equilibrium for $t \geq a$.

We first solve the ODE. The homogeneous solution is
$$y_h(t) = c_1 \cos(\omega_0 t) + c_2 \sin(\omega_0 t) = A\cos(\omega_0 t - \phi).$$

We will use the Laplace transform to find the particular solution that satisfies the initial condition $y_p(0) = y_p'(0) = 0$. Since $y(0) = y'(0) = 0$ (as in section 4.3.2), the general solution being the sum of the homogeneous and particular solutions means that for the initial value problem,
$$0 = y(0) = c_1 + y_p(0), \quad 0 = y'(0) = \omega_0 c_2 + y_p'(0),$$
i.e., $c_1 = c_2 = 0$. In other words, the solution to the IVP will be $y(t) = y_p(t)$.

In order to simplify the calculations, it will henceforth be assumed that the natural frequency satisfies $\omega_0 = 1$ (the natural period is 2π). Applying the Laplace transform to both sides yields that the Laplace transform of the particular solution is given by
$$(s^2+1)\mathscr{L}[y](s) = \mathscr{L}[f](s) \quad \leadsto \quad \mathscr{L}[y](s) = \frac{\mathscr{L}[f](s)}{s^2+1}. \tag{5.4.2}$$

Using the form of the solution given in (5.4.2) and the fact that
$$\mathscr{L}[f](s) = f_0 e^{-s} + f_1 e^{-as},$$
we have
$$\mathscr{L}[y](s) = f_0 e^{-s} \frac{1}{s^2+1} + f_1 e^{-as} \frac{1}{s^2+1}.$$

Since $\mathscr{L}[\sin(t)](s) = 1/(s^2+1)$, inverting using the result of Proposition 5.10 yields that the solution is
$$y(t) = f_0 \sin(t-1) H(t-1) + f_1 \sin(t-\omega_1) H(t-a)$$
$$= \begin{cases} 0, & 0 \leq t < 1, \\ f_0 \sin(t-1), & 1 \leq t < a, \\ f_0 \sin(t-1) + f_1 \sin(t-a), & a \leq t. \end{cases}$$

In order to determine the parameter values for which $y(t) \equiv 0$ for $t > a$, we need to first rewrite the solution. Doing so yields
$$y(t) = -[f_0 \sin(1) + f_1 \sin(a)]\cos(t) + [f_0 \cos(1) + f_1 \cos(a)]\sin(t)$$
$$= \sqrt{f_0^2 + f_1^2 + 2f_0 f_1 \cos(a-1)} \cos(t - \phi),$$

5.4. Case studies

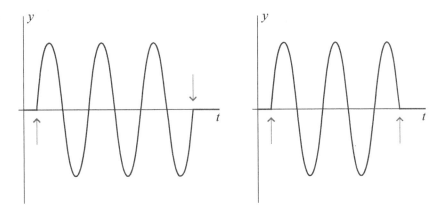

Figure 5.2. *Cartoon plots of the solution curve for $f_1 = -f_0$ (left figure) and $f_1 = f_0$ (right figure). The time a is chosen in each figure so that there is no motion for $t > a$. In the left panel, $a = 1 + 4\pi$, while in the right panel, $a = 1 + 3\pi$. The arrows depict the direction of the forcing at $t = 1$ and $t = a$.*

where the exact expression for the phase shift ϕ is irrelevant here. The second line follows after some algebraic and trigonometric simplifications (e.g., see the first example of this in section 2.6.1). Defining the square of the amplitude,

$$A(f_1, a) := f_0^2 + f_1^2 + 2f_0 f_1 \cos(a-1),$$

it is clearly the case that $y(t) \equiv 0$ for $t > a$ if and only if $A(f_1, a) = 0$.

Since the function $A(f_1, a)$ represents an amplitude and is therefore always non-negative, the desired values of f_1 and a represent points for which the amplitude is minimized. In other words, in order to answer the question, we need to find the critical points of $A(f_1, a)$. The critical points are found by solving

$$0 = \frac{\partial A}{\partial f_1} = 2f_1 + 2f_0 \cos(a-1), \quad 0 = \frac{\partial A}{\partial a} = -2f_0 f_1 \sin(a-1).$$

Since $f_0, f_1 \neq 0$, from the equation on the right we have that $a = 1 + k\pi$ for $k = 1, 2, \ldots$. From the equation on the left, we then get

$$f_1 = \begin{cases} f_0, & a = 1 + (2k-1)\pi, \ k = 1, 2, \ldots, \\ -f_0, & a = 1 + 2k\pi, \ k = 1, 2, \ldots. \end{cases}$$

Physically speaking, in order to stop the motion, we must apply at $t = a$ a force equal in magnitude to that applied when $t = 1$. Furthermore, we must apply the force at a time that is some integer multiple of the natural period after which the force was initially applied. This is precisely the time when the mass is passing through its rest position. If $a - 1$ is an even multiple of π, then the force at $t = a$ is applied in the direction opposite to the original force (see the left panel in Figure 5.2). On the other hand, if $a - 1$ is an odd multiple of π, then the force is applied in the same direction as when it was applied at $t = 1$ (see the right panel in Figure 5.2).

5.4.3 ▪ One-tank mixing problem

Recalling the discussion in section 2.6.1, we know that a one-tank mixing problem in which the initial concentration of brine in the tank is zero is modeled by the first-order ODE

$$y' + ay = aV_0 c(t), \quad y(0) = 0. \tag{5.4.3}$$

Here $y(t)$ is the amount of salt in the tank, $a > 0$ is the ratio of the outflow rate of briny solution to the total amount of solution in the tank (V_0), and $c(t)$ is the incoming concentration of briny solution. As was the case in the study presented in section 2.6.1, the incoming concentration will be assumed to vary periodically. However, this time, we will assume an injection strategy. If ω is the period, then a briny concentration will be instantaneously injected at $t = j\omega$ for $j = 1, 2, \ldots$; otherwise, the incoming solution is freshwater.

A mathematical model for this injection strategy is

$$c(t) = c_0 \omega \sum_{j=1}^{\infty} \delta(t - j\omega), \quad \omega > 0,$$

where c_0 represents the mean concentration of the injected briny solution at $t = \omega, 2\omega, \ldots$. In order to verify this statement regarding c_0, first note that calculating the mean of a periodic function over one period does not fix the interval over which the integral takes place. Thus, we have that the mean is

$$\frac{1}{\omega} \int_{\omega/2}^{3\omega/2} c(t)\,dt = \frac{1}{\omega} \int_{\omega/2}^{3\omega/2} c_0 \omega \delta(t - \omega)\,dt = c_0,$$

where the last equality follows from the definition of the delta function. Although the mean is fixed, the *magnitude* of the forcing, $c_0 \omega$, is dependent on the period. Indeed, for a fixed mean concentration, the magnitude will increase linearly as a function of the period.

We find the solution that satisfies $y(0) = 0$ using Laplace transform techniques. Applying the Laplace transform to both sides and using the table in (5.2.1) yields

$$(s+a)\mathscr{L}[y](s) = f_0 \sum_{j=1}^{\infty} e^{-j\omega s} \quad \leadsto \quad \mathscr{L}[y](s) = aV_0 c_0 \omega \sum_{j=1}^{\infty} \frac{e^{-j\omega s}}{s+a}.$$

Inverting gives the solution to be

$$y(t) = aV_0 c_0 \omega \sum_{j=1}^{\infty} e^{-a(t-j\omega)} H(t - j\omega).$$

The concentration in the tank is

$$c_{\text{tank}}(t) = \frac{y(t)}{V_0} = ac_0 \omega \sum_{j=1}^{\infty} e^{-a(t-j\omega)} H(t - j\omega).$$

Unfortunately, from this form of the solution, it is not yet clear as to what the solution looks like for $t \gg 0$.

In order to better understand the solution for large t, consider the following representation of the particular solution. By the definition of the unit step function,

5.4. Case studies

we can rewrite the solution as

$$c_{\text{tank}}(t) = ac_0\omega \begin{cases} 0, & 0 \leq t < \omega, \\ e^{-a(t-\omega)}, & \omega \leq t < 2\omega, \\ e^{-a(t-\omega)} + e^{-a(t-2\omega)}, & 2\omega \leq t < 3\omega, \\ e^{-a(t-\omega)} + e^{-a(t-2\omega)} + e^{-a(t-3\omega)}, & 3\omega \leq t < 4\omega \\ \vdots & \vdots \\ \sum_{j=1}^{n} e^{-a(t-j\omega)}, & n\omega \leq t < (n+1)\omega. \end{cases}$$

On noting that

$$e^{-a(t-j\omega)} = e^{-at}e^{aj\omega} = e^{-at}(e^{a\omega})^j$$

and using the geometric series formula

$$\sum_{j=1}^{n} r^j = r\frac{r^n - 1}{r - 1},$$

we have with $r = e^{a\omega}$

$$\sum_{j=1}^{n} e^{-a(t-j\omega)} = e^{-at} \sum_{j=1}^{n} e^{aj\omega}$$

$$= \frac{e^{an\omega} - 1}{e^{a\omega} - 1} e^{-a(t-\omega)} \qquad (5.4.4)$$

$$= \frac{e^{a\omega}}{e^{a\omega} - 1} e^{-a(t-n\omega)} - \frac{e^{a\omega}}{e^{a\omega} - 1} e^{-at}.$$

Consequently, we can write the concentration in the tank as

$$c_{\text{tank}}(t) = c_0 \frac{a\omega e^{a\omega}}{e^{a\omega} - 1} \begin{cases} 0, & 0 \leq t < \omega, \\ e^{-a(t-n\omega)} - e^{-at}, & n\omega \leq t < (n+1)\omega, \ n = 1, 2, \ldots. \end{cases}$$
(5.4.5)

For large t, the concentration in the tank is

$$c_{\text{tank}}(t) \sim c_0 \frac{a\omega e^{a\omega}}{e^{a\omega} - 1} e^{-a(t-n\omega)}, \quad n\omega \leq t < (n+1)\omega,$$

where n is large enough so that $an\omega \gg 0$, i.e., $n \gg (a\omega)^{-1}$. This is an ω-periodic sawtooth function and is illustrated in Figure 5.3 for $t > 3$ (dashed curve) and $t > 4$ (solid curve).

Now that we have a concise expression for the concentration in the tank, we can do an analysis. The mean of the concentration in the tank is

$$\frac{1}{\omega} \int_{n\omega}^{(n+1)\omega} c_{\text{tank}}(t)\,dt = \frac{c_0}{\omega} \frac{a\omega e^{a\omega}}{e^{a\omega} - 1} \int_{n\omega}^{(n+1)\omega} e^{-a(t-n\omega)}\,dt = c_0, \qquad (5.4.6)$$

so for large time, we get the physically realistic answer that the mean concentration in the tank is the same as the incoming concentration.

The variation of the tank concentration about the mean satisfies the bounds

$$-c_0 \frac{1 + a\omega - e^{a\omega}}{e^{a\omega} - 1} \leq c_{\text{tank}}(t) - c_0 \leq c_0 \frac{1 + a\omega e^{a\omega} - e^{a\omega}}{e^{a\omega} - 1}.$$

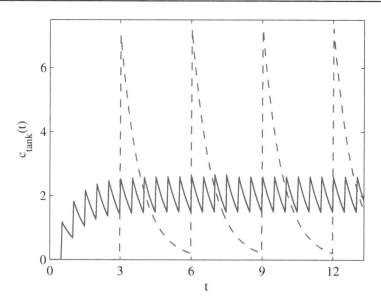

Figure 5.3. *Plots of the concentration in the tank with $a = 1.2$ and $c_0 = 2$. The solid curve is the solution when $\omega = 0.5$, and the dashed curve is the solution when $\omega = 3.0$. We see that for a desired mean concentration in the tank (in this case the desired mean concentration is $c_0 = 2$), if there is to be less of a variation about the mean, the injection period must be relatively small.*

Regarding the variation about the mean, we have two distinguished limits. First, suppose that $0 < \omega \ll 1$ is small (the solid solution curve), which means that the frequency of the variation of the incoming concentration is large. Using the approximation

$$e^{a\omega} \sim 1 + a\omega + a^2\omega^2/2$$

means that

$$\frac{1 + a\omega - e^{a\omega}}{e^{a\omega} - 1} \sim \frac{a^2\omega^2/2}{a\omega} = \frac{1}{2}a\omega$$

and that

$$\frac{1 + a\omega e^{a\omega} - e^{a\omega})}{e^{a\omega} - 1} \sim \frac{1 + a\omega(1 + a\omega) - (1 - a\omega - a^2\omega^2/2)}{a\omega} = \frac{1}{2}a\omega.$$

Thus, for large time, the variation about the mean has the approximate bounds

$$-\frac{1}{2}a\omega c_0 \leq c_{\text{tank}}(t) - c_0 \leq \frac{1}{2}a\omega c_0 \quad \rightsquigarrow \quad \left(1 - \frac{1}{2}a\omega\right)c_0 \leq c_{\text{tank}}(t) \leq \left(1 + \frac{1}{2}a\omega\right)c_0.$$

Now suppose that ω is large (the dashed solution curve); i.e., the frequency of the variation of the incoming concentration is small. We now have the approximations

$$\frac{1 + a\omega - e^{a\omega}}{e^{a\omega} - 1} \sim -1$$

and

$$\frac{1 + a\omega e^{a\omega} - e^{a\omega}}{e^{a\omega} - 1} \sim a\omega.$$

5.4. Case studies

Consequently, for large time, the variation about the mean has the approximate bounds

$$-c_0 \leq c_{\text{tank}}(t) - c_0 \leq a\omega c_0 \quad \rightsquigarrow \quad 0 \leq c_{\text{tank}}(t) \leq (1+a\omega)c_0.$$

The maximum of the concentration in the tank is large, which is consistent with the magnitude of the incoming concentration being large.

5.4.4 ▪ Strongly damped mass-spring system

Using the mathematical equivalence between RLC circuits and damped mass-spring systems (see section 3.1.5), the analysis presented here is also valid for a circuit in which there is a large resistance.

We originally saw in section 3.1.3 that the damped and forced mass-spring system can be written in the form

$$y'' + 2by' + \omega_0^2 y = f(t). \tag{5.4.7}$$

Here we will assume that the damping is strong, i.e., $b \gg 0$. Our first task is to show that under this assumption, the model equation effectively becomes a variant of the first-order equation (5.4.3). In other words, for large damping, the force due to acceleration is negligible. Such a situation occurs, e.g., in the case that the ODE is being used to model the effect of a shock absorber. After this has been accomplished, we can then use the analysis presented in section 5.4.3 to understand the solution behavior.

In order to see why the acceleration term can be ignored, it will be convenient to rewrite the equation in a different form. Rescale time via

$$\tau = \frac{t}{2b}$$

and use the chain rule to get

$$\frac{d}{dt} = \frac{1}{2b}\frac{d}{d\tau}, \quad \frac{d^2}{dt^2} = \frac{1}{4b^2}\frac{d^2}{d\tau^2}.$$

After some simplification, the ODE can be rewritten as

$$\frac{1}{4b^2}\frac{d^2 y}{d\tau^2} + \frac{dy}{d\tau} + \omega_0^2 y = f(2b\tau).$$

The primary effect of the rescaling has been to move the coefficient associated with the damping term to the coefficient associated with the acceleration term. Since $1/b^2 \ll 1$, we can (formally) drop the acceleration (first) term in the equation to get the approximate system

$$\frac{dy}{d\tau} + \omega_0^2 y = f(2b\tau).$$

In terms of the original time variable, since

$$t = 2b\tau \quad \rightsquigarrow \quad \frac{d}{d\tau} = 2b\frac{d}{dt},$$

on dividing through by $1/2b$, we can rewrite the approximate system as

$$y' + \frac{\omega_0^2}{2b}y = \frac{1}{2b}f(t), \quad b \gg 0. \tag{5.4.8}$$

Note that one can also formally derive the approximation (5.4.8) by first dividing each term in (5.4.7),

$$\frac{1}{2b}y'' + y + \frac{\omega_0^2}{2b}y = \frac{1}{2b}f(t),$$

and then ignoring the acceleration term. However, by taking this approach, it is not clear why the acceleration term can be ignored, while the other seemingly equally small terms must be kept.

Let us briefly consider one implication of using the approximate ODE (5.4.8) instead of the full ODE (5.4.7). Consider the initial value problem associated with each ODE. The full system requires both an initial position and an initial velocity, whereas the second requires only an initial position. Consequently, in using (5.4.8), we are implicitly saying that the initial velocity does not have a significant impact on the solution behavior.

Returning back to our approximation (5.4.8), this system is precisely the form of (5.4.3) with $a = \omega_0^2/2b$. Consequently, the ideas behind the analysis presented for the one-tank mixing problem are (formally) applicable to the study of the dynamics for the strongly damped mass-spring system. Let us now assume that the forcing function has the form

$$f(t) = f_0 \sum_{j=1}^{\infty} \delta(t - j\omega), \tag{5.4.9}$$

so the approximate system has the form

$$y' + \frac{\omega_0^2}{2b}y = \frac{f_0}{2b}\sum_{j=1}^{\infty}\delta(t-j\omega).$$

The forcing is ω-periodic, and, as we saw in section 5.4.3, it has mean f_0/ω. This ODE was solved in section 5.4.3 in the context of the one-tank mixing problem. Using the relationships

$$a = \frac{\omega_0^2}{2b} \ll 1, \quad aV_0c_0\omega \mapsto \frac{f_0}{2b},$$

we know that for $t > 0$ sufficiently large, the solution is approximately

$$y(t) \sim \frac{f_0}{2ab\omega}\frac{a\omega e^{a\omega}}{e^{a\omega}-1}e^{-a(t-n\omega)}, \quad n\omega \le t < (n+1)\omega.$$

Here sufficiently large means $t > N\omega$ with N satisfying

$$aN\omega \gg 0 \quad \leadsto \quad N \gg \frac{2b}{\omega_0^2}\frac{1}{\omega} \gg 1.$$

Noting that

$$ab = \frac{1}{2}\omega_0^2,$$

we can rewrite the large time solution as

$$y(t) \sim \frac{f_0}{\omega \omega_0^2}\frac{a\omega e^{a\omega}}{e^{a\omega}-1}e^{-a(t-n\omega)}, \quad n\omega \le t < (n+1)\omega.$$

We now describe the behavior of the large-time solution. Using the calculation of (5.4.6) we see that the mean of this solution is $f_0/(\omega\omega_0^2)$ and is independent of the strength of damping. The variation about the mean is as what it was for the one-tank mixing problem:

$$\frac{f_0}{\omega\omega_0^2}\frac{1+a\omega-e^{a\omega}}{e^{a\omega}-1} \leq y(t) - \frac{f_0}{\omega\omega_0^2} \leq \frac{f_0}{\omega\omega_0^2}\frac{1+a\omega e^{a\omega}-e^{a\omega}}{e^{a\omega}-1}.$$

Since $a > 0$ is small, we have the approximations

$$\frac{1+a\omega-e^{a\omega}}{e^{a\omega}-1} \sim -\frac{1}{2}a\omega$$

and

$$\frac{1+a\omega e^{a\omega}-e^{a\omega}}{e^{a\omega}-1} \sim \frac{1}{2}a\omega.$$

Consequently, on using

$$\frac{1}{2}a\omega = \frac{\omega\omega_0^2}{4b},$$

we have the approximate solution bounds

$$-\frac{f_0}{4b} \leq y(t) - \frac{f_0}{\omega\omega_0^2} \leq \frac{f_0}{4b}.$$

In conclusion, for large time, the solution is a sawtooth ω-periodic function with mean $f_0/(\omega\omega_0^2)$. Moreover, the variation about the mean is small and of $\mathcal{O}(1/b)$.

Remark 5.22. *The only worrisome part in the analysis is that we dropped the second-order derivative term; in other words, we assumed that for large damping, the effect of acceleration is negligible. For the problem at hand, this assumption is valid, as the piece that is being missed by the analysis decays exponentially fast to zero as $t \to +\infty$ (see Figure 5.4). However, it need not be the case that the assumption is valid for other problems. The interested reader should consult Holmes [21] for an introduction to problems of this type, which are known as singular perturbation problems.*

Exercises

Exercise 5.4.1. *Consider a 5-gallon tank that is initially filled with distilled water. A briny solution with a concentration of $c(t)$ lb/gal, where $c(t)$ is given by*

$$c(t) = \begin{cases} 0, & 0 \leq t < 1, \\ 10, & 1 \leq t < 4, \\ 3, & 4 \leq t, \end{cases}$$

flows into the tank at a rate of 2 gal/min. The uniformly mixed solution exits the tank at the same rate, so that the amount of solution in the tank remains constant.

(a) *What is the concentration of the briny solution in the tank, say, $c_{\text{tank}}(t)$, at time t?*

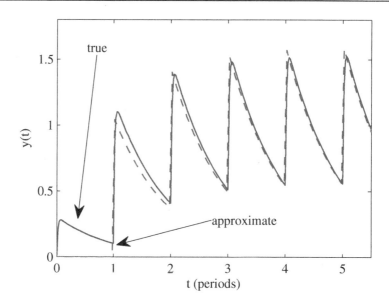

Figure 5.4. *Two solution plots for the strongly damped mass-spring system when $b = 15$ and $\omega_0 = 4$. The forcing function is given by the periodic delta-function model of (5.4.9) with $f_0 = 32$ and $\omega = 2$. The period of the horizontal axis refers to the period of the forcing function. Both solution plots start with the initial position $y(0) = 0$. The plot of the solution to the second-order ODE (the true model) with initial velocity $y'(0) = 8$ is given by the solid curve. The plot of the solution to the first-order ODE (the approximate model) is given by the dashed curve. Note that after three forcing periods, the two solutions almost coincide.*

(b) For which value(s) is $c_{\text{tank}}(t)$ maximized? What is the maximum value?

(c) What happens to the concentration in the tank as $t \to +\infty$?

Exercise 5.4.2. *Consider the forced oscillator that is modeled by the equation*

$$y'' + 4y = 2\delta(t-1) + f_0 \delta(t-a), \quad a > 1.$$

Further suppose that the initial condition is given by $y(0) = y'(0) = 0$. Determine value(s) f_0 and a so that the amplitude of the solution for $t > a$ is two.

Exercise 5.4.3. *Consider the forced oscillator that is modeled by the equation*

$$y'' + 4y = \begin{cases} 0, & 0 \leq t < 1, \\ 8, & 1 \leq t < a, \\ 0, & a \leq t. \end{cases}$$

Further suppose that the initial condition is given by $y(0) = y'(0) = 0$. Determine value(s) of a so that the amplitude of the solution for $t > a$ is minimized.

Exercise 5.4.4. *Suppose that a bank account pays a 4% interest rate (recall the discussion in section 2.6.2). Suppose that the account initially contains $10,000. Further suppose that $500 is deposited into the account in one lump sum in the middle of each year for 20 consecutive years.*

(a) *Letting $x(t)$ denote the amount of money in the account at time t, make a convincing argument that $x(t)$ solves the ODE*

$$x' = 0.04x + 500 \sum_{j=1}^{20} \delta(t - j + 1/2), \quad x(0) = 10{,}000.$$

(b) *How much money is in the account after 21 years?*

Exercise 5.4.5. *Reconsider the one-tank mixing problem of section 5.4.3. Instead of supposing that the mean concentration over one period is c_0, now suppose that the concentration at the time of injection is c_0. In other words, now suppose that*

$$c(t) = c_0 \sum_{j=1}^{\infty} \delta(t - j\omega).$$

(a) *What is the mean concentration over one period?*

(b) *What is the approximate concentration in the tank for $t \gg 0$?*

(c) *What is the mean concentration in the tank for $t \gg 0$?*

(d) *If the period is large, i.e., $\omega \gg 0$, what is the maximum concentration in the tank?*

(e) *If the period is small, i.e., $0 < \omega \ll 1$, what are the lower and upper bounds on the variation of the concentration in the tank about its mean?*

5.5 ▪ The transfer function, convolutions, and variation of parameters

We will conclude our study of the Laplace transform by looking at the solution through a different lens. In doing so, we will be able to construct a variation of parameters solution that is valid for an ODE of any order.

Consider the nth-order ODE given by

$$y^{(n)} + a_{n-1} y^{(n-1)} + \cdots + a_1 y' + a_0 y = f(t). \tag{5.5.1}$$

On setting

$$Y_p(s) := \mathscr{L}[y_p](s), \quad F(s) := \mathscr{L}[f](s),$$

we know from Exercise 5.3.3 and the linearity of the Laplace transform that the transform of the particular solution with zero initial data is

$$(s^n + a_{n-1} s^{n-1} + \cdots + a_1 s + a_0) Y_p(s) = F(s) \quad \rightsquigarrow \quad Y_p(s) = H(s) F(s).$$

Here

$$H(s) := \frac{1}{s^n + a_{n-1} s^{n-1} + \cdots + a_1 s + a_0}$$

is known as the *transfer function* for the ODE (5.5.1). In the engineering vernacular, the transfer function represents the physical system that is being modeled by an

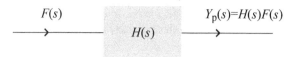

Figure 5.5. *The input $F(s)$ is the Laplace transform of the forcing function $f(t)$, and the output $Y_p(s)$ is the Laplace transform of the particular solution $y_p(t)$. The physical system being modeled is described by the transfer function $H(s)$.*

ODE. See Figure 5.5 for a typical graphical representation of a system response (the solution $y(t)$) to a given input (the forcing function $f(t)$).

Given the transfer function, it turns out that we can use the fact that the Laplace transform of the solution to the ODE (5.5.1) is the product of the transfer function and the Laplace transform of the forcing function to immediately write down the solution. Denote by $h(t)$ the inverse Laplace transform of the transfer function, i.e., $\mathscr{L}[h](s) = H(s)$. Defining the *convolution* of $h(t)$ and $f(t)$ to be

$$h \star f(t) := \int_0^t h(t-u)f(u)\,du,$$

we have the following:

Lemma 5.23. *Consider the ODE (5.5.1). Letting $h(t)$ denote the inverse Laplace transform of the transfer function $H(s)$, the particular solution with zero initial data is*

$$y_p(t) = h \star f(t) \quad \left(= \int_0^t h(t-u)f(u)\,du\right).$$

In other words,

$$Y_p(s) = H(s)F(s) \quad \leadsto \quad y_p(t) = h \star f(t).$$

Proof. We have already seen that the Laplace transform of the particular solution is given by
$$Y_p(s) = H(s)F(s).$$
Given that we are proposing that the solution is the convolution of $h(t)$ and $f(t)$, in order to prove the result, we must then show that the Laplace transform of a convolution of two functions is precisely the product of the Laplace transform of each function.

By definition, the Laplace transform of the convolution is

$$\mathscr{L}[h \star f](s) = \int_0^{+\infty} e^{-st} h \star f(t)\,dt = \int_0^{+\infty}\int_0^t e^{-st} h(t-u)f(u)\,du\,dt.$$

Reversing the order of integration yields

$$\int_0^{+\infty}\int_0^t e^{-st} h(t-u)f(u)\,du\,dt = \int_0^{+\infty}\int_u^{+\infty} e^{-st} h(t-u)f(u)\,dt\,du,$$

5.5. The transfer function, convolutions, and variation of parameters

and applying the change of variables $t = u + r$ means that

$$\int_0^{+\infty}\int_u^{+\infty} e^{-st} h(t-u) f(u)\, dt\, du = \int_0^{+\infty}\int_0^{+\infty} e^{-s(u+r)} h(r) f(u)\, dr\, du.$$

We conclude by noting that

$$\int_0^{+\infty}\int_0^{+\infty} e^{-s(u+r)} h(r) f(u)\, dr\, du = \int_0^{+\infty}\int_0^{+\infty} e^{-sr} h(r) e^{-su} f(u)\, dr\, du$$

$$= \left(\int_0^{+\infty} e^{-sr} h(r)\, dr\right)\left(\int_0^{+\infty} e^{-su} f(u)\, du\right),$$

which is simply the product of $H(s)$ and $F(s)$. $\square$

While Lemma 5.23 gives the variation of parameters formulation for the particular solution to the nth-order ODE (5.5.1), it is not yet clear as to how the inverse of the transfer function relates to the ODE. Consider the homogeneous IVP

$$\begin{aligned} & y^{(n)} + a_{n-1} y^{(n-1)} + \cdots + a_1 y' + a_0 y = 0, \\ & y(0) = y'(0) = \cdots = y^{(n-2)}(0) = 0,\ y^{(n-1)}(0) = 1. \end{aligned} \quad (5.5.2)$$

On using the result of Exercise 5.3.3 we have for these initial data

$$\mathscr{L}[y^{(j)}](s) = s^j \mathscr{L}[y](s),\ j = 0,\ldots, n-1;\quad \mathscr{L}[y^{(n)}](s) = s^n \mathscr{L}[y](s) - 1.$$

Thus, on using the Laplace transform, we see that the transform of the solution to the IVP is

$$(s^n + a_{n-1} s^{n-1} + \cdots + a_1 s + a_0)\mathscr{L}[y](s) - 1 = 0 \quad \leadsto \quad \mathscr{L}[y](s) = H(s).$$

Inverting yields that $h(t)$ is the solution to the homogeneous IVP (5.5.2). We have consequently proven the following:

Corollary 5.24. *Consider the ODE (5.5.1). The particular solution with zero initial data is*

$$y_p(t) = h \star f(t) \quad \left(= \int_0^t h(t-u) f(u)\, du\right),$$

where $h(t)$ is the solution of the homogeneous IVP

$$\begin{aligned} & y^{(n)} + a_{n-1} y^{(n-1)} + \cdots + a_1 y' + a_0 y = 0, \\ & y(0) = y'(0) = \cdots = y^{(n-2)}(0) = 0,\ y^{(n-1)}(0) = 1. \end{aligned}$$

Alternatively, we have that $h(t)$ is the solution to the nonhomogeneous IVP with zero initial data:

$$\begin{aligned} & y^{(n)} + a_{n-1} y^{(n-1)} + \cdots + a_1 y' + a_0 y = \delta(t), \\ & y(0) = y'(0) = \cdots = y^{(n-1)}(0) = 0. \end{aligned}$$

In this context, $h(t)$ can be thought of as the response of the system to an impulse applied at $t = 0$, so the solution is referred to as the *impulse response*. From this perspective, the particular solution with zero initial data is given as the convolution of the impulse response and the external forcing function. This is all a generalization of the discussion in section 5.4.1.

Example 5.25. Let us find the particular solution with zero initial data for the forced but undamped mass-spring system modeled by

$$y'' + \omega_0^2 y = f(t).$$

The transfer function for this system is given by

$$H(s) = \frac{1}{s^2 + \omega_0^2} = \frac{1}{\omega_0} \frac{\omega_0}{s^2 + \omega_0^2},$$

so that the Laplace transform of the particular solution is

$$Y_p(s) = H(s)F(s) = \frac{1}{\omega_0} \frac{\omega_0}{s^2 + \omega_0^2} F(s).$$

By Corollary 5.24, the inverse of the transfer function, $h(t)$, is the solution to the homogeneous IVP with initial data $y(0) = 0$, $y'(0) = 1$. Using the table in (5.2.1),

$$h(t) = \frac{1}{\omega_0} \sin(\omega_0 t),$$

so by Lemma 5.23, the particular solution is

$$y_p(t) = \frac{1}{\omega_0} \int_0^t \sin(\omega_0(t-u)) f(u) \, du.$$

On using the identity

$$\sin(\omega_0(t-u)) = \sin(\omega_0 t)\cos(\omega_0 u) - \cos(\omega_0 t)\sin(\omega_0 u),$$

we can rewrite the solution as

$$y_p(t) = \frac{1}{\omega_0} \sin(\omega_0 t) \int_0^t \cos(\omega_0 u) f(u) \, du - \frac{1}{\omega_0} \cos(\omega_0 t) \int_0^t \sin(\omega_0 u) f(u) \, du.$$

While we will not go into the confirming details, this is precisely the form of the answer that is gotten by using the variation of parameters formulation of Theorem 4.9, under the assumption that the particular solution satisfies $y_p(0) = y_p'(0) = 0$. ∎

Example 5.26. Consider the third-order ODE

$$(D+1)(D+2)(D+3)y = f(t).$$

We wish to find the particular solution with zero initial data. The transfer function is

$$H(s) = \frac{1}{(s+1)(s+2)(s+3)} = \frac{1}{2}\frac{1}{s+1} - \frac{1}{s+2} + \frac{1}{2}\frac{1}{s+3},$$

5.5. The transfer function, convolutions, and variation of parameters

and using the table in (5.2.1), the inverse is

$$h(t) = \frac{1}{2}e^{-t} - e^{-2t} + \frac{1}{2}e^{-3t}.$$

By Corollary 5.24, the function $h(t)$ is the solution to the homogeneous IVP with initial data $y(0) = y'(0) = 0$ and $y''(0) = -1$. On using Lemma 5.23, the desired particular solution is

$$y_p(t) = \int_0^t h(t-u)f(u)\,du$$
$$= \frac{1}{2}e^{-t}\int_0^t e^u f(u)\,du - e^{-2t}\int_0^t e^{2u} f(u)\,du + \frac{1}{2}e^{-3t}\int_0^t e^{3u} f(u)\,du.$$

The simplified form was found using linearity and

$$\int_0^t e^{\alpha(t-u)} f(u)\,du = \int_0^t e^{\alpha t} e^{-\alpha u} f(u)\,du = e^{\alpha t}\int_0^t e^{-\alpha u} f(u)\,du. \quad\blacksquare$$

Let us conclude our discussion of the transfer function by considering what it tells us about the underlying physical system. There are (at least) three important things to note about the transfer function:

(a) it is the inverse of the characteristic polynomial for the ODE (5.5.1);

(b) the *poles*, i.e., those s-values for which the function $1/H(s)$ has a zero, are the zeros of the characteristic polynomial); and

(c) the inverse is a homogeneous solution that satisfies a certain initial condition (by Corollary 5.24).

Regarding the homogeneous solution to (5.5.1), we know that the location of the zeros of the characteristic polynomial determine the solution behavior. If all of the zeros have negative real part, then the homogeneous solution will be transient; i.e., it will decay to zero exponentially fast as $t \to +\infty$: the zero solution is (asymptotically) stable. If (at least) one of the zeros has positive real part, then there will be some solutions that grow exponentially fast as $t \to +\infty$: the zero solution will be unstable. If some of the zeros are purely imaginary (but simple), then there will be solutions that are uniformly bounded but that do not decay: the zero solution is stable. Thus, the poles of the transfer function determine the solution behavior. Figure 5.6 gives two cartoons in which the potential poles of a transfer function are plotted.

As we saw in Chapter 4, if for a given ODE we are able to factor the corresponding characteristic polynomial, then we can write down the homogeneous solution. Conversely, if we are given the zeros of the characteristic polynomial, then we can re-create the corresponding homogeneous ODE. Since the transfer function is the inverse of the characteristic polynomial, the same statement applies regarding the location of the poles.

Example 5.27. Suppose that the poles of the transfer function are given by

$$s_1 = -2 + i4, \quad s_2 = -2 - i4.$$

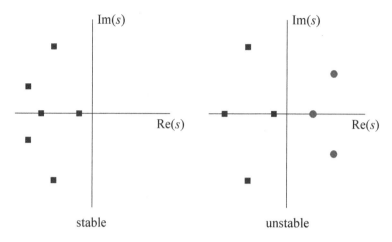

Figure 5.6. *The poles of the transfer function are denoted by squares and circles. Recall that if $s = a + ib$, then $\mathrm{Re}(s) = a$ and $\mathrm{Im}(s) = b$. In the left figure, all of the poles have negative real part, so that the homogeneous solution is a transient solution. In the right figure, some of the poles (the circles) have positive real part, so that the zero solution is unstable.*

Since both poles have negative real part, the homogeneous solution is transient. Furthermore, since

$$H(s) = \frac{1}{[s-(-2+i4)][s-(-2-i4)]} = \frac{1}{s^2+4s+20},$$

the characteristic polynomial for the ODE is

$$p(s) = s^2 + 4s + 20.$$

Thus, the original nonhomogeneous ODE is

$$y'' + 4y' + 20y = f(t). \quad \blacksquare$$

We can use the location of the poles to determine the relevant parameters—natural frequency and damping—associated with a physical system. Recall that the homogeneous problem is

$$y'' + 2by' + \omega_0^2 y = 0,$$

where the parameters in terms of the physical constants are

$$2b = \frac{c}{m}, \quad \omega_0^2 = \frac{k}{m}.$$

We can rewrite these equations to say that

$$k = \omega_0^2 m, \quad c = 2bm;$$

in other words, we can write the spring constant and damping constant as some multiplicative multiple of the mass. Thus, given the location of the poles, we can re-create the homogeneous ODE as in Example 5.27 and then use the coefficients to determine the physical constants.

Remark 5.28. In electrical circuit theory, we have
$$2b = \frac{R}{L}, \quad \omega_0^2 = \frac{1}{LC},$$
so in terms of the inductance, the capacitance and resistance can be written as
$$\frac{1}{C} = \omega_0^2 L, \quad R = 2bL.$$

Example 5.29. Suppose that the poles of the transfer function are given by
$$s_1 = -5 + i3, \quad s_2 = -5 - i3.$$
Since
$$H(s) = \frac{1}{[s-(-5+i3)][s-(-5-i3)]} = \frac{1}{s^2 + 10s + 34},$$
the characteristic polynomial for the ODE is
$$p(s) = s^2 + 10s + 34,$$
and the associated homogeneous ODE is
$$y'' + 10y' + 34y = f(t).$$
Thus,
$$2b = 10, \quad \omega_0^2 = 34,$$
so in terms of the mass, the other physical quantities are
$$k = 34m, \quad c = 10m. \quad \blacksquare$$

Exercises

Exercise 5.5.1. *Find the transfer function associated with the following ODEs and state whether or not the solution of the homogeneous problem is transient.*

(a) $y'' + 10y' + 25y = f(t)$

(b) $y'' + 8y' + 17y = f(t)$

(c) $y'' + 8y' + 12y = f(t)$

(d) $y'' - 5y' + 6y = f(t)$

(e) $(D^2 + 4)(D+3)^2(D+6)y = f(t)$

Exercise 5.5.2. *For the following ODEs, find the particular solution with zero initial data using a convolution integral.*

(a) $y'' + 10y' + 25y = f(t)$

(b) $y'' + 8y' + 17y = f(t)$

(c) $y'' + 8y' + 12y = f(t)$

(d) $y'' - 5y' + 6y = f(t)$

(e) $(D^2 + 4)(D+3)^2(D+6)y = f(t)$.

Exercise 5.5.3. *For the following transfer functions, (a) write down the corresponding nonhomogeneous ODE, (b) determine if the homogeneous solution is transient, and (c) write down the general solution to the nonhomogeneous ODE by writing the particular solution as a convolution integral.*

(a) $H(s) = [s^2 + 12s + 35]^{-1}$

(b) $H(s) = [s^2 + 8s + 32]^{-1}$

(c) $H(s) = [s^2 + 64]^{-1}$

(d) $H(s) = [(s+2)(s+6)^2(s^2+6s+34)]^{-1}$

Exercise 5.5.4. *For the following problems, the poles of a transfer function are given. Write down the corresponding nonhomogeneous ODE.*

(a) $s_1 = 2, s_2 = -5$

(b) $s_1 = -1 + i2, s_2 = -1 - i2$

(c) $s_1 = 3, s_2 = 3$

(d) $s_1 = 2, s_2 = 5, s_3 = 8$

(e) $s_1 = -4, s_2 = -1, s_3 = i5, s_4 = -i5$

Exercise 5.5.5. *For the following problems, the poles of a transfer function are given. If the corresponding physical system is that of a damped mass-spring system, determine the damping constant, c, and spring constant, k, in terms of the mass, m.*

(a) $s_1 = -2, s_2 = -7$

(b) $s_1 = -2 + i4, s_2 = -2 - i4$

(c) $s_1 = -5, s_2 = -5$

(d) $s_1 = -3 + i8, s_2 = -3 - i8$

(e) $s_1 = i3, s_2 = -i3$

Exercise 5.5.6. *For the following problems, the poles of a transfer function are given. If the corresponding physical system is that of an RLC circuit, determine the resistance, R, and capacitance, C, in terms of the inductance, L.*

(a) $s_1 = -4, s_2 = -9$

(b) $s_1 = -1 + i6$, $s_2 = -1 - i6$

(c) $s_1 = -7$, $s_2 = -7$

(d) $s_1 = -8 + i$, $s_2 = -8 - i$

(e) $s_1 = i3$, $s_2 = -i3$

Group projects

5.1. Consider the forced oscillator that is modeled by the equation

$$y'' + 4y = f_0 \omega \sum_{j=1}^{\infty} \delta(t - j\omega), \quad \omega > 0.$$

Further suppose that the initial condition is given by $y(0) = y'(0) = 0$.

(a) What is the average of the forcing function over one period?

(b) Is it possible for there to be resonant forcing (see section 4.3.2 for the case where the forcing function is modeled by $f_0 \cos(\omega t)$)?

(c) If there is resonant forcing, for which value(s) of ω does it occur?

(d) If there is resonant forcing for $\omega = \omega^*$, describe the solution behavior for $\omega \sim \omega^*$. Compare your answer with that described in Figure 4.4.

(e) What is the solution behavior when $0 < \omega \ll 1$?

(f) What is the solution behavior when $\omega \gg 0$? Compare your answer with that described in Figure 4.3.

5.2. Suppose that the incoming concentration for the one-tank mixing problem ODE (5.4.3) in section 5.4.3,

$$y' + ay = aV_0 c(t), \quad y(0) = 0,$$

is of the form

$$c(t) = 2c_0 \sum_{j=0}^{\infty} [H(t - j\omega) - H(t - (j + 1/2)\omega)].$$

(a) Plot a graph of $c(t)$. What is its period?

(b) Show that the mean of $c(t)$ is c_0.

(c) Find the approximate concentration for $t \gg 0$.

(d) What is the mean of the concentration for $t \gg 0$?

(e) What is the variation of the concentration about its mean for $t \gg 0$?

(f) Is there a phase shift between the concentration and the forcing for $t \gg 0$?

(g) Assume that $t \gg 0$. Carefully describe the solution behavior in the two distinguished limits, $0 < \omega \ll 1$ and $\omega \gg 1$.

5.3. Recall that in section 3.1.2, we provided the linear Ackerman model for the concentrations of glucose and insulin in the blood:

$$G' = aG + bI + J(t), \quad G(0) = G_0,$$
$$I' = cG + dI, \quad I(0) = I_0.$$

We will suppose that the ingestion of insulin is modeled by

$$J(t) = J_0 \omega \sum_{j=1}^{\infty} \delta(t - j\omega), \quad \omega > 0.$$

This form of $J(t)$ models the ingestion of $J_0 \omega$ units of insulin when $t = \omega, 2\omega, 3\omega, \ldots$.

(a) What is the average amount of insulin being ingested over one period?

(b) Show that the concentration of insulin satisfies the second-order linear ODE

$$I'' - (a+d)I' + (ad - bc)I = cJ(t); \quad I(0) = I_0, \ I'(0) = cG_0 + dI_0.$$

(c) If $I(t)$ solves the ODE in (b), how do we find $G(t)$ without solving another ODE?

(d) Under the model assumptions

$$a, b, c < 0, \quad d > 0,$$

show that the concentration of insulin in the blood for $t \gg 0$ does not depend on the initial amount of insulin and glucose in the body.

(e) Show that for $t \gg 0$, the concentration of insulin in the body is modeled by an ω-periodic function.

(f) Now use the experimentally determined coefficient values:

$$a = -2.92, \quad b = -4.34, \quad c = 0.21, \quad d = -0.78.$$

Assume that $t \gg 0$. What is the average concentration of insulin that is present in the body between successive injections?

(g) Assume that $t \gg 0$. Carefully describe the solution behavior in the two distinguished limits, $0 < \omega \ll 1$ and $\omega \gg 1$. Use amplitude and phase diagrams (if appropriate).

5.4. Consider the two-tank mixing problem of section 3.7.1. The governing equations are

$$x_1' = -2ax_1 + ax_2 + aV_0 c(t), \quad x_1(0) = x_{1,0},$$
$$x_2' = ax_1 - 2ax_2, \quad x_2(0) = x_{2,0}.$$

We will suppose that the incoming concentration is modeled by

$$c(t) = c_0 \omega \sum_{j=1}^{\infty} \delta(t - j\omega), \quad \omega > 0.$$

This form of $c(t)$ models the injection of $c_0 \omega$ units of briny solution every $t = \omega, 2\omega, 3\omega, \ldots$.

(a) What is the average concentration being injected over one period?

(b) Show that the amount of salt in the second tank satisfies the second-order linear ODE

$$x_2'' + 4ax_2' + 3a^2 x_2 = a^2 V_0 c(t); \quad x_2(0) = x_{2,0}, \; x_2'(0) = ax_{1,0} - 2ax_{2,0}.$$

(c) If $x_2(t)$ solves the ODE in (b), how do we find $x_1(t)$ without solving another ODE?

(d) Show that the amount of salt in the second tank for $t \gg 0$ does not depend on the initial amount of salt in the either tank.

(e) Show that for $t \gg 0$, the concentration of salt in the second tank is modeled by an ω-periodic function.

(f) Assume that $t \gg 0$. What is the average concentration of salt in the second tank that is present between successive injections?

(g) Assume that $t \gg 0$. Carefully describe the solution behavior in the two distinguished limits, $0 < \omega \ll 1$ and $\omega \gg 1$.

Chapter 6
Odds and ends

> *We are usually convinced more easily by reasons we have found ourselves than by those which occurred to others.*
>
> — Blaise Pascal

The purpose of this chapter is to briefly discuss three classical topics that have not yet been touched on. We will not cover any of them in extensive detail. Instead, we will focus on a few key examples associated with each topic that will point the reader towards a basic understanding of the associated material. It is expected that if mastery of the material is desired, then the reader will consult another text (such as [9, 25, 37]) to get those additional details. It will be assumed throughout this chapter that the ODE being considered will have unique solutions over some time interval.

6.1 • Separation of variables

In Chapter 2, we studied linear scalar ODEs of the form

$$x' = a(t)x + f(t)$$

and showed that the general solution is of the form

$$x(t) = \Phi(t)c + \Phi(t)\int \Phi(t)^{-1} f(t)\,dt, \quad \Phi(t) = e^{\int a(t)\,dt}. \quad (6.1.1)$$

In section 2.2.1, we discussed the existence and uniqueness theory for the general problem

$$x' = g(t,x), \quad x(t_0) = x_0,$$

and stated a theorem that says that as long as $g(t,x)$ and $\partial_x g(t,x)$ are continuous, then the IVP will have a unique solution. However, we did not state what is that solution; indeed, in most cases, it is not possible to have an explicit solution formula such as that given in (6.1.1). If the function $g(t,x)$ is *separable*, i.e., it can be written as a product of a function of t and a function of x, then it will be possible to provide an *implicit* solution formula. In some cases, this implicit formula can be made into an explicit formula.

The implicit formula follows from the following argument. We start by assuming that $g(t,x)$ is separable, i.e.,

$$g(t,x) = h(t)f(x) \quad \leadsto \quad x' = h(t)f(x). \tag{6.1.2}$$

In order to solve this ODE, we first rewrite it in the form

$$\frac{1}{f(x)}\frac{dx}{dt} = h(t)$$

and then integrate both sides with respect to t:

$$\int \frac{1}{f(x)}\frac{dx}{dt}\,dt = \int h(t)\,dt. \tag{6.1.3}$$

The right-hand side of the above equation makes sense. As for the left-hand side, there is the implicit assumption that $x = x(t)$, so we should write it as

$$\int \frac{1}{f(x)}\frac{dx}{dt}\,dt = \int \frac{1}{f(x(t))} x'(t)\,dt.$$

But if we set $u = x(t)$, then by the integration method of u-substitution, we get

$$\int \frac{1}{f(u)}\,du = \int \frac{1}{f(x(t))} x'(t)\,dt.$$

Thus, we can rewrite (6.1.3) as

$$\int \frac{1}{f(u)}\,du = \int h(t)\,dt.$$

Since u is a dummy integration variable for the integral on the left-hand side, we can replace it with x. Furthermore, since the integrals on both sides are indefinite integrals, there is an arbitrary integration constant associated with the implicit solution. In conclusion, a general solution to the separable ODE (6.1.2) is

$$\int \frac{1}{f(x)}\,dx = \int h(t)\,dt + c, \tag{6.1.4}$$

where $c \in \mathbb{R}$ is arbitrary. If we set

$$F(x) := \int \frac{1}{f(x)}\,dx, \quad H(t) := \int h(t)\,dt,$$

then the solution can be written as

$$F(x) = H(t) + c.$$

The solution is implicit because it is not clear that we can explicitly invert $F(x)$.

6.1. Separation of variables

Example 6.1. Consider the logistic equation

$$x' = rx\left(1 - \frac{x}{k}\right), \quad r, k > 0.$$

The biological significance of this equation will be discussed in more detail in section 6.3.1. In order to simplify our analysis, we first rewrite the ODE as

$$x' = \frac{r}{k} x(k - x).$$

Writing

$$h(t) = \frac{r}{k}, \quad f(x) = x(k - x),$$

we know by (6.1.4) that the implicit general solution is given by

$$\int \frac{1}{x(k-x)} dx = \int \frac{r}{k} dt + c.$$

The method of partial fractions yields

$$\frac{1}{x(k-x)} = \frac{1}{k}\left(\frac{1}{x} + \frac{1}{k-x}\right),$$

so on integrating, we have the implicit solution

$$\frac{1}{k}[\ln(x) - \ln(k - x)] = \frac{r}{k} t + c.$$

We now make the solution explicit. First, we have

$$\ln\left(\frac{x}{k-x}\right) = rt + c,$$

where we have relabeled the arbitrary constant ck as c. Exponentiating both sides and using the law of exponents gives

$$\frac{x}{k-x} = ce^{rt}.$$

Here we again relabeled the arbitrary constant e^c as c. We now have an algebraic equation in x. Multiplying through by $k - x$ gives

$$x = ce^{rt}(k - x) \quad \rightsquigarrow \quad (1 + ce^{rt})x = cke^{rt}.$$

We can no longer relabel the arbitrary constant because it appears on both sides of the equation. We finally simplify to get the explicit solution

$$x(t) = \frac{cke^{rt}}{1 + ce^{rt}}.$$

Now consider the IVP

$$x' = \frac{r}{k} x(k - x), \quad x(0) = x_0.$$

From our solution formula, we have

$$x_0 = x(0) = \frac{ck}{1+c} \quad \leadsto \quad c = \frac{x_0}{k-x_0}.$$

Plugging this value of c into the solution formula and simplifying gives

$$x(t) = \frac{kx_0 e^{rt}}{k - x_0 + x_0 e^{rt}}.$$

From this solution formula, we get the asymptotic behavior

$$x_0 < 0: \lim_{t \to T^-} x(t) = -\infty, \quad T := \frac{1}{r}\ln\left(\frac{x_0 - k}{x_0}\right),$$

$$x_0 > 0: \lim_{t \to +\infty} x(t) = k.$$

This differentiation in asymptotic behaviour based on the initial data is an important feature of the model and will be considered more closely in section 6.3.1. ∎

Exercises

Exercise 6.1.1. *Find an explicit general solution for the following ODEs.*

(a) $x' = 6\sin(3t)x^2$

(b) $x' = 2t\cot(x)$

(c) $x' = e^{t-x}$

6.2 • Nonlinear scalar autonomous ODEs and the phase line

A nonlinear scalar autonomous ODE is a special case of the separable ODE with $h(t) \equiv 1$ and will be written as

$$x' = f(x). \tag{6.2.1}$$

We know that the implicit solution is given by

$$\int \frac{1}{f(x)}\,dx = t + c.$$

In special cases, it may be possible to invert to get an explicit solution (such as in Example 6.1), but in general either

(a) it may not be possible to derive an explicit solution or

(b) the explicit solution is so complicated that it seemingly hides solution behavior.

For an example of case (a), consider the ODE

$$x' = \frac{e^{-x}}{x} \quad \leadsto \quad xe^x - x = t + c.$$

6.2. Nonlinear scalar autonomous ODEs and the phase line   277

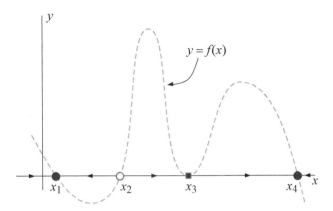

Figure 6.1. *The phase line for a given function $f(x)$. The plot of $y = f(x)$ is given by the dashed curve. The phase line is presented on the x-axis. Unstable equilibrium solutions are shown as open circles, stable equilibrium solutions are shown as filled circles, and semistable equilibrium solutions are represented by filled squares.*

For an example of case (b), consider

$$x' = \sec(x) \quad \leadsto \quad x = \cos^{-1}(t+c).$$

In this section, we will not focus on finding implicit (or explicit) solutions for the ODE. Instead, we will determine the asymptotic solution behavior from a graphical perspective. This approach should be reminiscent of the phase-plane analysis presented in section 3.5.

Suppose that the graph of $f(x)$ is as given in Figure 6.1. The idea behind drawing the phase line is quite straightforward. We know from Calculus I that for a differentiable function $x(t)$,

(a) $x'(t) < 0$ means the function is decreasing, and

(b) $x'(t) > 0$ means the function is increasing.

For the ODE, we do not know the solution $x(t)$; however, we do know the derivative $x'(t)$ through the ODE (6.2.1). Consequently, wherever $f(x) > 0$, the solution will be increasing, and wherever $f(x) < 0$, the solution will be decreasing. We will represent this behavior by appropriately pointing arrows on the x-axis of the graph of $f(x)$, and in this context the x-axis will be called the *phase line*. The points where $f(x) = 0$ are the equilibrium solutions for the ODE, as in this case $x' \equiv 0$, means that $x(t) \equiv x^*$ for some x^*.

For an example, suppose that $f(x)$ has a plot as given in Figure 6.1. Here there are four equilibrium solutions, i.e., four points where $f(x) = 0$. First, consider the solution x_1. Since

$$x < 0 : f(x) > 0 \quad \leadsto \quad x'(t) > 0,$$
$$x_1 < x < x_2 : f(x) < 0 \quad \leadsto \quad x'(t) < 0,$$

the solution x_1 is stable. In particular, we have

$$x(0) \in (-\infty, x_1) \cup (x_1, x_2) \quad \leadsto \quad \lim_{t \to +\infty} x(t) = x_1.$$

The same can be said about the solution x_4. Since

$$x_3 < x < x_4 : f(x) > 0 \rightsquigarrow x'(t) > 0,$$
$$x_4 < x : f(x) < 0 \rightsquigarrow x'(t) < 0,$$

the solution x_4 is stable. In particular,

$$x(0) \in (x_3, x_4) \cup (x_4, +\infty) \rightsquigarrow \lim_{t \to +\infty} x(t) = x_4.$$

Now consider the solution x_2. Since

$$x_1 < x < x_2 : f(x) < 0 \rightsquigarrow x'(t) < 0,$$
$$x_2 < x < x_3 : f(x) > 0 \rightsquigarrow x'(t) > 0,$$

it will be the case that no matter how close $x(0)$ is to x_2, the solution $x(t)$ will not stay "near" x_2 as t increases. In particular,

$$x(0) \in (x_1, x_2) \rightsquigarrow \lim_{t \to +\infty} x(t) = x_1$$

and

$$x(0) \in (x_2, x_3) \rightsquigarrow \lim_{t \to +\infty} x(t) = x_3.$$

The equilibrium solution x_2 is unstable.

As for the solution x_3, we have $f(x) > 0$ both to the right and to the left of this equilibrium solution. Consequently, $x'(t) > 0$ both to the right and to the left. We say that this solution is semistable. In particular, for initial data to the left of x_3,

$$x(0) \in (x_2, x_3) \rightsquigarrow \lim_{t \to +\infty} x(t) = x_3,$$

but for initial data to the right of x_3,

$$x(0) \in (x_3, x_4) \rightsquigarrow \lim_{t \to +\infty} x(t) = x_4.$$

If the initial data is to the left, then the solution acts as if x_3 was stable, while if the initial data is to the right, the solution acts as if the equilibrium solution is unstable.

Exercises

Exercise 6.2.1. *Draw the phase line for each of the following autonomous scalar ODEs. Identify the equilibrium solutions on your phase line and label each one as stable, unstable, or semistable.*

(a) $x' = 3x(1-x)^2(4-x)$

(b) $x' = \sin(x)$

(c) $x' = \cos(2x)$

(d) $x' = -2x^3(1+x)(1-x)(3-x)^4$

(e) $x' = 4x^2(2+x)(4-x)(7-x)^3$

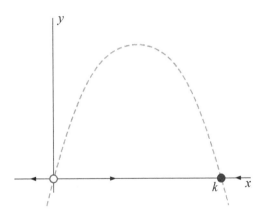

Figure 6.2. *The phase line for the logistic equation* (6.3.1). *The plot of* $y = rx(1-x/k)$ *is given by the dashed curve. The phase line is presented on the x-axis. The unstable equilibrium solution* $x = 0$ *is shown as an open circle, and the stable equilibrium solution* $x = k$ *is shown as a filled circle.*

6.3 ▪ Case studies

6.3.1 ▪ The logistic equation

Consider the logistic equation

$$x' = rx\left(1 - \frac{x}{k}\right), \tag{6.3.1}$$

which was explicitly solved in Example 6.1. In the setting of population dynamics, the variable x represents the total population and is assumed to be nonnegative.

Let us begin by considering the biological meaning of the constants. When x is small, the ODE is approximately linear:

$$x' \sim rx.$$

This linear ODE has the solution $x(t) = ce^{rt}$. The parameter r then describes the exponential growth rate for small populations. This is consistent with the analysis in Example 6.1, where it was shown that $x = 0$ is an unstable equilibrium solution. As for the constant k, we saw in Example 6.1 that $x = k$ is a stable equilibrium solution. The parameter k then represents the *carrying capacity* of the environment.

We now solve the ODE (6.3.1) through a phase-line analysis. The plot of $f(x)$ and the associated phase line are given in Figure 6.2. Here there are two equilibrium solutions. First, consider the point $x = 0$. Since $f(x) < 0$ for $x < 0$ (which means $x'(t) < 0$ for $x < 0$) and $f(x) > 0$ for $0 < x < k$ (which means $x'(t) > 0$ for $0 < x < k$), the equilibrium solution $x = 0$ is unstable. Now consider the point $x = k$. Since $f(x) > 0$ for $0 < x < k$ and $f(x) < 0$ for $x > k$ (which means $x'(t) < 0$ for $x > k$), the equilibrium solution $x = k$ is stable. This asymptotic behavior not only is seen by studying the graph of $f(x)$ but was demonstrated in Example 6.1 via a careful study of the explicit solution. The phase-line analysis corroborates the analysis found by studying the explicit solution.

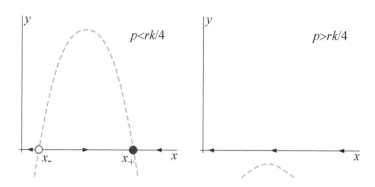

Figure 6.3. *In both panels, the plot of the function $f(x)$ is given by the dashed curve for the case of constant predation. The phase line is presented on the x-axis. In the left panel, there are two equilibrium solutions, while in the right panel, there are no equilibrium solutions.*

6.3.2 ▪ Budworm population dynamics

This problem is considered in Strogatz [37, Chapter 3.7]. Let $x(t) \geq 0$ represent the budworm population at time t. Under the assumptions of no predation and a finite environment, a mathematical model that describes the budworm population is the logistic model

$$x' = rx\left(1 - \frac{x}{k}\right).$$

Now suppose that the budworm population experiences a constant predation (primarily by birds). This additional assumption means that the logistic model should read

$$x' = rx\left(1 - \frac{x}{k}\right) - p,$$

where $p > 0$ corresponds to the rate of predation. A phase-line analysis of this ODE shows two possibilities. If $0 < p < rk/4$, then there will be two equilibrium solutions, $0 < x_- < x_+ < k$ (see the left panel of Figure 6.3). The solution x_- is unstable, and the solution x_+ is stable. If the initial population satisfies $x(0) < x_-$, then there will be a finite time such that $x(t) = 0$; in other words, the budworms will go extinct. On the other hand, if $x(0) > x_-$, then $x(t) \to x_+$ as $t \to +\infty$, so that the budworm population will stabilize at some level below the carrying capacity of the environment. On the other hand, if $p > rk/4$, then there will be no equilibrium solutions, and for any initial population $x(0)$, there will be a finite time such that $x(t) = 0$ (see the right panel of Figure 6.3). For predation rates above this threshold, the budworms will always become extinct.

It is probably unrealistic to assume that the predation rate does not depend on the population of budworms. It is more likely to be the case that the rate will be small if the population is small and will achieve some maximal level as the size of the population increases. In nondimensionalized form, one possible model is

$$x' = rx\left(1 - \frac{x}{k}\right) - \frac{x^2}{1 + x^2}.$$

For this model, the predation function is given by

$$p(x) = \frac{x^2}{1 + x^2},$$

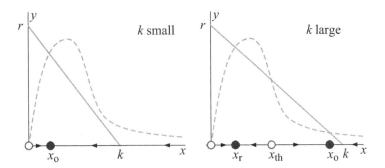

Figure 6.4. *In both panels, the plot of the two pieces of the function $f(x)$ are given by the solid curve ($y = r(1 - x/k)$) and dashed curve $y = x/(1 + x^2)$ for the case where the predation depends on the population. The intersection point(s) of these two curves correspond to equilibrium solutions. The phase line for the full problem is presented on the x-axis. The stability of each equilibrium solution follows from the fact that (a) $x = 0$ is unstable and (b) the stability of each equilibrium solution must alternate between stable and unstable. In the left panel, there is one nonzero equilibrium solution, while in the right panel, there are three nonzero equilibrium solutions.*

and it has the biologically relevant properties that $p(0) = 0$, $p'(x) > 0$, and $p(x) \to 1$ as $x \to +\infty$. In other words, there is no predation when there are no budworms, the rate of predation increases as the population of budworms increases, and the rate of predation has a maximal value.

For this model, there is always an unstable fixed point at $x = 0$. Since the right-hand side can be rewritten as

$$rx\left(1 - \frac{x}{k}\right) - \frac{x^2}{1+x^2} = x\left[r\left(1 - \frac{x}{k}\right) - \frac{x}{1+x^2}\right],$$

the other fixed point(s) are found by solving

$$r\left(1 - \frac{x}{k}\right) - \frac{x}{1+x^2} = 0 \quad \rightsquigarrow \quad r\left(1 - \frac{x}{k}\right) = \frac{x}{1+x^2}.$$

Two possibilities are detailed in Figure 6.4. Let $r > 0$ be fixed. If $k > 0$ is not too large, then there will be precisely one solution to this equation (see the left panel of Figure 6.4). This equilibrium solution will be stable.

On the other hand, if k is sufficiently large, there will be three solutions, say, $0 < x_r < x_{th} < x_o$ (see the right panel of Figure 6.4). The stable equilibrium solution x_r is known as the refuge level of the population, and the stable equilibrium solution x_o is known as the outbreak level. The unstable equilibrium solution x_{th} is the threshold solution. If the initial budworm population satisfies $x(0) < x_{th}$, then the budworm population will approach the refuge level, whereas if $x(0) > x_{th}$, the budworm population will approach the outbreak level. From the perspective of pest management, one would like the population to approach the refuge level.

Although we will not prove this fact here, for fixed r there is a value of k, say, k^*, such that if $k < k^*$, there is one additional equilibrium solution, whereas if $k > k^*$, there are three additional equilibrium solutions. This change in the number of equilibrium solutions at a particular parameter value is known as a *bifurcation*, and it is a fascinating and important area of study. The interested reader should consult the wonderful book of Strogatz [37] for an introduction to this topic.

6.3.3 • Snowball Earth

A simple ODE used to model the balance of global energy can be written in the form

$$x' = [1 - \alpha(x)]Q - \sigma x^4. \tag{6.3.2}$$

See Kaper and Engler [22, Chapter 2]. Here $x > 0$ is the ratio of the effective surface temperature of the Earth and the heat capacity of the Earth, $\sigma > 0$ is Stefan's constant, $Q > 0$ is proportional to the solar constant (an indication of how much solar energy is captured by the Earth per unit time), and $\alpha(x) > 0$ is the temperature-dependent albedo of the Earth. The term σx^4 in (6.3.2) follows from the Stefan-Boltzmann law, which governs the energy radiated by a black body. Since snow and ice have a much higher albedo than open water, the function $\alpha(x)$ should have the asymptotics

$$\alpha(x) \sim \begin{cases} \alpha_i, & x < x_-, \\ \alpha_w, & x > x_+. \end{cases}$$

Here $\alpha_i > \alpha_w$ and $x_- < x_+$. The term $\alpha_i \sim 0.7$ refers to the albedo of ice, and $\alpha_w \sim 0.3$ is the albedo of water. The value $(x_- + x_+)/2$ corresponds to the freezing temperature of water. For the sake of discussion, we will assume that

$$x_- = 250, \quad x_+ = 280,$$

and set

$$\alpha(x) = \frac{1}{2} - \frac{1}{5}\tanh(x - 265) \quad \leadsto \quad 1 - \alpha(x) = \frac{1}{2} + \frac{1}{5}\tanh(x - 265).$$

We have various possibilities for the phase line as a function of the parameter Q. As was the case for the budworm population dynamics model, each scenario depends on how Q relates to two critical values, $Q_- < Q_+$. The critical points for the ODE are found by solving the nonlinear equation

$$1 - \alpha(x) = \frac{\sigma}{Q}x^4, \quad x > 0.$$

If $Q < Q_-$, then there will be only one solution that will have a value less than the freezing temperature of water, and it will correspond to a stable critical point. If $Q_- < Q < Q_+$, then there will be three critical points: the outer ones will be stable, and the inner one will be unstable. Finally, if $Q > Q_+$, there will again be one stable critical point; however, the value will be greater than the freezing temperature of water (see Figure 6.5).

We have the following physical interpretation of the solution, assuming that the heat capacity of the Earth is fixed. If Q is too small, i.e., the Earth does not capture a sufficient amount of solar energy (e.g., there is a large amount of ice covering the surface of the Earth), then the equilibrium temperature is below freezing, and the Earth is stuck in an ice age. If Q is sufficiently large (e.g., the sun heats up or more radiation is trapped by the atmosphere), then the equilibrium temperature is well above freezing. Finally, if $Q_- < Q < Q_+$, then it is possible to have either scenario. It simply depends on the initial temperature of the Earth. If it is low enough, then there will be an ice age, whereas if it is high enough, there will be an ice-free planet.

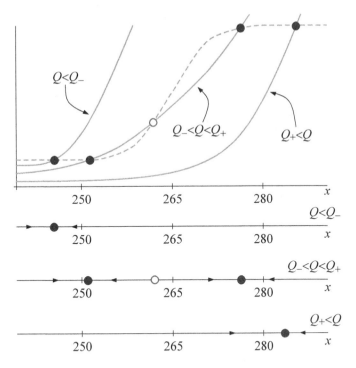

Figure 6.5. *In the top panel, there is the plot of the two pieces of the function $f(x)$: the solid curves are $y = \sigma x^4/Q$ for various values of Q, and the dashed curve is $y = 1 - \alpha(x)$. The intersection point(s) of these two curves correspond to equilibrium solutions. The phase line for each representative value of Q is presented below the graphs of the functions.*

6.4 ▪ Power series solutions

The goal here is to derive explicit solutions for nonautonomous nth-ODEs:

$$y^{(n)} + a_{n-1}(t)y^{(n-1)} + \cdots + a_1(t)y' + a_0(t)y = f(t). \tag{6.4.1}$$

Our assumption is that all of the coefficients $a_j(t)$ for $j = 1, \ldots, n-1$, as well as the forcing function $f(t)$, can each be represented by a power (Taylor) series on some common interval $|t - t_0| < R$ for some $R > 0$. If this is the case, then the idea is that the general solution to the ODE can also be represented by a convergent power series, i.e.,

$$y(t) = \sum_{j=0}^{\infty} a_j (t - t_0)^j.$$

For second-order ODEs the first two coefficients are determined by the initial conditions:

$$y(t_0) = a_0, \quad y'(t_0) = a_1.$$

The remaining coefficients associated with the series solution will solve some recurrence relationship.

We will not attempt to develop a general theory in this text. Instead, we will focus on just a couple of examples so that the main ideas associated with generating and solving the recurrence relationship are clear. In these examples, it will always be assumed for the sake of clarity that $t_0 = 0$.

Example 6.2. Let us find the power series for the general solution to

$$y' - y = 0.$$

In the language of (6.4.1), we have $a_0(t) \equiv 1$, so our expectation is that a power series solution to this ODE should converge for all t. We already know that the general solution is $y(t) = ce^t$, so it should be the case that the power series solution can be recognized as the exponential solution.

We assume that the general solution has the power series representation

$$y(t) = \sum_{j=0}^{\infty} a_j t^j,$$

where the coefficients need to be determined by the ODE. Since the derivative of a power series is determined through term-by-term differentiation, we have

$$y'(t) = \sum_{j=0}^{\infty} j a_j t^{j-1} = \sum_{j=1}^{\infty} j a_j t^{j-1} = \sum_{j=0}^{\infty} (j+1) a_{j+1} t^j.$$

Substitution of these power series representations into the ODE gives

$$\sum_{j=0}^{\infty} (j+1) a_{j+1} t^j - \sum_{j=0}^{\infty} a_j t^j = 0,$$

which on combining the sums gives

$$\sum_{j=0}^{\infty} \left[(j+1) a_{j+1} - a_j \right] t^j = 0.$$

In order for a power series to be zero for all t, each of the coefficients must be equal to zero. This fact means that for $j = 0, 1 \ldots$,

$$(j+1) a_{j+1} - a_j = 0 \quad \leadsto \quad a_{j+1} = \frac{1}{j+1} a_j.$$

We now have a recurrence relationship for the coefficients. The coefficient a_0 is arbitrary. For the next coefficients, we have

$$a_1 = a_0,$$
$$a_2 = \frac{1}{2} a_1 = \frac{1}{2} a_0,$$
$$a_3 = \frac{1}{3} a_2 = \frac{1}{2 \cdot 3} a_0,$$
$$a_4 = \frac{1}{4} a_3 = \frac{1}{2 \cdot 3 \cdot 4} a_0.$$

The pattern is clear, and by an induction argument, we can conclude that

$$a_j = \frac{1}{j!} a_0.$$

6.4. Power series solutions

The power series for the solution is then given by

$$y(t) = \sum_{j=0}^{\infty} \frac{1}{j!} a_0 t^j = a_0 \sum_{j=0}^{\infty} \frac{1}{j!} t^j.$$

An application of the ratio test shows that this series converges for all t. Moreover, we recognize that

$$e^t = \sum_{j=0}^{\infty} \frac{1}{j!} t^j;$$

hence, from the approach of finding the solution via a power series, we recover the expected solution $y(t) = a_0 e^t$, where a_0 is an arbitrary constant. ∎

Example 6.3. Let us find the power series solution to

$$y'' + y = 0.$$

We know that the general solution is

$$y(t) = c_1 \cos(t) + c_2 \sin(t),$$

so it should be the case that the power series solution can be recognized as the sum of these two trigonometric solutions.

We assume that the general solution has the power series representation

$$y(t) = \sum_{j=0}^{\infty} a_j t^j,$$

where the coefficients need to be determined by the ODE. Since the derivative of a power series is determined through term-by-term differentiation, we have

$$y'(t) = \sum_{j=0}^{\infty} j a_j t^{j-1} = \sum_{j=1}^{\infty} j a_j t^{j-1} = \sum_{j=0}^{\infty} (j+1) a_{j+1} t^j.$$

As for the second derivative, we differentiate the above expression to get

$$y''(t) = \sum_{j=0}^{\infty} j(j+1) a_{j+1} t^{j-1} = \sum_{j=1}^{\infty}\sum_{j=0}^{\infty} j(j+1) a_{j+1} t^{j-1} = \sum_{j=0}^{\infty} (j+1)(j+2) a_{j+2} t^j.$$

Substitution of these power series representations into the ODE gives

$$\sum_{j=0}^{\infty} (j+1)(j+2) a_{j+2} t^j + \sum_{j=0}^{\infty} a_j t^j = 0,$$

which on combining the sums gives

$$\sum_{j=0}^{\infty} \left[(j+1)(j+2) a_{j+2} + a_j \right] t^j = 0.$$

In order for a power series to be zero for all t, each of the coefficients must be equal to zero. This fact means that for $j = 0, 1 \ldots$,

$$(j+1)(j+2)a_{j+2} + a_j = 0 \quad \leadsto \quad a_{j+2} = -\frac{1}{(j+1)(j+2)}a_j.$$

We now have a recurrence relationship for the coefficients. The coefficients a_0 and a_1 are arbitrary. For the next coefficients, we have

$$a_2 = -\frac{1}{1\cdot 2}a_0,$$

$$a_3 = -\frac{1}{2\cdot 3}a_1,$$

$$a_4 = -\frac{1}{3\cdot 4}a_2 = \frac{1}{1\cdot 2\cdot 3\cdot 4}a_0,$$

$$a_5 = -\frac{1}{4\cdot 5}a_3 = \frac{1}{2\cdot 3\cdot 4\cdot 5}a_1,$$

$$a_6 = -\frac{1}{5\cdot 6}a_4 = -\frac{1}{1\cdot 2\cdot 3\cdot 4\cdot 5\cdot 6}a_0,$$

$$a_7 = -\frac{1}{6\cdot 7}a_5 = -\frac{1}{2\cdot 3\cdot 4\cdot 5\cdot 6\cdot 7}a_1.$$

The pattern is clear, and by an induction argument, we can conclude for $j = 1, 2, \ldots$, that

$$a_{2j} = (-1)^j \frac{1}{(2j)!}a_0, \quad a_{2j+1} = (-1)^j \frac{1}{(2j+1)!}a_1.$$

The power series for the solution is then

$$y(t) = \sum_{j=0}^{\infty} \left((-1)^j \frac{1}{(2j)!} a_0 t^{2j} + (-1)^j \frac{1}{(2j+1)!} a_1 t^{2j+1} \right)$$

$$= a_0 \sum_{j=0}^{\infty} (-1)^j \frac{1}{(2j)!} t^{2j} + a_1 \sum_{j=0}^{\infty} (-1)^j \frac{1}{(2j+1)!} t^{2j+1}.$$

An application of the ratio test shows that this series converges for all t. Moreover, we recognize that the first sum is the Maclaurin series for $\cos(t)$ and that the second sum is the Maclaurin series for $\sin(t)$. Hence, from the approach of finding the solution via a power series, we recover the expected solution. ∎

Exercises

Exercise 6.4.1. *Find a power series representation centered at $t = 0$, i.e., $y(t) = \sum a_j t^j$, for the following ODEs.*

(a) $y' = t^2 y$

(b) $y' = y + \dfrac{8}{1-t}$ *(hint: use the power series formulation for $(1-t)^{-1}$)*

(c) $t^2 y'' - 6t y' + 10 y = 0$

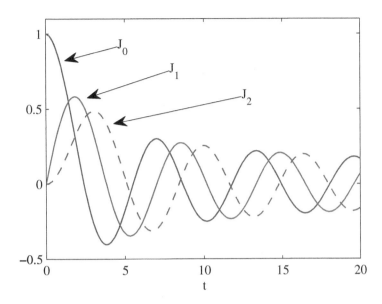

Figure 6.6. *A plot of the first three Bessel functions $J_0(t)$ (solid curve with $J_0(0) = 1$), $J_1(t)$ (solid curve with $J_1(0) = 0$), and $J_2(t)$ (dashed curve).*

6.5 ▪ Case studies

Now that we have seen the mechanics for using power series to solve a simple first-order ODE, we will consider two nonautonomous second-order ODEs. Each of these is of physical interest. Indeed, many equations of interest in mathematical physics—Legendre's equation and Hermite's equation are two classical examples—can be solved through finding the power series of the solution.

6.5.1 ▪ Bessel's equation

In the study of the vibrations of a circular membrane, one is led to studying the ODE

$$t^2 y'' + t y' + (t^2 - n^2) y = 0.$$

Here $n \in \mathbb{N}_0$ is an nonnegative integer. The equation is known as *Bessel's equation*. In standard form, the equation would be written as

$$y'' + \frac{1}{t} y' + \left(1 - \frac{n^2}{t^2}\right) y = 0,$$

and since the coefficients are not continuous as $t = 0$, it is not clear that there is a solution that is continuous at $t = 0$. However, such a solution does exist: it is known as a *Bessel function of nth order* and is denoted by $J_n(t)$ [**13**, Chapter V.5] (see Figure 6.6 for plots of the first three Bessel functions). Our goal is to construct $J_n(t)$ via a power series for each n and to then show that the Bessel function is well defined for all t.

In order to simplify the problem, let us assume that $n = 0$, so the equation to be studied is
$$t^2 y'' + t y' + t^2 y = 0.$$
We assume that the desired solution has the power series representation
$$y(t) = \sum_{j=0}^{\infty} a_j t^j;$$
thus, we are implicitly assuming that the solution is continuous at $t = 0$. After differentiating term by term,
$$y'(t) = \sum_{j=0}^{\infty} j a_j t^{j-1} = \sum_{j=0}^{\infty} (j+1) a_{j+1} t^j,$$
and
$$y''(t) = \sum_{j=0}^{\infty} j(j+1) a_{j+1} t^{j-1} = \sum_{j=0}^{\infty} (j+1)(j+2) a_{j+2} t^j.$$
Since
$$t^2 y'' = \sum_{j=0}^{\infty} (j+1)(j+2) a_{j+2} t^{j+2},$$
$$t y' = \sum_{j=0}^{\infty} (j+1) a_{j+1} t^{j+1} = a_1 t + \sum_{j=0}^{\infty} (j+2) a_{j+2} t^{j+2},$$
$$t^2 y = \sum_{j=0}^{\infty} a_j t^{j+2},$$
on substitution into Bessel's equation, we have
$$\sum_{j=0}^{\infty} (j+1)(j+2) a_{j+2} t^{j+2} + a_1 t + \sum_{j=0}^{\infty} (j+2) a_{j+2} t^{j+2} + \sum_{j=0}^{\infty} a_j t^{j+2} = 0.$$
Combining the sums gives
$$a_1 t + \sum_{j=0}^{\infty} \left[(j+1)(j+2) a_{j+2} + (j+2) a_{j+2} + a_j \right] t^{j+2} = 0,$$
which after simplification yields
$$a_1 t + \sum_{j=0}^{\infty} \left[(j+2)^2 a_{j+2} + a_j \right] t^{j+2} = 0.$$
All the coefficients must be zero, which means that
$$a_1 = 0$$
$$(j+2)^2 a_{j+2} = -a_j, \quad j = 0, 1, \ldots.$$

As a side remark, the assumption that the solution be continuous at $t = 0$ implies that instead of finding two linearly independent solutions to the second-order

6.5. Case studies

ODE, we will find only one solution. The other solution must not be continuous at $t = 0$; indeed, it can be shown that the other solution, say, $y_2(t)$, has the series representation

$$y_2(t) = \ln\left(\frac{t}{2}\right) \sum_{j=0}^{\infty} b_j t^j.$$

The recurrence relationship for the coefficients is $a_1 = 0$ with

$$a_{j+2} = -\frac{1}{(j+2)^2} a_j, \quad j = 0, 1, \ldots.$$

We now solve this relationship. Suppose first that j is odd. In addition to $a_1 = 0$, we have

$$a_3 = -\frac{1}{9} a_1 \quad \leadsto \quad a_3 = 0,$$
$$a_5 = -\frac{1}{25} a_3 \quad \leadsto \quad a_5 = 0,$$
$$a_7 = -\frac{1}{49} a_5 \quad \leadsto \quad a_7 = 0.$$

The pattern is clear, and from an induction argument, we have that if j is odd, the coefficient $a_j = 0$. Now suppose that j is even. The coefficient a_0 is arbitrary, and

$$a_2 = -\frac{1}{4} a_0,$$
$$a_4 = -\frac{1}{16} a_2 = \frac{1}{4 \cdot 16} a_0,$$
$$a_6 = -\frac{1}{36} a_4 = -\frac{1}{4 \cdot 16 \cdot 36} a_0,$$
$$a_8 = -\frac{1}{64} a_6 = \frac{1}{4 \cdot 16 \cdot 36 \cdot 64} a_0.$$

The pattern is now clear. On noting that

$$\frac{1}{4 \cdot 16} = \frac{1}{2^{2 \cdot 2}} \frac{1}{2 \cdot 2} \frac{1}{2},$$
$$\frac{1}{4 \cdot 16 \cdot 36} = \frac{1}{2^{2 \cdot 3}} \frac{1}{2 \cdot 3} \frac{1}{2 \cdot 3},$$
$$\frac{1}{4 \cdot 16 \cdot 36 \cdot 64} = \frac{1}{2^{2 \cdot 4}} \frac{1}{2 \cdot 3 \cdot 4} \frac{1}{2 \cdot 3 \cdot 4},$$

after applying an induction argument, we can conclude that the even coefficients are given by

$$a_{2j} = (-1)^j \frac{1}{2^{2j}} \frac{1}{j! j!} a_0.$$

The power series for the solution is then

$$y(t) = a_0 \sum_{j=0}^{\infty} (-1)^j \frac{1}{2^{2j}} \frac{1}{j! j!} t^{2j} = a_0 \sum_{j=0}^{\infty} \frac{(-1)^j}{j! j!} \left(\frac{t}{2}\right)^{2j}.$$

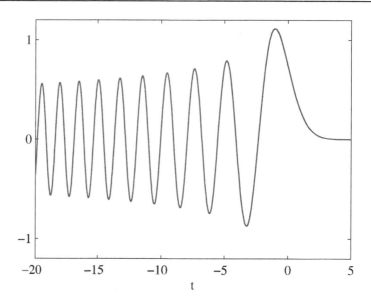

Figure 6.7. *A plot of the solution to Airy's equation with the initial condition* $y(0) = 1/\Gamma(2/3)$ *and* $y'(0) = -3^{1/3}/\Gamma(1/3)$. *Here* $\Gamma(t)$ *is the gamma function.*

It is clear that $y(0) = a_0$, so that this power series solution is bounded at $t = 0$. An application of the ratio test further shows that this solution is well defined for all t. In conclusion, we write the Bessel function of 0th-order as

$$J_0(t) = \sum_{j=0}^{\infty} \frac{(-1)^j}{j!j!} \left(\frac{t}{2}\right)^{2j},$$

where without loss of generality we set $a_0 = 1$ in our series solution. The construction of $J_n(t)$ is left as Exercise 6.5.1.

6.5.2 ▪ Airy's equation

In wave propagation problems, one is led to studying *Airy's equation*:

$$y'' - ty = 0.$$

A solution to Airy's equation, known as the *Airy function*, models the transition of wave behavior that decays as $t \to +\infty$ and oscillates at $t \to -\infty$—"ripples" behind a moving wave [1, Chapter 4.6] (see Figure 6.7). The Airy function is also the solution to Schrödinger's equation for a particle confined within a triangular potential well and for a particle in a one-dimensional constant force field and serves as a model for the diffraction of light.

We assume that the desired solution to Airy's equation has the power series representation

$$y(t) = \sum_{j=0}^{\infty} a_j t^j.$$

After differentiating term by term,

$$y'(t) = \sum_{j=0}^{\infty} j a_j t^{j-1} = \sum_{j=0}^{\infty} (j+1) a_{j+1} t^j,$$

6.5. Case studies

and
$$y''(t) = \sum_{j=0}^{\infty} j(j+1)a_{j+1}t^{j-1} = \sum_{j=0}^{\infty}(j+1)(j+2)a_{j+2}t^j.$$

Since
$$ty = \sum_{j=0}^{\infty} a_j t^{j+1} = \sum_{j=1}^{\infty} a_{j-1} t^j,$$

on substitution into Airy's equation, we have
$$\sum_{j=0}^{\infty}(j+1)(j+2)a_{j+2}t^j - \sum_{j=1}^{\infty} a_{j-1} t^j = 0.$$

Combining the sums and noting the difference in where the index of each sum begins, we have
$$2a_2 + \sum_{j=1}^{\infty}\left[(j+1)(j+2)a_{j+2} - a_{j-1}\right]t^j = 0.$$

All the coefficients must be zero, which means that
$$a_2 = 0$$
$$(j+1)(j+2)a_{j+2} = a_{j-1}, \quad j = 1, 2, \ldots.$$

It will be convenient to note that the second relationship can be rewritten as
$$(j+2)(j+3)a_{j+3} = a_j, \quad j = 0, 1, \ldots.$$

We now have our recurrence relationship for the coefficients. In order to solve it, we need to consider three different subsets of the positive integers. First, suppose that j is a multiple of three. We have

$$a_3 = \frac{1}{2 \cdot 3} a_0,$$
$$a_6 = \frac{1}{5 \cdot 6} a_3 = \frac{1}{2 \cdot 3 \cdot 5 \cdot 6} a_0,$$
$$a_9 = \frac{1}{8 \cdot 9} a_6 = \frac{1}{2 \cdot 3 \cdot 5 \cdot 6 \cdot 8 \cdot 9} a_0.$$

The pattern is clear, and from an induction argument, we have that if $j = 3\ell$ for some integer ℓ,
$$a_{3\ell} = \frac{1}{(2 \cdot 3)(5 \cdot 6) \cdots ((3\ell - 1)(3\ell))} a_0.$$

Now suppose that j is a multiple of three plus one. We have

$$a_4 = \frac{1}{3 \cdot 4} a_1,$$
$$a_7 = \frac{1}{6 \cdot 7} a_4 = \frac{1}{3 \cdot 4 \cdot 6 \cdot 7} a_1,$$
$$a_{10} = \frac{1}{9 \cdot 10} a_7 = \frac{1}{3 \cdot 4 \cdot 6 \cdot 7 \cdot 9 \cdot 10} a_1.$$

The pattern is clear, and from an induction argument, we have that if $j = 3\ell + 1$ for some integer ℓ,

$$a_{3\ell+1} = \frac{1}{(3 \cdot 4)(6 \cdot 7) \cdots ((3\ell)(3\ell+1))} a_1.$$

Finally, suppose that j is a multiple of three plus two. Recalling that $a_2 = 0$, we have

$$a_5 = \frac{1}{4 \cdot 5} a_2 \rightsquigarrow a_5 = 0,$$

$$a_8 = \frac{1}{7 \cdot 8} a_5 \rightsquigarrow a_8 = 0,$$

$$a_{11} = \frac{1}{10 \cdot 11} a_8 \rightsquigarrow a_{11} = 0.$$

The pattern is clear, and from an induction argument, we have that if $j = 3\ell + 2$ for some integer ℓ, then $a_{3\ell+2} = 0$.

On collecting terms, the general solution to Airy's equation is

$$y(t) = a_0 \left(1 + \sum_{\ell=1}^{\infty} \frac{1}{(2 \cdot 3)(5 \cdot 6) \cdots ((3\ell-1)(3\ell))} t^{3\ell} \right)$$

$$+ a_1 \left(t + \sum_{\ell=1}^{\infty} \frac{1}{(3 \cdot 4)(6 \cdot 7) \cdots ((3\ell)(3\ell+1))} t^{3\ell+1} \right).$$

An application of the ratio test further shows that each sum is well defined for all t, which implies the same for the full general solution.

The Airy function itself is the solution to Airy's equation with initial condition

$$y(0) = \frac{1}{\Gamma(2/3)}, \quad y'(0) = -\frac{3^{1/3}}{\Gamma(1/3)}.$$

Here $\Gamma(t)$ is the celebrated *gamma function* (see Markushevich [28, Chapter II.10] for many details about this function).

Exercises

Exercise 6.5.1. *Consider Bessel's equation as written in section 6.5.1 for $n \geq 1$. By assuming that the power series for the solution that is continuous at $t = 0$ is given by*

$$y(t) = t^k \sum_{j=0}^{\infty} a_j t^j$$

for some integer $k > 0$, show that the Bessel function of nth order is given by

$$J_n(t) = \left(\frac{t}{2}\right)^n \sum_{j=0}^{\infty} \frac{(-1)^j}{j!(j+n)!} \left(\frac{t}{2}\right)^{2j}.$$

Further show that $J_n(t)$ is well defined for all t. (Hint: The ratio test may be useful.)

Exercise 6.5.2. *Hermite's equation is given by*

$$y'' - 2ty' + 2\lambda y = 0,$$

where λ is real valued. Assume that there is a power series solution:

$$y(t) = \sum_{j=0}^{\infty} a_j t^j.$$

(a) *Show that the recursion relationship is*

$$a_{j+2} = \frac{2(j-\lambda)}{(j+1)(j+2)} a_j, \quad j = 0, 1, \ldots.$$

(b) *Show that if there are an infinite number of nonzero coefficients, then the power series does not converge for all values of t.*

(c) *Determine the values of λ for which there only a finite number of nonzero coefficients. The associated solutions are known as the Hermite polynomials.*

Exercise 6.5.3. *Legendre's equation is*

$$(1-t^2)y'' - 2ty' + \lambda(\lambda+1)y = 0,$$

where λ is real valued. Assume that there is a power series solution:

$$y(t) = \sum_{j=0}^{\infty} a_j t^j.$$

(a) *Show that the recursion relationship is*

$$a_{j+2} = \frac{(j-\lambda)(j+\lambda+1)}{(j+1)(j+2)} a_j, \quad j = 0, 1, \ldots.$$

(b) *Show that if there are an infinite number of nonzero coefficients, then the power series does not converge for all values of t.*

(c) *Determine the values of λ for which there only a finite number of nonzero coefficients. The associated solutions are known as the Legendre polynomials.*

Bibliography

[1] M. Ablowitz and A. Fokas. *Complex Variables: Introduction and Applications.* Cambridge Texts in Applied Mathematics. Cambridge University Press, 2nd edition, 2003. (Cited on p. 290)

[2] E. Ackerman, J. Rosevear, and W. McGuckin. A mathematical model of the glucose tolerance test. *Phys. Med. Biol.*, 9:202–213, 1964. (Cited on p. 137)

[3] E. Ackerman, L. Gatewood, J. Rosevear, and G. Molnar. Blood glucose regulation and diabetes. In F. Heinmets, editor, *Concepts and Models of Biomathematics*, Chapter 4, pages 131–156. Marcel Dekker, 1969. (Cited on p. 137)

[4] K. Billah and R. Scanlan. Resonance, Tacoma Narrows bridge failure, and undergraduate physics textbooks. *Amer. J. Phys.*, 59:118–124, 1991. (Cited on p. 224)

[5] P. Blanchard, R. Devaney, and G. Hall. *Differential Equations.* Brooks/Cole, Fourth edition, 2012. (Cited on p. x)

[6] M. Boelkins, M. Potter, and J. Goldberg. *Differential Equations with Linear Algebra.* Oxford University Press, 2009. (Cited on p. vii)

[7] V. Bolie. Coefficients of normal blood glucose regulation. *J. Appl. Physiol.*, 16(5):783–788, 1961. (Cited on p. 137)

[8] R. Borelli and C. Coleman. *Differential Equations: A Modeling Approach.* Prentice Hall, 1987. (Cited on p. 189)

[9] W. Boyce and R. DiPrima. *Elementary Differential Equations.* John Wiley & Sons, Eighth edition, 2005. (Cited on pp. vii, 273)

[10] R. Bronson and G. Costa. *Schaum's Outline of Differential Equations.* Schaum's Outlines. McGraw-Hill, Third edition, 2009. (Cited on p. x)

[11] J. Brown and R. Churchill. *Complex Variables and Applications.* McGraw-Hill, 7th edition, 2004. (Cited on p. 160)

[12] B. Conrad. *Differential Equations: A Systems Approach.* Pearson Education, 2003. (Cited on p. x)

[13] R. Courant and D. Hilbert. *Methods of Mathematical Physics*, volume 1. Interscience Publishers, 1953. (Cited on p. 287)

[14] P. Drazin and R. Johnson. *Solitons: An Introduction.* Cambridge University Press, Cambridge, 1989. (Cited on p. 201)

[15] C. Edwards and D. Penney. *Differential Equations and Linear Algebra.* Pearson, 3rd edition, 2008. (Cited on p. vii)

[16] H. Eves. *Elementary Matrix Theory*. Dover Publications, 1980. (Cited on p. 62)

[17] T. Fay and S. Graham. Coupled spring equations. *Int. J. Math. Educ. Sci. Technol.*, 34(1):65–79, 2003. (Cited on p. 228)

[18] S. Goode and S. Annin. *Differential Equations and Linear Algebra*. Prentice Hall, 2007. (Cited on p. x)

[19] D. Griffiths and D. Higham. *Numerical Methods for Ordinary Differential Equations: Initial Value Problems*. Undergraduate Mathematics Series. Springer, 2010. (Cited on p. 110)

[20] J. Hefferon. *Linear Algebra*. Virginia Commonwealth University Mathematics, 2009. http://joshua.smcvt.edu/linalg.html/. (Cited on p. viii)

[21] M. Holmes. *Introduction to Perturbation Methods*, volume 20 of *Texts in Applied Mathematics*. Springer-Verlag, 2nd edition, 2013. (Cited on p. 259)

[22] H. Kaper and H. Engler. *Mathematics and Climate*. SIAM, 2014. (Cited on p. 282)

[23] R. Lamberson, R. McKelvey, R. Noon, and C. Voss. A dynamic analysis of the viability of the Northern Spotted Owl in a fragmented forest environment. *Conservation Biology*, 6:505–512, 1992. (Cited on p. 92)

[24] D. Lay. *Linear Algebra and its Applications*. Pearson, 4th edition, 2012. (Cited on p. 92)

[25] N. Lebedev. *Special Functions and Their Applications*. Dover Publications, New York, 1972. (Cited on p. 273)

[26] S. Lipschutz and M. Lipson. *Schaum's Outline of Linear Algebra*. Schaum's Outlines. McGraw-Hill, Fifth edition, 2012. (Cited on p. x)

[27] W.-S. Lu, H.-P. Wand, and A. Antoniou. An efficient method for the evaluation of the controllability and observability gramians of 2-D digital filters and systems. *IEEE Trans. Circ. Sys.*, 39(10):695–704, 1992. (Cited on p. 97)

[28] A. Markushevich. *Theory of Functions*. Chelsea Publishing Co., New York, 1985. (Cited on p. 292)

[29] G. McBride and N. French. A new linear analytical SIR model for age-dependent susceptibility and occupation-dependent immune status. In *SMO'06 Proceedings of the 6th WSEAS International Conference on Simulation, Modelling and Optimization*, pages 222–226. WSEAS, 2006. (Cited on pp. 192, 193)

[30] L. Meirovitch. *Analytical Methods in Vibrations*. MacMillan Series in Applied Mechanics. The MacMillan Company, New York, 1967. (Cited on p. 201)

[31] C. Meyer. *Matrix Analysis and Applied Linear Algebra*. SIAM, 2000. (Cited on pp. 80, 202)

[32] R. Nagle, E. Saff, and A. Snider. *Fundamentals of Differential Equations*. Addison-Wesley, Fifth edition, 2000. (Cited on p. x)

[33] F. Paladini, I. Renna, and L. Renna. A discrete SIRS model with kicked loss of immunity and infection probability. *J. Phys.: Conf. Ser.*, 285:012018, 2011. (Cited on pp. 90, 91)

[34] L. Perko. *Differential Equations and Dynamical Systems*. Springer-Verlag, 2nd edition, 1996. (Cited on p. 168)

[35] J. Polking. *Ordinary Differential Equations Using MATLAB*. Pearson, 3rd edition, 2003. (Cited on p. 110)

[36] J. Smith. *Introduction to Digital Filters: With Audio Applications*. W3K Publishing, 2007. (Cited on p. 85)

[37] S. Strogatz. *Nonlinear Dynamics and Chaos with Applications to Physics, Biology, Chemistry, and Engineering*. Westview Press, 2015. (Cited on pp. ix, 232, 273, 280, 281)

[38] R. Vein and P. Dale. *Determinants and Their Applications in Mathematical Physics*, volume 134 of *Applied Mathematics Sciences*. Springer, 1999. (Cited on p. 62)

[39] T. Yuster. The reduced row echelon form of a matrix is unique: A simple proof. *Math. Magazine*, 57(2):93–94, 1984. (Cited on p. 7)

Index

Abel's formula, 150
Airy function, 290
Airy's equation, 290

basis, 47–51, 55, 69, 72, 73, 150, 151, 166
beating, 223, 230
Bessel equation, 287
Bessel function, 287
bifurcation, 281

carrying capacity, 279
Cauchy–Euler equation, 216
center, 162
characteristic polynomial, 72, 202
complex number
  complex conjugate, 66
  imaginary part, 65
  magnitude, 66
  real part, 65
convolution, 262
critical point, 152

damping
  critical, 219
  over, 218
  under, 219
DC motor, 199
delta function, 237, 251
determinant, 60, 61, 72, 86
diabetes model, 137, 234, 270
dimension, 50, 69
direction field, 104
dynamical system, 87, 90, 92

eigenspace, 72, 73
eigenvalue, 71, 72, 79, 91, 150, 166, 190
  algebraic multiplicity, 73
  geometric multiplicity, 73
  multiple, 73

  simple, 73
eigenvector, 71, 72, 91, 150, 190
equilibrium solution, 152
Euler's formula, 67

filter
  controllable, 85
  observable, 85, 97
forced pendulum, 139, 232, 233
Fourier coefficient, 80, 228
Fourier expansion, 80, 150, 228
free variable, 9, 25, 33, 37, 50, 57, 58, 73

Gaussian elimination, 8
Google PageRank algorithm, 97

Heaviside function, 237
Hermite polynomial, 293
Hermite's equation, 292
homogeneous linear system, 24, 26, 33, 42
homogeneous solution, 24–26, 57, 113, 117, 126, 147, 151, 207, 210

impulse response, 251, 264
initial value problem (IVP), 102

Laplace transform, 240, 248, 251, 261
  inverse, 244, 249, 252, 254, 262
lead ingestion model, 189, 198
leading entry, 7
Legendre equation, 293
Legendre polynomial, 293
linear
  combination, 13
  dependence, 33, 36
  independence, 33, 57, 151
logistic equation, 279

Markov process, 88
mass-spring system
  coupled, 227, 235
  damped, 138, 224, 257
  undamped, 219, 251, 269
matrix
  augmented, 8, 21, 24, 55, 170
  companion, 202
  identity, 6, 19, 37, 50, 58, 180
  inverse, 21, 58, 79, 151
  spectral decomposition, 80
matrix-valued solution, 146, 149, 151, 176
matrix/matrix multiplication, 18, 21, 69
matrix/vector multiplication, 15, 19, 26, 30, 41, 55, 79, 80, 88, 145, 160, 169, 210
method of undetermined coefficients, 120, 127, 130, 212, 220, 224, 232
mixing problem
  one tank, 126, 253, 269
  three tanks, 198
  two tanks, 135, 183, 197, 270
Mythbusters, 226

natural frequency, 219, 221, 228, 230, 232, 235, 252, 266
numerical method
  Dormand–Prince, 110
  Euler, 108
  improved Euler, 109
  Runge–Kutta, 109

phase line, 277
phase plane, 154
pivot column, 7, 33, 48, 50, 57
pivot position, 7, 56
poles, 265

rank, 50, 56, 58

resonant forcing, 223
row-reduced echelon form, 7

saddle point, 155
separable, 273
singular perturbations, 259
space
  column, 43, 47, 50, 69
  null, 42, 56, 58, 69
  sub-, 40, 69
  vector, 40

span, 30, 41, 44, 47, 52, 55, 56
spanning set, 32, 47, 50, 58
stable node, 157
stable spiral, 162
state-space model, 97
susceptible-infected-recovered (SIR) model
  continuous, 192, 200
  discrete, 89

Tacoma Narrows Bridge, 224

transfer function, 261
transient, 219, 224

unstable node, 159
unstable spiral, 162

variation of parameters, 118, 176, 210
vector transpose, 81

Wronskian, 150